Christiane Naumann

"Where We Used to Plough"

KÖLNER ETHNOLOGISCHE STUDIEN

herausgegeben von

Martin Rössler

Band 37

LIT

Christiane Naumann

"Where We Used to Plough"

Environmental Governance, Rural Livelihoods and Social-Ecological Change in South Africa

LIT

Cover image: Dam that served the former Grobler Irrigation Scheme in Sediba (Photograph: C. Naumann)

This book is printed on acid-free paper.

Bibliographic information published by the Deutsche Nationalbibliothek
The Deutsche Nationalbibliothek lists this publication in the Deutsche Nationalbibliografie; detailed bibliographic data are available on the Internet at http://dnb.d-nb.de.

ISBN 978-3-643-90844-5 (pb)
ISBN 978-3-643-95844-0 (PDF)
Zugl.: Köln, Univ., Diss., 2015

A catalogue record for this book is available from the British Library

Klosbachstr. 107
CH-8032 Zürich
Tel. +41 (0) 44-251 75 05
E-Mail: zuerich@lit-verlag.ch http://www.lit-verlag.ch
Distribution:
In the UK: Global Book Marketing, e-mail: mo@centralbooks.com
In North America: International Specialized Book Services, e-mail: orders@isbs.com
In Germany: LIT Verlag Fresnostr. 2, D-48159 Münster
Tel. +49 (0) 2 51-620 32 22, Fax +49 (0) 2 51-922 60 99, e-mail: vertrieb@lit-verlag.de
In Austria: Medienlogistik Pichler-ÖBZ, e-mail: mlo@medien-logistik.at
e-books are available at www.litwebshop.de

To my parents and brother

Acknowledgements

This dissertation would never have been possible without the support and help of many people. First and foremost, I would like to thank all headmen and residents of the four research sites in Thaba Nchu (Middeldeel, Talla, Rooifontein, and Sediba) for welcoming me and for taking the time to talk and answer my numerous, sometimes repetitive questions. Moreover, I thank the farmers of the commercial and land reform farms for giving me insight into their land management. I am grateful that I found wonderful people who assisted me during the whole research process: Tsametse Pohe, Portia Segalo, Michael Nyelele and Kedikilwe Mihi did a great job in explaining, translating, transcribing and collecting data. Sandile van Staden and Matthews Zimema assisted me particularly in the initial research phase.

I thank Philip Henning and Matsidiso Kgware, who offered me a place to stay on the farm and in Sediba, respectively. Moreover, I thank the officers of the Thaba Nchu and Botshabelo extension office of the Department of Agriculture and Rural Development for establishing contacts with the communities and providing information about the area.

In Bloemfontein, I thank our project partners at the University of the Free State, particularly Chris Du Preez and Rida van Heerden, who were both always concerned with my well-being, and Elmarie Kotzé, who was not only a colleague but also became a friend. I appreciated the revealing conversations and discussions I had with other present (and former) UFS scholars: Petro Esterhuyse, Hermanus Fouché, Robert Gordon and Godfrey Kundhlande. Bahedile Letlala taught me Setswana and helped with the translation of some interviews.

In Germany, I thank Michael Bollig for his pragmatic advice and valuable comments on methodological and theoretical problems during the research and writing process. Thomas Widlok made very useful suggestions for the first draft of this dissertation, and Hartmut Lang helped me with ‘cracking’ some ethnodemographic ‘nuts’. I thank all colleagues who supported me academically and practically during the doctorate, particularly Clemens Greiner. Pax and Emma Welter proofread the thesis. Monika Feinen made an illustrative map of the Thaba Nchu area.

The dissertation emerged in the context of the research project *Resilience, Collapse and Reorganisation in East- and South Africa's Savannahs* (RCR) funded by the *Deutsche Forschungsgemeinschaft* (DFG). Thanks to all colleagues of the RCR project, particularly Sebastian Rasch for the pleasant cooperation and our (sometimes almost spiritual) discussions of the resilience concept; Julia Tischler and Katharina Löber for copying some archival sources; Roloef Oomen, Anja Linstädter and Alexandra Sandhage-Hoffman for explaining the findings of the natural sciences; Katharina Brüser for taking care of the purchase of aerial photography; and Werner Schuck for organisational support.

Writing a dissertation can be a tedious process, with ups and downs as well as minor and major challenges. Knowing that I was not alone with my questions and doubts made the difficult times of the research and writing process easier. This was first and foremost due to the lively exchange with fellow PhD students. I thank all *Ethnodocs* and colleagues in Marienburg for the academic and emotional support. Especially, I would like to thank Anne-Christina Achterberg-Boness, Anja Becker, Franziska Bedorf, Sara de Wit, Anna Grumblies, Thekla Kelbert, Diego Menestrey-Schwieger, Ulrike Wesch and Bea Wittger for our invaluable conversations, 'writing camps' and inspiring brainstorming and discussion sessions.

Last but not least, I would like to thank my friends and relatives in Leipzig, Hamburg, Cologne and other parts of Germany; my parents, Sabine and Volker Naumann, and my brother, Christoph Naumann, for their unlimited support; and Niels and Bent Anders Dreber for their encouragement and love, and for cheering me up whenever it was necessary.

Table of Contents

List of Figures

List of Tables

List of Acronyms

ANC	African National Congress
ARC	Agricultural Research Council
BAD	Department of Bantu Administration and Development (previously: Department of Native Affairs)
BNP	Bophuthatswana National Party
CPA	Communal Property Association
IRWH	Infield Rainwater Harvesting
LRAD	Land Redistribution for Agricultural Development
LSU	Large stock unit
NAD	Department of Native Affairs (later: Department of Bantu Administration and Development)
NGO	Non-governmental organisation
ORC	Orange River Colony
PTO	Permission to occupy
RDP	Reconstruction and Development Programme (RDP)
SADT	South African Development Trust (previously: South African Native Trust)
SANT	South African Native Trust (later: South African Development Trust)
SLAG	Settlement Land Acquisition Grant
SSU	Small stock unit
TLRRP	Thaba Nchu Land Reform Research Project
UFS	University of the Free State

Glossary Setswana – English

bogadi (pl. magadi)	bridewealth
dijo tsa naga	wild edible fruits
go le monate	it was nice
kgomo (pl. dikgomo)	cattle
kgosana (pl. dikgosana)	headman
kgosi (pl. dikgosi)	chief
kgotla (pl. dikgotla)	assembly, court
korong (pl. dikorong)	wheat
leuba (pl. mauba)	drought, famine
lelapa (pl. malapa)	household, homestead, house of a married woman, agnatic family group
lengope (pl. mangope)	donga
lerole le le hibidi	red dust
letamo (pl. matamo)	dam
mabele (pl. lebele)	sorghum
mmidi	maize
mokete (pl. mekete)	celebration, feast
morogo (pl. merogo)	vegetable, wild food plants
mpho ya badimo	gift for the ancestors
muthi	herbal medicine, magic medicine
nawa (pl. dinawa)	bean
phitlho (pl. diphiltho)	funeral
pitsa (pl. dipitsa)	pot
setlamo	a united group
tsie (pl. ditsie)	locust
tsimo (pl. masimo)	field

1. INTRODUCTION

Prologue

When I visited Thaba Nchu for the first time, I had a clear research proposition in mind. Being part of a research unit that studies resilience in social-ecological systems, I wanted to investigate how people use land, how they adapt to environmental crises and whether they agree or disagree with the way in which some scientists describe communal rangelands, i.e. that they are prone to degradation because of mismanagement and common pool resource management dilemmas. In the first months, still somewhat blindly following my research proposal, I tried to gain initial insights into local perceptions of rangeland change. However, whenever I asked questions about the condition of the vegetation in the past, people told me about former fields and the maize and wheat they had cultivated. I realised that the rangelands were not their major concern. Instead, they wanted to tell a different story: a story of a time when the people of Thaba Nchu still produced much of their own food and made a living from the land. They wanted to speak about state officials and agricultural extension officers who came to show farmers new machinery and introduced 'modern' cultivation methods, and who later took everything the people had: their livestock, their fields, and eventually their homes. But my informants also talked about the present and about how much has changed in the new South Africa; about what they had gained and what they had lost after the end of apartheid; and about how important social transfers, particularly pensions, are for their livelihoods. It was striking that the recent conditions found in Thaba Nchu were strongly connected to the past; and that a better understanding of the issue of land use could not be achieved by focussing only on the present time or the local level, but required a much broader perspective, both temporally and spatially.

1.1 Scope and Aims of the Thesis

This dissertation focusses on environmental governance and institutional development, as well as agro-economic change and current livelihood strategies of rural households in Thaba Nchu. Situated in the Free State province in central South Africa, the region's history is characterised by numerous shifts of political-economic context. In 1833, Batswana groups settled in a centralised village at the foot of the 'Black Mountain', *Thaba Nchu* in Setswana, the language of the Batswana people. The village soon developed into a large settlement surrounded by fields and cattle posts where young men guarded the livestock. Settlement patterns changed in the 1870s when family groups established new villages in the hinterland of Thaba Nchu. In the late nineteenth century, the area was annexed by the Orange Free State, which shortly afterwards defined two parts of the region as 'native reserves', i.e. areas where only black[1] people were allowed to live. 'Native reserves' existed in various areas throughout the country. They soon developed into reservoirs for labour migrants who became increasingly roped into the capitalistic economy. By tying black people to rural areas and enabling them to earn a supplementary income from land use, wages paid in the industrial sector were kept low.

The reserve areas were small. Although land was added later during the establishing of so-called 'homelands', in total only 13.7% of the country's land was designated for occupation by black South Africans. The 'native areas' were overpopulated and their natural resources placed under high pressure. The ecological conditions in the 'native reserves' prompted concern on the part of government officials, and the need for conservation methods was increasingly articulated. Thaba Nchu was one of the first districts where the so-called 'betterment' schemes were enforced – a programme which aimed at a rehabilitation of the natural resources and land use practices. During the mid-twentieth century, the regulation and control of land use became increasingly violent and finally culminated in the resettlement of hundreds of households in order to form grid-patterned villages. In the late 1970s, Thaba Nchu became part of the Bophuthatswana homeland, a 'pseudo-nation state' (Ferguson 2006) that was never internationally recognised. The Bophuthatswana government, seeking to achieve economic independence from the South African state, heavily subsidised agricultural production, yet it could not stop the agricultural decline that had started in the beginning of the decade. After 1994, Thaba Nchu was

---

[1]. A study that deals for a large part with South African history faces the problem of racial categories constructed by apartheid ideologists. Officially four distinct 'races' were distinguished – namely 'whites' or 'Europeans', 'Coloureds', 'Indians', and 'Africans' or 'blacks'. A critical, historical analysis of South African politics – and also post-apartheid transformations (Posel 2001) – has to work with these categories of racial difference which are still today used by many South African citizens and even the state. Nevertheless, I wish to stress that by using these categories I do not intend to ignore or downplay the fact that they are constructed and arbitrary and based on a racist ideology.

reincorporated into South Africa, which again marked a significant shift, particularly with regard to agriculture. Many rural households are still engaged in pastoral production, gardening and other forms of land use. However, their main incomes are derived from wage labour and social grants paid by the South African government.

Against this historical background the book addresses several topics and questions. Some of them refer to broader theoretical concerns related to social-ecological systems and resilience, which are formulated in Chapter 1.2. Others relate specifically to the Thaba Nchu context and are briefly presented here. The overarching theme highlighted throughout the entire time period under consideration is 'environmental governance'. This includes examinations both of environmental governance from an ideological and discursive perspective and of technical interventions constituting attempts at environmental governance on the local level. More precisely, I will reconstruct how officials in the past invented narratives of an ecological crisis and used these as a legitimation for intervention with regard to land use practices. Furthermore, I will explore conservation and rehabilitation measures and their effects on both the socio-economic organisation of rural communities and the natural resource base. In that regard, it is critical to examine the role that formal institutions played. Implemented top-down by the state, these institutions not only altered existing land use practices, but also undermined socially embedded institutions that regulated mutual cooperation between households. However, such transformations were not only instigated by state intervention. Social and political changes on the local level, as well as regional economic dynamics, also impacted resource use and land tenure.

Today, Thaba Nchu's rural communities are best described as post-agrarian societies characterised by low agricultural production, diversification of non-farm incomes and declining population numbers. In addition, the region experienced mass layoffs in formerly typical occupation spheres with the end of apartheid. Thus, households are currently faced with very few opportunities to engage in wage employment. How households try to maintain their livelihoods under these conditions is therefore another critical question here. This question refers, on the one hand, to the diversification of livelihood strategies and, on the other, to forms of cooperation and mutual support.

1.2 Theoretical Approaches to Human-Environment Interactions

In the following subchapters, I will situate the study within a broader theoretical context. As the first step, I will introduce the concept of social-ecological systems, which constitutes the analytical framework guiding the book. After giving a summary of the central assumptions of the concept, I will outline some criticism of the social-ecological system approach, which was directed *inter alia* against the concept's insufficient inclusion of historical and political conditions. In the second step, I argue for an integration of the approaches of political ecology and environmental history in the analysis of social-ecological systems in order to avoid some of the pitfalls addressed by academic critics. Finally, as the third step, I will elaborate upon the main theoretical foci of analysis of social-ecological dynamics, namely processes of coupling and decoupling, institutional change and social resilience.

1.2.1 The Analytical Framework: Social-Ecological Systems

Contemporary scholars from the social and natural sciences (for instance Anderies et al. 2004; Walker and Salt 2006; Berkes, et al. 2008; Chapin, III, et al. 2009; Crane 2010; Bollig 2012; Glaser et al. 2012a; Widlok et al. 2012; Leslie and McCabe 2013; Linstädter et al. 2016) understand human-environment relations in terms of coupled social-ecological systems in which, in simple terms, people depend on natural resources and ecosystem services while the environment is influenced by human activities (Chapin, III, et al. 2009:6). The interdisciplinary nature of the concept makes it a 'boundary object' (Becker 2012:38) that helps to connect research from various disciplines. Social-ecological systems are understood as complex adaptive systems (Holland 1992) that are 'composed of many heterogeneous components that interact with each other in parallel' (Abel and Stepp 2003:n. pag.). Accordingly, the systems are not characterised by linearity, as is typical for equilibrium systems, but rather show a high degree of non-linear and unpredictable dynamics (Glaser et al. 2012b:7–8). The components of social-ecological systems encompass bio-physical (e.g. climate, vegetation, fauna, soil and geology), socio-political (e.g. institutions, markets, policies and national economies) and cultural dimensions (e.g. discourses, identities, norms, knowledge and behaviour). These different elements of a social-ecological system are coupled across various spatial and temporal scales with mutual feedbacks existing among and between the elements and scales. For instance, policies decided at the national level influence land use practices of resource users at the regional level, which in turn may have long-term impacts on the resource base. A study that looks through the lens of a social-ecological system approach must identify the temporal and the spatial scales which are of most interest (Widlok et al. 2012:270). For analytical purposes, the size and boundary of a social-ecological system is determined by the objectives and aims of the study (Chapin, III, et al. 2009:9) or, in other words, by 'the problem at hand' (Leslie and McCabe 2013:116).

The rather arbitrary determination of boundaries and scales of interest proposed by social-ecological systems scholars is problematic. In fact, in its empirical praxis social-ecological research tends to focus on the local scale only, e.g. land use in a certain catchment area, and defines boundaries along merely geographical lines (Widgren 2012:104). However, neither ecosystems nor social systems have 'clear cut [sic] boundaries' (Moran 1990:22). Still, I will not solve the 'boundary problem' (Leslie and McCabe 2013:116), as a social-ecological system approach requires the definition of boundaries. Spatially, the book focusses on the Thaba Nchu area that was part of the former homeland Bophuthatswana (Figure 1). Environmental policies, laws and institutions implemented in this area differed significantly from those applied to the adjoining white-owned farmland. Hence, from an analytical point of view, this boundary definition makes sense for the thematic focus of this dissertation. Yet, the problem at hand – how state regulation affected the human-environment relationship in this region – also requires the consideration of influences that may have evolved outside the system's boundaries. Of course, these include not only government interventions, but also issues, such as droughts or animal diseases, that were not restricted to the Thaba Nchu area and affected the region as a whole.

Even if a definition of system boundaries is meaningful from an analytical perspective, it does not preclude the possibility that the unit of analysis may be highly heterogeneous. Whereas ecological systems may be characterised by 'patchiness' (Moran 1990:23), social systems are stratified and shaped by differences in terms of power, gender, class and ethnicity (Widgren 2012:104). Thus, they constitute 'social fields' in the Bourdieusian sense, in which actors have different dispositions, are endowed with various types of capital and struggle to achieve or maintain a dominant position within the social field (Papilloud 2003:31–38; Schwingel

2009:82–102). According to the critique, the heterogeneous social positions of actors are often ignored by social-ecological studies, as they tend to emphasise shared norms and strategies of groups rather than the different interests of various actor groups (Widgren 2012:104). Another point of criticism is that of lack of historical depth. Despite the acknowledgment of temporal cross-scale linkages, social-ecological system research often has a narrow focus on present conditions (ibid. 103). Although some recent studies take a historical perspective on social-ecological dynamics (e.g. Costanza et al. 2007; Li and Li 2012; Widlok et al. 2012; Butzer 2012; Bollig 2013; Nayak 2014; Rotarangi and Stephenson 2014), a leading journal of the social-ecological system school of thought, *Ecology and Society*, has so far published only a small number of historical case studies (Ommer and Paterson 2014:n. pag.). However, a study that seeks to understand why the properties of a social-ecological system are the way they are must look beyond recent processes and explain the past dynamics that shaped the system's identity.

Despite the justified criticism – I will offer some possible solutions to the problems highlighted in the next section – the social-ecological systems approach provides a meaningful theoretical framework for this dissertation. First, this study emerged in an interdisciplinary context that seeks to explain the dynamics of human-environment relationships. In such contexts, the search for a boundary object is necessary in order to find a common basis, a starting point for the analysis. The concept of coupled social-ecological systems is an appropriate tool for the 'study of the manifold [...] interdependencies among natural and social processes' (Becker 2012:44). Moreover, the emphasis on coupling and complex dynamics avoids an oversimplistic conception of human-environment relations. Second, the existence of linkages between various temporal and spatial scales is a proposition of the social-ecological system approach. Hence, the definition of the system boundaries for the interdisciplinary research does not prohibit *per se* the consideration of past and present processes at the regional, national and global level.

Politicising the Analytical Framework: Approaches of Environmental History and Political Ecology

In contrast to many social-ecological system scholars, some others have emphasised the importance of the historical and political context for an understanding of past and present human-environment interactions. Particularly significant are studies published in the tradition of 'political ecology' and the overlapping field of 'environmental history'. Neither field is characterised by a single coherent theory or methodology (Carruthers 2003:4; Robbins 2012:5); both can rather be described as 'interdisciplinary arenas' (Dovers 2000:132; Carruthers 2003:4). Environmental history emerged as a new strand of American and French historiography in the 1970s (Worster 1988:290–91). In the analysis of accounts written by settlers, missionaries, colonial administrators and other actors, the long-term perspective of ecological change became the centre of attention. Environmental histories contributed *inter alia* to a new understanding of apparently stable ecosystems by showing that they had been subject to continual fluctuations and transformations in the past. Moreover, they challenged studies that offered oversimplistic explanations for recent ecological change (Neumann 2005:52–53; Robbins 2012:66–67). With a focus on sub-Saharan Africa, a growing body of case studies (for instance Beinart 1984, 1989; Carruthers 1989; Grove 1989; Showers 1989; Jacobs 1999; Potts 2000; McCann 2001; Anderson 2002; Carswell 2003; Jacobs 2003b, 2003a) and essay collections (for instance Anderson and Grove 1987; McCann 1999; Beinart and McGregor 2003; Dovers et al. 2003; Beinart and Hughes 2007; Beinart 2008) emerged. A common feature of these is that they reconstruct how 'western' ideas of conservation, soil erosion and resource degradation shaped colonial politics. Moreover, they deconstruct mainstream narratives that depicted the land use practices of peasants and pastoralists as deteriorative *per se*.

Studies subsumed under the term 'political ecology' emphasise that ecological systems are the 'product of political process[es]' (Robbins 2012:19–20) and explore environmental changes with consideration of power relations and conflicts, stratification and inequalities, and local-global linkages. Comprehensive overviews of the major contributions to this field have been offered by several authors in the past 15 years (e.g. Scoones 1999; Neumann 2005; Biersack 2006; Blaikie 2008; Robbins 2012; for a critical discussion of political ecology see Vayda 2009). Here, I will focus only on one theme of political ecology research, namely environmental governance. This term refers to

> 'the broad range of political, economic, and social structures and processes that shape and constrain actors' behaviour towards the environment. [….] It implies rule creation, institution-building,

> and monitoring and enforcement. But it also implies a soft infrastructure of norms, expectations, and social understandings of acceptable behaviour towards the environment, in processes that engage the participation of a broad range of stakeholders.' (Levy and Newell 2005:2–3)

The overarching theme of this dissertation is the history of 'environmental governance' and its effects on rural livelihoods in the long-term. From a political ecology perspective, it is critical to ask how powerful accounts of unsustainability and degradation were invented, backed up by 'tragedy narratives' (Robbins 2012:191), and used as legitimation for 'conservation', 'development' and 'control' (ibid. 176–98). With regard to the latter, a specialised strand of literature has emerged that explores the effects of the 'betterment' schemes in South African homelands since the 1940s (for instance De Wet 1989; Hendricks 1989; McAllister 1989; Hendricks 1990; De Wet 1995; Fay 2012). These studies deconstruct the rationale for 'betterment' interventions by showing that they were motivated by political concerns. However, the authors mostly focus on the social, economic and political outcomes of the state interventions, while addressing environmental issues only marginally if at all – De Wet's (1995) work being a notable exception in that regard.

The book seeks to integrate environmental history, political ecology and the social-ecological system approach. Thereby, I hope to avoid some of the pitfalls addressed by the criticism elaborated above. By adopting a longitudinal perspective on social-ecological dynamics covering a period of more than a century, the linkages between several temporal scales will become clearer. On the other hand, an emphasis on politics, power relations and social stratification will reflect the heterogeneities of the social system.

1.2.2 Coupling and Decoupling in Social-Ecological Systems

Whereas the term 'coupling' describes the mutual feedbacks that exist between the components of social-ecological systems, the term 'decoupling' is defined by Hoole and Berkes (2010) as the process in which the feedbacks between different subsystems are cut off (ibid. 305). Social-ecological coupling has been conceptualised by several scholars (e.g. Chapin, III, et al. 2009; Stafford Smith et al. 2009; Butzer 2012). Decoupling, on the other hand, is only addressed by a small number of authors and still requires clarification. Hoole and Berkes (2010) illustrate in their study the decoupling of Herero communities from the land that constitutes today's

Etosha National park in Namibia, and argue for a recoupling of the social and ecological subsystems by re-establishing the linkages between them (ibid. 305):

> 'They [the Herero] were forcibly ousted from the area after the Park was established, and have remained alienated from their traditional territory and places inside the Park. We characterise this as a "decoupling" of people from their environment. The empowerment of the Herero through CBNRM and the development of local institutions for community-based conservation offer prospects for recoupling people and the land, that is, restoring connections with the land inside Etosha National Park.' (Hoole and Berkes 2010:304)

The case study by Hoole and Berkes portrays a good example of a complete decoupling process, but also a very obvious one. The Herero people had been removed from their ancestral lands. After the fencing in of the Etosha National Park in 1973, the Herero became entirely detached from their specific environment to which they had become accustomed, as they were denied access to the conservation area and its natural resources. I argue here that social-ecological decoupling can also be a less literal process as it does not necessarily require a physical detachment of humans from their environment. Decoupling not only occurs when legislations explicitly prohibit resource use; it can also take place when people are still living in a certain area, as the case of Thaba Nchu suggests.

Another study that addresses decoupling in social-ecological systems is Li and Li's (2012) article on rangeland management in Mongolia. The authors illustrate that China's interventions regarding the improvement of livestock breeds and pastoral production had negative effects on the social ecology of the rangeland system in Inner Mongolia. Until the 1950s the pastoral system was characterised by little government intervention, low agricultural inputs, diverse breeds that were adapted to the harsh environment, and extensive networks for animal exchange, which enhanced the quality of livestock. With the implementation of a centrally planned economy during China's collective period, pastoral production underwent significant transformations, ideologically underpinned with ideas of modernisation and the need to support the national economy. In practice, the interventions encompassed the restructuring of institutions of resource use, the introduction of new sheep breeds, and the increase of agricultural inputs, e.g. artificial insemination. With the transition towards a market economy after 1983, the pastoral economy experienced further transformations driven by economic incentives aimed at commercialisation. Besides modernisation, the state increasingly argued that ecological deterioration made an intervention necessary (ibid.). As a result, the

> 'philosophy and norms of the herders are increasingly adapting to those of the external world instead of the needs of their local rangeland SES. Consequently, the rangeland SES is transforming from a closely coupled system in which the social system adapted to the natural system, to an approach in which the social system is separating from and trying to conquer the natural system. On the other hand, the decoupling of the natural and social components of the local SES is being accompanied by increased connectivity with the external world, especially in the social sphere, as herders become more fully involved with the market and better adapted to strong governmental interference with local practices.' (Li and Li 2012:n. pag.)

Li and Li's descriptions – both are environmental scientists – of the coupling and decoupling processes are in some parts problematic. They tend to romanticise the pre-collective-period pastoral system in Inner Mongolia by calling it an 'isolated and self-sufficient natural economy' (ibid. n. pag.). The term 'decoupling' is used synonymously with 'increasing external contacts'. Accordingly, most – if not all – pastoralists and agro-pastoralists in the world are decoupled from their environment, since regional, national and global forces are omnipresent in rural areas today. Moreover, their account reproduces the narrative that so-called 'traditional societies' lived in harmony with nature, which is a similar oversimplification to that found in tragedy narratives that describe pre-colonial societies as destructive *per se*. Still, Li and Li show that decoupling processes occur in more subtle ways than in Hoole and Berkes' (2010) case suggested. Comprehensive government interventions were initiated and justified in debates on the modernisation of farming methods and the degradation of natural resources. The new institutions implemented by state authorities undermined existing locally established institutions and eventually resulted in a gradual decoupling of some social and ecological components of the system. As I will show, similar mechanisms were at work in Thaba Nchu. One insight that can be drawn from the case study of Mongolia is that institutions are critical to the coupling and decoupling in social-ecological systems. Thus, the role of institutions will be briefly discussed in the following section.

1.2.3 Institutional Development

Institutions are broadly defined as the 'rules of the game in a society or, more formally [...] the humanly devised constraints that shape human interaction' (North 1990:3). These rules play an important role in social-ecological systems, as they are historically and politically evolved structures that guide humans' engagement with their environments and are thus central to the coupling of the system components. The broad definition of institutions refers to both formalised and informal rules, or, to use Cleaver's (2002) terms, 'bureaucratic' and 'socially embedded' institutions (ibid:13). Whereas the former refers to contracts and rights that are implemented by the state or other authorities, the latter describes arrangements 'based on culture, social organisation and daily practice' (ibid.). However, this differentiation does not imply that these two types of institutions are always clearly distinguishable from one another. Rather, they shape each other, are 'interwoven' (Engels and Moss 2003:368), can be either 'complementary or in conflict with each other' (Widlok 2003:212) and often have a multipurpose character (Cleaver 2002:11).

In this study, I will deal with both socially embedded and bureaucratic institutions that couple the social and ecological components of the Thaba Nchu system. With regard to the institutions based on daily cultural practices, I am particularly interested in two arrangements that have formerly played a central role in rural agro-pastoral production, namely the lending of cattle (*mahisa*) in patron-client relationships and the working groups that cooperated during harvesting (*letsema*). I will attempt to trace the ways in which these institutions have changed over time and whether their purposes are reflected in recent codes of conduct. When dealing with socially embedded institutions, the concept of 'bargaining power' (Knight 1995; Ensminger and Knight 1997) seems useful in order to understand how power asymmetries influence the emergence and transformation of social norms. In short, the concept assumes that institutions are 'a by-product of strategic conflicts' (Ensminger and Knight 1997:4) between actors that differ in terms of resource access and thus do not possess equal bargaining power in conflicts over social norms. For instance, actors who have more resources are less vulnerable to sanctions that follow norm deviation than actors who are marginalised. Thus, the former might be able to challenge established institutions and pursue the creation of further advantageous rules (ibid. 4-5).

The concept of bargaining power does explain some of the transformations that took place in the present case study. However, in a context such as Thaba Nchu, which is characterised by massive top-down implementation of bureaucratic institutions, the idea of bargaining power is not sufficient for an understanding of all of these changes. Thus, the broader political context also has to be considered (Juul 2001:57), and the role and power of the state as an institution-crafting agent must be explored (Moberg 1994; Lund 2006). With regard to bureaucratic institutions, I will focus upon formal acts and proclamations (e.g. the Natives Land Act of 1913 and 'Betterment' Proclamation of 1939) and some of the specific rules included within these regulations, such as the carrying capacity and rotational grazing management of communal rangelands, the differentiation of certain land types (i.e. residential, grazing and arable land) and the culling of livestock species. Strikingly, these top-down-implemented rules have a multipurpose character, as they were officially aimed at optimising land use systems but were in fact an attempt to exercise political control. As a by-product, some of these laws and regulations undermined socially embedded institutions central in rural Thaba Nchu, created a new coupling between social and ecological components of the system or accelerated their decoupling. Each of these transformations will be addressed in the thesis. Moreover, I will examine the endurance of these institutions: what happened to them after the end of apartheid?

1.2.4 Resilience

Social-ecological systems theory is strongly connected with 'resilience thinking', as Walker and Salt (2006) put it. In the last two decades, much has been published on the resilience concept and its application to the issue of sustainable management of ecosystems and social-ecological systems (for instance Holling 1973; Carpenter et al. 2001; Holling and Gunderson 2002; Folke 2006; Walker and Salt 2006; Berkes et al. 2008; Chapin, III, et al. 2009; Folke et al. 2010). It is beyond the scope of this thesis to give a comprehensive overview of the recent debates. Instead, I will briefly describe the origin of the term, its application to the analysis of social-ecological and social systems, and the limitations of the concept as emphasised by social scientists.

The concept of resilience emerged in ecology in the late 1960s and early 1970s (Folke 2006:254). It was defined by Holling as 'a measure of the persistence of systems and of their ability to absorb change and disturbance and still maintain the same relationships between populations or state variables' (Holling 1973:14). Thereafter, many ecologists in the 1980s and 1990s applied resilience as a measure of how fast a system can return to an equilibrium state after disturbance (Allen et al. 2010:5). In later works, Holling called this measure of return 'engineering resilience' and explicitly contrasted it with an 'ecosystem resilience' that 'emphasises conditions far from any equilibrium steady state, where instabilities can flip a system into another regime of behaviour' (Holling 1996 reprinted in Gunderson et al. 2010:53). Thus, ecological resilience is strongly connected with the idea of complex adaptive systems characterised by multiple states, interacting components, non-linearity and unpredictability (Holland 1992).

In the past decades, numerous amendments and redefinitions of the concept of resilience have emerged, and the term has become increasingly used in other disciplines (for an overview see Brand and Jax 2007). Proponents of the social-ecological systems approach have argued that the concept is critical for understanding complex human-environment interactions. In this context resilience is defined as 'the capacity of a social-ecological system to absorb a spectrum of shocks or perturbations and to sustain and develop its fundamental function, structure, identity, and feedbacks through either recovery or reorganisation in a new context' (Chapin, III, et al. 2009:24). Whereas ecosystem resilience emphasised robustness and persistence, social-ecological resilience additionally refers to the system's capacity for adaptation, transformation, self-organisation, learning and innovation (Carpenter et al. 2001; Folke 2006).

The increasing use of the concept of resilience as a 'boundary object' (Brand and Jax 2007:n.pag.) to describe and analyse social-ecological dynamics evoked harsh critique by some ecologists, who argued that the broadening of the descriptive, ecological concept developed by Holling resulted in ambiguity, and thus a loss of analytical value (ibid). Social scientists joined in the critique, particularly after the *Resilience Alliance* (www.resalliance.org), a research network that focusses on the study of human-environment dynamics, suggested a definition of resilience of social-ecological systems (Detlef Müller-Mahn, personal communication). Cannon and Müller-Mahn (2010) argue that the transfer of a concept that originated in ecology to social systems can be misleading and even dangerous because it leads to a depoliticisation of social phenomena. Moreover, failure or collapse in social-ecological systems is easily blamed on humans when thinking in resilience terms that emphasise the system's capacity for self-organisation (ibid. 623, 632). For instance, what about cases where people have no means to act otherwise because of economic or political constraints? People's coping and adaptive capacities depend on their ability to make their own decisions, and are influenced by the degree to which their voices are heard in a wider political arena (Voss 2008:41; Keck and Sakdapolrak 2013:9–10). However, these capacities are not equally distributed within a household, community, region, and so forth. People and households that live within the same community may have different levels of resilience (Cannon 2008:10). Accordingly, the collapse of a system or parts of it might only affect a few, e.g. the most vulnerable.

The criticism of the depoliticised character of resilience is an outcome of the concept's strong connection to ecology and the social-ecological systems framework. Like the latter, the concept of resilience is a boundary object that facilitates interdisciplinary communication. However, from an anthropological perspective the concept is too vague as it focusses on the recovery and reorganisation of system functions, structures, and identities. This is problematic in that, to the extent that it ignores the historical origins of the structures of the system, it tends to ignore the different interests of social actors, and it runs the risk of reducing national and global political-economic processes to the status of mere outside factors. Thus, for the purpose of this dissertation a more specific definition of resilience must be found that considers – as political ecology and environmental history do – past and present power relations and politics.

Since the 2000s, social scientists have used the resilience concept to understand social dynamics. Adger (2000), one of the first authors to suggest a definition of social resilience, characterised it as the 'ability of communities to withstand external shocks to their social infrastructure' (ibid. 361). This rather vague definition has been specified and extended over the past 15 years. Common to all definitions is that they refer to social entities, which may be households, communities or organisations, and their ability 'to tolerate, absorb, cope with and adjust to environmental and social threats of various kinds' (Keck and Sakdapolrak 2013:8).

The resilience of a social entity is reflected in its coping, adaptive and transformative capacities. The first of these refers to actors' short-term responses to disturbances and threats, while the second addresses their capacity to learn from past experiences and to adopt strategies in order to secure future wellbeing. Transformative capacity refers to actors' 'ability to access assets and assistance from the wider socio-political arena [...] to participate in decision-making processes, and to craft institutions that both improve their individual welfare and foster societal robustness towards future crisis' (Keck and Sakdapolrak 2013:11). Whereas the first two capacities deal with rather technical and socio-economic aspects of resilience, it is particularly the concept of transformative capacity that facilitates a stronger inclusion of notions of politics and power. Moreover, the dimension of 'access to assets' acknowledges stratification and an unequal distribution and accumulation of various kinds of capital. Following Bourdieu (1983), capital can be distinguished into three main types: economic, cultural and social. Economic capital includes quantifiable goods that are directly convertible into money and institutionalised in property rights (ibid. 186). Cultural capital refers, *inter alia*, to education and is institutionalised in certificates and academic qualifications (ibid. 190-192).[2] Social capital is determined by an actor's social networks that can be effectively mobilised, as well as by the economic, cultural and social capital possessed by the people to whom the actor is related (ibid. 192-193).

How can social resilience be measured? Leslie and McCabe (2013) argue that a direct measurement of resilience is difficult. Instead, one should focus on surrogates of resilience, i.e. 'attributes [...] that have been identified as enhancing resilience and the adaptive capacity' (ibid. 116). For their comparison of Turkana and Simanjiro pastoralists in Kenya and Tanzania, respectively, they chose several surrogates, including diversity in herd management and the extent of exchange networks (ibid. 126). I will not discuss Leslie and McCabe's choices in detail here, but instead will identify surrogates that are important to the present case study. A first indicator is livelihood diversification. In a post-agrarian society such as Thaba Nchu, diversified livelihood options are presumably more important for households in general than the diversification of pastoral management strategies, which is only relevant to those who keep livestock. A second surrogate, I suggest, is the accumulation of different types of capital; by this, I mean social and cultural capital in addition to economic capital. Assuming that the rural communities are stratified, an important question that must be answered is this: are different kinds of actors endowed with different kinds

---

[2] Bourdieu's (1983) concept of cultural capital refers not only to institutionalised forms, but also to embodied capital (e.g. cultural knowledge) and objectified capital, such as paintings, books and monuments (ibid. 187–90). However, I focus mainly on cultural capital in terms of formal education and qualifications.

of capital? Connected to the aspect of social capital, but more important on the village level, is a third surrogate of resilience: namely, the support networks and institutions of cooperation. Bearing in mind the past state interventions that have been indicated only briefly so far, a pressing question is whether such networks and institutions still exist and, if so, how they have changed over time. By identifying these surrogates, it has become clear that I address resilience first and foremost on two levels: the household and the community. Whereas I have used mainly ethnographic data collected during field research to examine the former level, the latter can be explored with the help of both historical and recent data.

1.3 Outline of the Dissertation

In the next section, I will elaborate on the research context and conditions, my role in the field, and challenges experienced. Moreover, I will illustrate in detail the sampling procedures and data collection methods that were applied during the research process, and address issues of data quality. In Chapter 2, I briefly introduce the region and take the reader on a 'tour' of the research area.

Chapter 3 leads back to the late nineteenth and early twentieth century. The focus of this chapter is on the local level. Based on archival documents and oral histories I will explore the continuities and stabilities in rural livelihoods in one of the three 'native reserves' of the Orange Free State, as well as its existing and growing interconnections with the regional and national level. The historical analysis will highlight the strong coupling of social and ecological components of the system, and emphasise that both stratification and institutions enforcing cooperation were important social structures that determined land use.

Chapter 4 covers, in part, the same time period as Chapter 3, but addresses the government's perspective on human-environment interactions in the Thaba Nchu area. I will elucidate how government officials used ecological concerns – which were at least partly exaggerated – as justification for taking increasing control of land use practices in Thaba Nchu. I will discuss the narratives that emerged in official debates on the one hand, and on the other hand the specific technical planning and implementation of conservation methods on the local level. Additionally, I shed light on the question how the different interest groups in Thaba Nchu reacted to these initial interventions.

Focussing on the mid-twentieth century, Chapter 5 deals with the core of forced transformations in Thaba Nchu's history, namely the resettlement of all rural households. I will elaborate on the political developments on the national and regional levels and elucidate their direct effects on the local households. Historical landscape changes will be visualised by analysing aerial photographs. Moreover, in this chapter I again address the issue of acceptance of intervention, but this time in a more detailed way. Attention is particularly given to the ways in which the people of Thaba Nchu tried to resist the more drastic measures, and to whether their resistance took the form of a unanimous, concerted action.

Chapter 6 then investigates the developments that took place between the foundation of the Bophuthatswana homeland in the 1970s and the official end of the apartheid state in 1994. In this section, it is argued that the social-ecological system had significantly changed since the mid-twentieth century. The new democratic government still intervened in terms of land use planning, but with a new political aim, i.e. the economic independence of the homeland. The chapter will illustrate how Thaba Nchu's suitability for agricultural production in the two decades was reassessed, and what this reassessment meant for the future agricultural development of the region.

Chapter 7 is concerned with post-apartheid South Africa and the question of how the state intervenes in Thaba Nchu today. I will explore the regional achievements of the land reform programme that was initiated in 1994 and elaborate on the new role the state has taken with regard to the regulation of land use patterns. It is argued that previously invented narratives (i.e. overpopulation and resource degradation) still shape debates about the potential development of the region. The chapter includes a critical examination of the main assumptions of these narratives and their validity by analysing population dynamics since the beginning of the twentieth century.

Chapter 8 changes the analytical focus and is concerned with the question of how rural households make a living in a post-agrarian setting. At the core of this chapter is an analysis of ethnographic data collected in 2010 and 2011. After illustrating new trends in demographic patterns and their possible causes, I describe the structure of households in the research sites today. Emphasis will be placed on household economy, socio-economic stratification and the various kinds of livelihood strategies that people from different wealth strata pursue. Moreover, I will address institutionalised forms of cooperation and mutual support and show how these have changed since the twentieth century.

In the last chapter of the dissertation I link the historical and ethnographic chapters with the theory (Chapter 9). Based on the specific case of rural Thaba Nchu, I will discuss what conclusions can be drawn regarding the applied concepts, and answer the leading research questions in the theory chapter.

1.4 Approaching the Field Methodologically

The study is embedded in the interdisciplinary DFG-funded research project *Resilience, Collapse and Reorganisation in Social-Ecological-Systems of East- and South Africa's Savannahs* (RCR) which focusses on two research sites in Kenya (Lake Naivasha and Lake Baringo) and two in South Africa (Kuruman and Thaba Nchu). During its first phase (2010-2013), the RCR project was divided in nine sub-projects, each of them having its own thematic focus. My study was part of sub-project A1, a cooperation between the Departments of Soil Science of the Universities of Bonn and the Free State and the Department of Social Anthropology of the University of Cologne. The major aim of the sub-project was to assess the sensitivity and extent of soil responses to different land use regimes in the semi-arid grasslands and savannahs of South Africa. Commercial farms, communal areas, and farms allocated in the course of the post-apartheid land reform programme were identified as land use regimes of interest. While the soil scientists worked in both Thaba Nchu and Kuruman, my research concentrated only on the former region. On the one hand, this was necessary because the holistic approach of social anthropology generally requires a longer research stay and the application of a variety of qualitative and quantitative methods. Research questions are open to change and are often only specified during the field research. Moreover, as the study often touches upon sensitive issues the researcher needs to build rapport (Bernard 2006:368). On the other hand, little was known about human-environment relations in the Thaba Nchu region. Although Colin Murray (1992) has written an excellent historical account of the political economy of the area, his work addressed social-ecological aspects only marginally. In contrast, the environmental history of Kuruman had already been elaborated in detail by Nancy Jacobs (Jacobs 2001, 2003b, 2003a).

1.4.1 Selection of the Field Sites

I joined the RCR project a few weeks after an initial kick-off trip to the South African research regions had been conducted by project colleagues.

Hence, the specific research sites – i.e. the commercial and land-reform farms as well as the communal villages – had been chosen before I became part of RCR. In hindsight, the communities that were part of my 'prescribed' sample were certainly not the worst choice, given my own research focus, for several reasons. First, there were two old communities that had been part of a 'native reserve' established in the late 19th century, on which I found rich material in the national archives in Pretoria. Second, the other two communities were relatively new and only developed during the 1960s. I therefore had research sites with very different histories. Third, one community was situated very close to Thaba Nchu town, and thus had a more peri-urban character than its counterparts. The only negative aspect of the site selection was that all four research sites were located in the northern part of Thaba Nchu. The southern parts are characterised by another soil and vegetation type and therefore from the natural scientists' perspective were not comparable with the northern part. From an anthropological point of view, the selection of at least one southern community would have been interesting in order to investigate whether the ecological differences might result in varying land management strategies. However, an additional site would have been too much, as four communities with more than 2000 inhabitants already constituted a large unit of study.

The commercial and the land-reform farms were not the main focus of my research. After the first field trip, it became clear that a direct comparison of small-scale land users in the communal areas and large-scale, commercial farmers was not reasonable, with regard to either land use strategies or household economy. I nevertheless conducted some interviews on these privately-owned farms as the information was necessary for the project counterparts. This enabled me to gain a broader picture of the region.

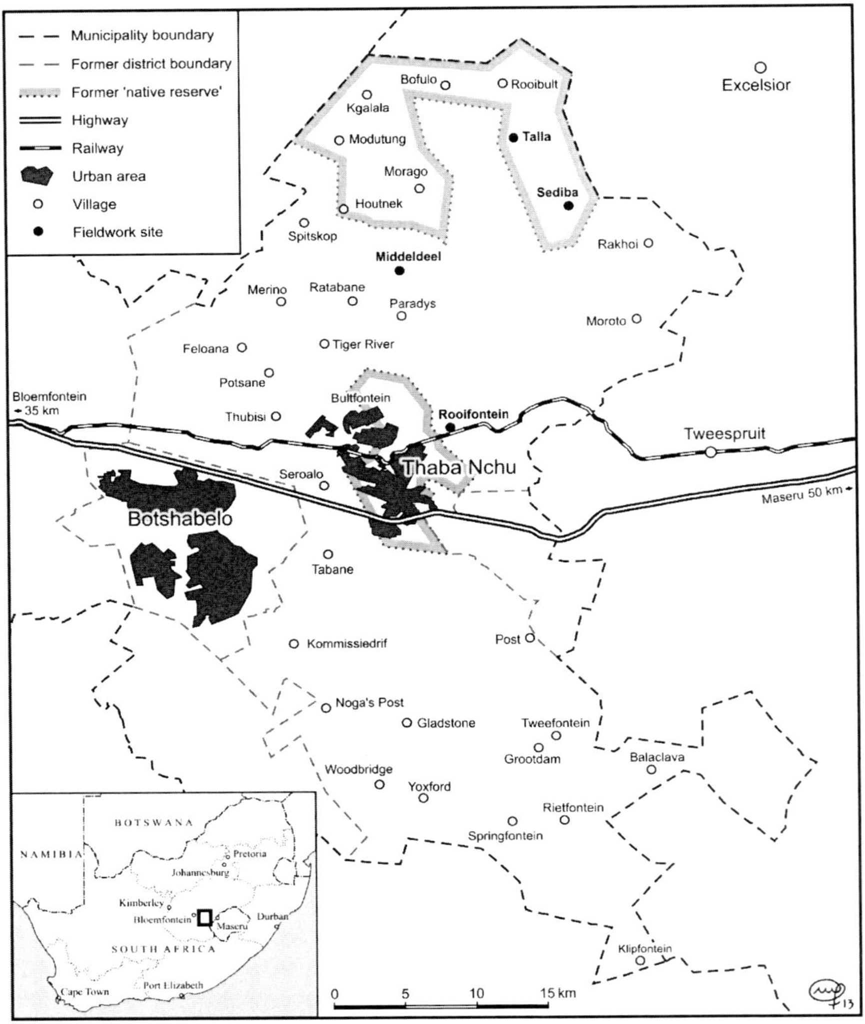

Figure 1 The research area (source: Naumann 2014)

In the course of the research it became apparent that it was not possible to spend the same amount of time in each of the four communities and to study all sites with the same intensity. The re-construction of the local histories, for instance, was time-intensive; not only because some interviews required more than one session, but also because it was necessary to talk to several people in one village in order to get a better idea of the heterogeneous perceptions, recollections and representations of certain events, e.g. the resettlement in the 1960s. Different families had different experiences and interests. To understand the historic dynamics and varieties in one community it was necessary to elucidate the history from different perspectives and standpoints. Thus, to find a balance between the varieties in the research field and the depth of the research topic, I decided with regard to certain aspects to concentrate on only two of the four communities. One of these was the village of Sediba, where I lived for seven months and gained the deepest insights and best contacts. The other community was Middeldeel, which is comparatively small and has a different history. In the following, I will briefly sketch the main phases of my research, reflect on my personal role in the field, and remark on some of the challenges I faced during the fieldwork.

1.4.2 Reflection on the Research Process

The research process was divided into two phases: an initial three-month period from mid-August 2010 until early November 2010 and a second nine-month period from mid-February 2011 until the end of November 2011. The first research stay had an explorative character, as I used the time to introduce myself and the proposed research to the residents of the four communities, government officials and other stakeholders. In the first days after my arrival, I talked to social anthropologists, agro-economists, and soil and vegetation scientists of the University of the Free State (UFS) who had worked in Thaba Nchu before. Many of them had been involved in the introduction and evaluation of an *Infield Rainwater-Harvesting* project (IFRH) in the area and were able to provide me with valuable information on the region. Moreover, I talked to officers of the Department of Agriculture and Rural Development in Bloemfontein and to the extension office in Thaba Nchu. The employers of the latter helped me to establish contact with the four research sites, namely Rooifontein, Sediba, Talla and Middeldeel. The officers introduced me to the respective headmen and arranged a meeting in each community, at which my assistant and I explained the research project. These meetings, which were attended by between twenty and fifty people, often developed in the same direction, as the following extracts from my field notes will illustrate:

> After we had introduced the research project, I suggested that people could ask questions. One of the women said that I must bring fodder for their animals as it was too dry and they would starve. I explained that I was part of a research project and that there was no money for drought relief. Then, a man asked whether I would bring jobs for them. I answered that I would need an assistant from the village for the second research phase, but that I could not offer additional jobs. Another man raised his hand and requested that I go from house to house and ask people about their problems. Moreover, he asked whether I could help with the compilation of a business plan. A third man noted that I should at least bring some money to the village when I finish my studies; whereupon the headman asked whether they could themselves decide what to do with the money or whether I would dictate the usage (field notes, 10 September 2010).

> Before I tried to gather some information on the village, I encouraged people to ask us questions. A woman said I should show them how to cultivate a field. I told her that I was not an agricultural expert. Then, she answered, I should at least bring the money for a plough. I explained again that the research project has no money for ploughs. After my assistant translated this, the woman just looked at the floor and did not talk anymore [...] When we were about to leave, some young men approached us and asked for money; first me, then my assistant. We both claimed that we had no money, and I started the car engine. One of the youngsters shouted that I should remember to bring KFC [Kentucky Fried Chicken] food for them next time (field notes, 8 September 2010).

Although I always tried to make clear that I was a PhD student from Germany, people regarded me as either government official or development worker; and because of that, they had clear expectations of how such meetings should proceed. Indeed, official meetings with representatives from departments followed a certain script, as I later observed. The compilation of wish lists for development needs was a standard method applied. In one case, I observed that government representatives brought a Kentucky Fried Chicken meal for each participant. However, in the long term the misconceptions with regard to my role were cleared up. In contrast to officials or NGOs I paid regular visits, came back after a longer absence in Germany, and finally – in the second field phase – even moved to a village. By that point at least, I was no longer perceived as a government official.

Data collection in the first research phase remained rather unsystematic. When visiting, I normally went to the headman's house and then did what Bernard (2006) calls 'hanging out' (ibid:368), i.e. spending time with people in order to build rapport. As my place of residence during this first research stay was Bloemfontein I had to drive between 1.5 and 2 hours to reach the villages. This was feasible in the exploratory phase but exhausting in the long term. In addition to these exploratory field visits, most of the archival data analysed in this dissertation were collected during my first research stay in 2010. In total, I spent about one month in the archives, one week in the Free State archives in Bloemfontein and three weeks in the national archives in Pretoria.

After spending three and a half months in Germany, during which time I took a look at the first collection of archival data and planned the second research phase, I returned to South Africa in February 2011. During the first weeks, I again lived in Bloemfontein, where I also attended a Setswana course, though for only eight hours. This was of course not enough to learn the language appropriately. I was able to make a bit of small talk, and I knew terms that were central to my research – e.g. 'cattle' (Setswana: *kgomo*) and 'fields' (Setswana *masimo*) – but I was unable to conduct longer conversations or interviews by myself. Some informants were fluent in English and Afrikaans – a language that I learned during my Masters' research in Namibia – and they switched to a language that I could understand. However, generally I was dependent on interpreters throughout the whole research process. Data collection during the second field phase was more systematic than in the first exploratory stay. Besides quantitative data that were gathered through mappings, rankings and surveys, I collected qualitative data on various topics. Moreover, I conducted a final period of archival research in the North West provincial archives in Mmabatho, where documents of the former Bophuthatswana homeland are filed. An overview of the entire research process is provided in Table 1.

After about two weeks in Bloemfontein I moved to a commercial farm east of the Thaba Nchu area. I could thereby reduce the driving time to the field. However, I still remained a visitor who would arrive at a village in the morning and leave in the evening. The first step of my second research stay was the search for local assistants in each village. The headmen of the communities helped me and suggested young people with sufficient English skills. Whereas the headmen of Sediba, Talla and Middeldeel chose women between 24 and 28 years of age, in Rooifontein I worked with a 36-year-old man. Employing local residents as field assistants had advantages and disadvantages. The headmen and villagers appreciated that I paid people from the village and thus gave something back to the community. The assistants had good knowledge of the households and often provided me with background information that shed new light on the interviews. Yet this closeness was sometimes also a restrictive factor, particularly when dealing with sensitive questions such as the household's monthly income. Moreover, the assistants were not trained interpreters, and the translation during narrative interviews was sometimes difficult, particularly in the beginning.

While I was staying at the commercial farm, I tried to find suitable accommodation in one of the research villages. Things developed the best in Sediba, mainly due to the efforts of my assistant, who was very communicative and did an invaluable job in introducing me to community members. After deciding that I would like to stay in Sediba, I talked to the headman, who immediately agreed. Before I could move in, a community meeting was held and my plans discussed with the residents. Nobody raised objections, and in May 2011 I moved to a small house in the middle of Sediba. Being a part of the village brought about a breakthrough in terms of my being accepted by the villagers, and helped to build trust. Residents of Sediba wanted me to participate in their daily activities and invitations to ancestral feasts, funerals and general visits were extended regularly. My daily presence in the village also generated better data. This is visible for instance in data on income, which are generally more reliable in Sediba than in the other villages. A further change to my role in the field took place when my partner visited me for about six weeks in June and July. Suddenly, the men were more open and even visited us in my house. Although it was clear that their attention was mainly directed to him, it nevertheless gave me more access to the 'male domain'. In interviews men were more communicative when my partner was present, even though I was the main conversation leader. The situation did not change completely after he left, but certainly the men were less open when only I and my female assistants conducted the interviews.

At the end of November 2011, I left Thaba Nchu and South Africa, but my ties to the field did not end with my departure. I visited South Africa again for a couple of weeks between July and September 2012. Although I stayed for most of the time in Potchefstroom in the North West province, I spent a weekend in my former research site. This 'coming back', though only for a short time, was a very important and positive aspect of my visit. On the one hand, it was evident – and also expressed by village inhabitants – that people were glad that I had not forgotten them. On the other hand, the short visit enabled me to gain some insights into recent developments, as a land claim made by the residents of the former Sediba reserve had just been settled.

1.4.3 Data Collection and Quality

In the following subchapters, I discuss in more detail the methods applied during the research and the quality (and weaknesses) of the data collected. Moreover, I will illustrate how informants in the field sites were selected.

Participant Observation, Informal and Narrative Interviews

Participant observation, the basis of social anthropological fieldwork (Bernard 2006:342), was an essential method that was applied during the whole research process, but whose real strength developed only when I moved to Sediba. Instead of talking only to selected informants and asking questions that were related to the research topic, participation meant spending time with people without being too focussed on specific aspects. Rather I tried to be open and to see what the day might bring. I often visited households spontaneously, took part when funerals (Setswana: *phitlho*), ancestral ceremonies (Setswana: *mpho ya badimo*, literally 'gift for the ancestors') and other celebrations (Setswana: *mokete*) were prepared, attended church services, or joined women when they did their washing or renovated the house. My language skills were a limitation to participant observation as I could only roughly understand the topics of conversations, but not the details. Thus, in most cases my assistants had to accompany me.

Table 1 Timetable of field research process and the applied methods

Month	Village Level	Household Level	Selected Informants	Miscellaneous
08/10				Literature research UFS Interviews with UFS researchers Commercial farms: Semi-structured interviews land use (n=2)
09/10	Community meetings and informal visits/interviews (all villages)			Interviews with officers of Department of Agriculture Land reform farms: Semi-structured interviews on land use (n=2) Archival research: Free State Repository (Bloemfontein)
10/10	Informal visits/interviews (all villages)		Oral History Village Elders (1 group interview; Sediba)	Archival research: National Archives Repository (Pretoria)
11/10–01/11	Germany Analysis of first data and planning of second research phase			
02/11				Language course UFS
03/11	Village mapping (all villages)			
04/11	Wealth rankings of sample of 70 households per village (13 interviews)			Development and translation of a census questionnaire
05/11		Ethnographic census and land use survey (n=45)		
06/11			Qualitative interviews on land use (n=10)	Transect walks
07/11	GPS mapping of rangeland and former fields (all villages)	Assistant: Land Use Monitoring (n=19)	Continued: Qualitative interviews on Land Use	Development land use monitoring questionnaire
08/11		Continued/Assistant: Land Use Monitoring	Oral History Interviews (n=21)	
09/11		Continued/Assistant: Land Use Monitoring	Continued: Oral History Qualitative Interviews on ecology and environmental change (n=6)	Development of ecological knowledge questionnaire Semi-structured interviews 'Excelsior-farms' owners (n=2)
10/11	Assistants: Livestock Census (Middeldeel, Rooifontein, Talla)	Continued/Assistant: Land Use Monitoring	Ecological knowledge survey (n=34)	Archival research: North West provincial archives (Mafikeng)
11/11		Continued/Assistant: Land Use Monitoring	Continued: Ecological knowledge survey Continued: Oral History Ranking of grasses and appearance (n=2)	Land reform farms: Semi-structured interviews on land use (n=1)

Informal interviews – conversations without a clear structure (Bernard 2006:209) – were another essential method that was applied throughout the research process. Observations and conversations, as well as methodological issues, were recorded in the daily field notes (DeWalt and DeWalt 2002:141). In addition to informal conversations I also used more structured techniques, conducting open-ended and semi-structured interviews (Schensul et al. 1999:121–48; Bernard 2006:210–50; de Munck 2009:139–58). Topics ranged from animal husbandry and gardening to ecological knowledge and perceptions of environmental change. Moreover, the collection of oral histories was a central part of the qualitative approach. Here, interviews focussed on life histories, past land use practices and household economy, as well as governmental interventions and their connected effects. All qualitative interviews were conducted with residents of Sediba and Middeldeel. To identify key informants for these topics I asked several individuals in the village: 'Imagine you want to know something about … [history of this village, animal husbandry, etc]; whom would you ask?' Normally, this generated a list of several names of possible informants. I interviewed those who were most frequently mentioned, and at the end of the interviews asked each of them a similar question. Narrative interviews were generally recorded, in cases where the informant was not against it. The most important and informative interviews were transcribed, many of them directly in the field. Others I listened to and partially transcribed back in Germany.

Archival Sources

I photographed a large number of historical sources stored in the archives in Bloemfontein, Mmabatho and particularly in Pretoria. Documents and reports there cover the period from the late nineteenth century until 1990.[3] Most important for the analysis were the minutes of the Sediba and Thaba Nchu 'native reserve' boards and the files of the Department of Native Affairs – later the Department of Bantu Administration and Development – both of which are located in the national archives in Pretoria. Unfortunately, I found only a few documents for the Bophuthatswana time period. The North West provincial archives had catalogued only small parts of

[3] Public records of governmental departments are normally transferred after 20 years to an archives repository. Hence, for the last two decades I do not have any historical documents.

their sources at the time of research. Moreover, some documents had not yet been transferred to the archives.

It is unnecessary to stress that the official sources are ideologically biased. Hence, the documents must be interpreted against the background of apartheid reasoning. However, there are some other characteristics of the historical sources that must be considered when writing a thesis that tries to reconstruct narratives of past human-environment relations in rural South Africa. For instance, the documents that I analysed in this study almost completely neglected women and their role in agriculture and the household economy. This 'female invisibility' (Bradford 1996:351) is not only true for primary sources[4] but was, for a long time, typical in African historiography (ibid.). Yet in 'their daily lives, through their families and their communities, in ritual and belief, in their travels, their struggles, and their travails, African women, as historical subjects, were active agents in the making of the colonial world' (Allman et al. 2002:1). Thus, a substantive, historical analysis must take into account gender biases in order to avoid what Bradford calls a 'sexual apartheid' (Bradford 1996:352).

The historical documents are biased with regard not only to gender but also to class and ethnicity. Some officials mentioned the stratification of rural societies. However, this was merely used as an argument for undermining the economic strength of the wealthier households but not paying attention to the lives of the poorest. Households without rights to land or livestock remained unheard. The case of ethnicity and ethnohistory is a bit different. Colonial officials addressed these issues and sometimes social anthropologists, such as P. L. Breutz, were commissioned to give an evaluation. However, in general the conceptualisations of ethnicity remained simplistic and essentialist (see also Chapter 2.1).

Another aspect complicating the historical analysis is that the sources vary considerably with regard to detail. This is evident for instance in the minutes of meetings of the reserve boards, which were compiled by the Native Commissioner. Some colonial officials wrote only brief comments, others pages-long records of the things discussed at such meetings. Moreover, one detects differences in these officials' abilities and interests. While some were more knowledgeable and concerned about environmental issues, others ignored ecological issues almost completely. This makes

---

[4] However, Bradford stresses that nineteenth-century accounts of the Cape and the Frontier did represent women as significant actors Bradford (1996:369).

it sometimes difficult to get a detailed picture of environmental dynamics over time. Still, the limitations of the archival sources can be compensated for by comparing them with other data. In the case of this dissertation these were gathered through narrative interviews of elders, transect walks and aerial photographs.

Exploring Household Economies and Land-Based Activities: Mapping, Ranking, Surveying and Monitoring

The topic of household economies was investigated with a range of methods that built on one another. To choose a number of households that were to be interviewed about their economic and land use activities, I first had to assess how many people and households lived in the research sites. This was done by a village mapping. Equipped with a Google Earth map (http://www.google.com/earth/), on which I could identify plots and houses, and a GPS device, my assistants and I went from homestead to homestead. Besides marking a 'waypoint' and assigning each household an identification number, I recorded preliminary information, e.g. name of the household head, number of members, duration of residency and whether they owned livestock or did gardening. In cases where we did not meet a household member, the information was provided by the assistant. During the mapping, it turned out that several houses on the map were empty. The inhabitants had moved to Thaba Nchu and visited their homes only sometimes, or had even left for good. I thus made a differentiation between households that were predominantly absent or present, i.e. that had their main place of residence in the village.

After collecting the basic information on about 700 households, I prepared a wealth ranking for each community. Information on appropriate sample sizes for ranking tasks varies. Whereas some scholars think that a ranking of 100-150 households is unproblematic (Grandin 1994), others suggest that samples should be smaller (Pretty et al. 1995). I decided to limit the sample size for each village to 70 households that I selected randomly from the mapping data. This proved to be the maximum manageable number, particularly for the older informants who sometimes needed up to three hours to complete the ranking task. In total, I conducted 13 individual interviews: four in Sediba and three in Middeldeel, Talla and Rooifontein respectively. In the interviews people were asked to sort the households according to their wealth. Typically, between 3 and 5 groups were made. Then, for each household a wealth score was calculated by dividing the ordinal number of the group by the total number of groups piled by the informants. The result was multiplied by 100; all the scores for each household were then added together, and finally divided by the number of ranking interviews done for the respective village (see also Grandin 1994:28). I then divided the households into four wealth groups per village, which I called 'rich', 'average', 'poor' and 'poorest'. Of each group, I randomly selected two to three households per community; these made up the sample for the ethnographic census (total sample: n=45; Sediba: n=13; Middeldeel: n=13; Rooifontein: n=11; Talla: n=8).

The census consisted of two parts. The first resembled a classical ethnographic census questionnaire (Lang and Pauli 2002). It included information on each household member, i.e. name, gender, relation to the household head, age, place of birth, ethnic group, religion, marital status, education, occupation, and membership in associations. Moreover, for each woman that had given birth to a child, a detailed birth history was recorded in order to allow an analysis of fertility dynamics. The second part of the questionnaire explored the household's economic and land use activities in more detail. The questionnaire was translated into Setswana. The interviews were mostly conducted by my assistants, but I was present in all cases in order to clarify ambiguities. The most issues raised in the questionnaire were unproblematic. The only exceptions were questions on income and child mortality (see also Chapter 8). A databank was created with Microsoft Access into which all data from the ethnographic census were entered. Data analysis was conducted with Microsoft Excel and SPSS.

As a next step, the land use activities of some of the 45 households were monitored for a period of four months (July-October) in order to assess these activities systematically. The monitoring questionnaires were developed from the findings from my observations, the ethnographic survey and semi-structured interviews. In addition to questions about the households' expenses and incomes from animal husbandry and gardening, the time spent on other land use activities, such as hunting and fishing or the collection of edible plants, dung and firewood, was included. For logistical reasons, I decided to interview only members of the households in Sediba and Middeldeel (n=19). The monitoring was conducted at the end of each month by my two assistants. In addition to the monthly monitoring, a livestock census was conducted in Talla, Rooifontein and Middeldeel. In Sediba my colleague from the Institute of Resource Economics of the University of Bonn did a complete assessment of all cattle-owning households.

Ecological Knowledge Survey

Before I started with the systematic collection of information about the ecological knowledge of residents of the rural communities, I conducted transect walks with elders who showed me some of the local plant species and soil types. After additional semi-structured interviews, I developed a semi-structured questionnaire to gather data about local ecological knowledge and perceptions of environmental change. The first part consisted of several free listing tasks (Bernard 2006:301–05; de Munck 2009:47–66) on various plant species, e.g. grasses, trees, medicinal plants, wild food plants, as well as on different soil types. The second part consisted of open-ended questions aimed to capture villagers' perceptions of rangeland and soil conditions and changes in vegetation, soil and climate, as well as possible reasons they perceived for any degradation of the natural resources. Informants for the ecological knowledge survey were selected through ethnographic sampling in which informants with different characteristics are chosen in order to explore the variety of answers to the research question (Werner and Bernard 1994; Schnegg and Lang 2008:20). For the study of ecological knowledge, I was particularly interested in the impact that informants' sex and age might have on the information given. In total, 34 people from Sediba and Middeldeel were interviewed. I also conducted interviews on certain grass species and their abundance (see also Bollig and Schulte 1999) with two informants who had proven to be knowledgeable in the survey. For these interviews, the names of the most common grass species were written on cards and presented to the informants. First, they were asked to rank the grasses according to what species cattle liked the most. In a second step, informants were asked to indicate for each species whether it had undergone changes in abundance patterns over time, i.e. whether the grasses increased, decreased or were indifferent compared to the past.

Analysing Land Use Change Spatially

Changes in the landscape and historical land use patterns were explored with the help of aerial photography provided by National Geo-Spatial Information (www.ngi.gov.za) and satellite images provided by Google Earth. Aerial photographs of the research area – or parts of it – are available for 1950, 1953, 1956, 1961, 1968, 1975, 1988, 1992, 1999, 2004 and 2007. The rich database indicates that the state had a huge interest in gathering precise information on land types and land use patterns which was necessary for intervention. In particular, before the resettlement of rural households that took place in the mid-1960s, frequent flights were conducted over the area. In order to interpret the spatial images correctly, I conducted several transect walks with informants on the ground. The locations of former settlements, fields and grazing camps were recorded with a GPS device. GPS waypoints were later imported into Google Earth, and maps of former land use patterns were compiled for Sediba, Talla and Middeldeel (see Appendix 11.1, 11.2, and 11.3). In the case of Rooifontein, the assessment was difficult, as the village residents were at odds with each other as to where the exact boundaries of the communal grazing areas and the former fields were located. As I found no surveyor documents of the Rooifontein area in the archives, I could not obtain these data from historical sources either.

2. ENTERING THE FIELD

2.1 The Research Area

The Thaba Nchu area constitutes the eastern part of the Mangaung metropolitan municipality, which is located in the Motheo district in the Free State province. Bloemfontein, the sixth largest city in South Africa and the capital of the Free State, is the economic hub of the metropolitan municipality. Between the province capital and Thaba Nchu lies the huge township of Botshabelo, which is the third urban area of the Mangaung municipality. In total, the administrative unit covers an area of 6,863 km². Bloemfontein is connected with other South African provinces by three highways (N1, N6 and N8), an east-west and south-north railway line, and a small national airport (Mangaung Metropolitan Municipality 2012:7).

According to the national census of 2011, 747,431 people were living in the Mangaung metropolitan municipality. Taking into account the last national census of 2001, when about 645,441 inhabitants were counted in Mangaung, this implies an average annual growth rate of about 1.5%, the second highest in the whole province (Statistics South Africa 2012a:48). In total, 231,921 households lived in the municipality in 2011, of which about 40.8% were female-headed (Statistics South Africa 2011). The Thaba Nchu region covers an area of approximately 1,290 km², encompassing an urban area that consists of several townships and a rural area with 37 villages. The total population of the whole area is 82,265. About 10,000 people are living in the rural communities.[5] The majority of the people in the Thaba Nchu area are Sesotho-speaking (Coplan 2010). At first glance this is astonishing, since the area was part of Bophuthatswana, the homeland of the Batswana people. However, historical sources indicate that already in the 1950s high numbers of Basotho had immigrated into the area, as it is very close to Basutoland, now Lesotho (Murray 1992:176). Still, the region became part of Bophuthatswana in the 1970s, which reveals the arbitrariness and superficiality of apartheid planners. Moreover, its designation as a 'Barolong area' ignores the fact that many Batswana who live there do not consider themselves to be Barolong. In my ethnographic census, slightly more than 10% of the people stated that they were

---

[5] Data are calculated from Frith (2013a) who provides the 2011 census data for the village level on his website.

Barolong. The biggest group of the people I interviewed for the survey called themselves Digoja (42%), a Setswana-speaking group that is related to the Barolong.[6] As the region is more heterogeneous than the common official accounts might suggest, I will in most parts of this dissertation refer to 'the people of Thaba Nchu' or 'the residents', or use other ethnic-neutral terms.

The research area belongs to the eastern part of the Highveld, with an altitude of about 1,400 – 1,600 m above sea level. Situated in the summer rainfall area, most precipitation occurs between October and April, with a peak between January and March. The mean annual precipitation is about 560 mm (Mucina and Rutherford 2006:382). Mean summer temperatures for Bloemfontein range from 15°C to 31°C. Winters are cold, with a mean temperature of -2°C to 16°C and about 60 days of frost. The biome is categorised as the Central Free State Grassland Biome whose landscape is characterised by undulating plains dominated by *Themeda triandra* when the vegetation is in good condition (Mucina and Rutherford 2006:381). On rangelands with a moderate condition *Eragrotis curvula* is dominant (ibid.), whereas in poor condition mainly *Cynodon dactylon* or *Aristida congesta* are found (Kotzé et al. 2013:222). The soils of the research sites showed no obvious signs of wind or water erosion, were clayey and classified as Lixisol (ibid. 221-222).

2.2 Thaba Nchu and the Research Sites: An Explorative Tour

When leaving Bloemfontein on highway N8 to the east, the Black Mountain rises on the horizon. Before approaching the lively town of Thaba Nchu, the road passes commercial farms, a few recently redistributed land-reform farms, and Botshabelo, a large township that was founded in the late 1970s to absorb the so-called 'surplus people' – the apartheid term for people evicted from white-owned farms, informal settlements in urban areas, or 'black spots' that had to be removed in order to consolidate 'homelands'. After passing Botshabelo, one enters the area that belonged to the former Bophuthatswana homeland. Before the main turnoff to Thaba Nchu

[6] Only scant information about the Digoja is found in the literature. Usually they are described as a 'historical group' that belonged to the first Batswana groups who migrated to Southern Africa. They settled in the area south-west of today's Lesotho. After their chief was killed in 1820 they scattered and affiliated with other Batswana groups (Lye and Murray 1980:26; Breutz 1989:4).

town is reached, one can leave the highway and drive to Selosesha. The township was developed on the drawing board in the 1960s. Selosesha was to be the new home for non-farming people who were expected to leave the rural villages during the 'betterment' (Chapter 5.4.3). However, landless families of the villages were not removed to the township because soon after its planning it was settled by numerous households engaged in wage labour in Bloemfontein. Today the main streets of Selosesha appear to belong to the 'better-off parts' of Thaba Nchu and are characterised by larger brick houses whose patios and garages testify to a certain wealth of the inhabitants.

The centre of Thaba Nchu consists of one main shopping street and a few smaller streets with more shops. Besides major chains such as the supermarket Shoprite, fast food restaurant Kentucky Fried Chicken, clothes store Ackermans and several banks, people can also buy at informal vegetable stalls or small shops run by Chinese or Bangladeshi immigrants. In the middle of the shopping street is a taxi rank where minibuses leave to Botshabelo and Bloemfontein every minute. Besides the Interstate bus line, they are the major means of transport for the inhabitants of the region. Apart from Sundays, the centre of Thaba Nchu is generally crowded. The peak of bustling activity, however, is during the first days of a new month, when people in Thaba Nchu town and the surrounding villages have just received their salaries and social benefits. Then, the streets are packed with people; women and children carrying large bags of maize meal and other groceries, street traders selling watches, clothes and other goods, and stores playing loud music to catch shoppers' attention. The monthly shopping of the rural population is normally done in the big supermarkets of Thaba Nchu town. But each research site had also at least one or two smaller shops (Figure 2) where people could purchase groceries, sanitary products and smaller household goods, albeit at comparatively high prices. Besides shops, transport facilities and a hospital, Thaba Nchu is home to several government offices, including the extension office of the Department of Agriculture and Rural Development.

Not far from the shopping streets are the premises of the Barolong Traditional Council. Here the chief (*kgosi*), his council and 42 headmen and headwomen (*dikgosana*) meet for their weekly assemblies. Today, the traditional council's main task is the allocation of land in the rural area and the resolution of conflicts. They also advise the Mangaung metropolitan municipality, but have no decision-making power in the municipality (Esterhuyse 2012:97). The Thaba Nchu area is divided into several wards: the three research sites in the north, namely Sediba, Talla and Middeldeel, belong to Ward 41, while Rooifontein belongs to Ward 43. Each ward is represented by a ward committee headed by a councillor who acts as a chairperson. These committees have ten additional members who live in the ward and represent the communities. Like the traditional council, ward committees have an advisory function in the municipality (Planact 2001:41).

Figure 2 Shop owner in her shop in Sediba (photograph: N. Dreber)

After passing the offices of the traditional council, the road leads to the northern rural area where my research was concentrated. Rooifontein is the first community reached after a few minutes' drive on the gravel road. Before entering the village one crosses a railway track, which connects Bloemfontein with Tweespruit. Whereas in the past passenger trains also stopped in Thaba Nchu, today only goods trains are passing. Rooifontein is strictly speaking not a village as since 1995 it counts as part of 'Thaba Nchu Urban' area and is a rather peri-urban settlement. It is the biggest community I studied. In 2011, I counted 265 households and 1,025 people. The settlement differs from the other villages in terms of infrastructure. For instance, it is the only research site where several taxi services exist. Whereas the residents in Sediba, Talla and Middeldeel depend on the Interstate buses that come a few times per day, Rooifontein inhabitants can make use of the minibuses that commute more or less every hour – perhaps even several times per hour – between Thaba Nchu and the peri-urban settlement. In contrast to the other villages, Rooifontein has overhead streetlights. The poles for such lights had also been built in Sediba, Talla and Middeldeel, even before I started my research in the area. However, by the time that I left, the municipality had not finished the construction and the villages were still pitch-dark after sunset. The streetlights are just one example that shows that in comparison to the other research sites Rooifontein is advantaged in terms of access to means of development. Housing is another example.

> **Excursion 1 – Residential Sites and Housing**: In all research sites, streets and residential plots, locally called 'stands', are laid out in a grid pattern, a structure that was established during the betterment schemes. When the villages were planned, the residential plots were of equal size, which could vary from community to community, but in the four research sites normally comprised about 0.2 ha. However, due to population growth several plots had to be subdivided in the past. Additionally, some new homesteads have emerged at the fringes of the villages. The younger informal village extensions have not yet been surveyed. As a consequence, these households cannot apply to have their homes connected to the local electricity network, or to have a pit toilet installed. Village inhabitants do not own their plots but possess a so-called PTO, a 'permission to occupy', which is a relic of the apartheid system. The PTOs were not common to all former 'homeland' areas, and in others they were abandoned after 1994 (Lahiff 2000:19). In Thaba Nchu, these permissions are registered at the office of the traditional council in town. In 2011 the applicant had to pay a fee of R250 for the registration.[7] However, not all families can afford to pay the money. Thus, strictly speaking they have no right to live on the plot. This can cause conflicts when the actual owner of the respective plot wants to remove the residing household. Possession of a PTO is necessary in order to apply for housing programmes, such as the Reconstruction and Development Programme (RDP).[8] Of the 11

---

[7] During my field research, the average currency rate was 1 Euro to 10 Rand.

[8] The Reconstruction and Development Programme is an integrative development programme that aims first and foremost at poverty alleviation and the reorganisation of the economic system. Part of the RDP strategy was the fulfilment of basic

> Rooifontein households that participated in the ethnographic census three had one of these popular RDP houses; all of those households were defined in the wealth ranking as 'poor' or 'poorest'. Rooifontein was the only community in my research sites where a couple of RDP houses had been built. With a few exceptions post-apartheid housing schemes focussed mainly on the semi-urban areas of Thaba Nchu, which often aroused the resentment of people living in the rural communities.[9] RDP houses are standardised in size and have one large room, which is often divided into a kitchen and a living room, as well as two bedrooms. Other non-subsidised houses vary significantly. One can find huge brick houses with a kitchen, living room, several sleeping rooms, and sometimes a dining room and a patio. On the other hand, there are medium-sized houses made of bricks or clay that have a kitchen and at least two other rooms. And finally, the poorest households stay in small clay houses with a total floor area of $20m^2$ at most, or in very simple corrugated iron shacks.

After passing Rooifontein the road leads towards the so-called Excelsior farms, which were purchased by the South African Development Trust (SADT) and incorporated into the 'homeland' in the 1980s (Chapter 6.3.2). During Bophuthatswana times, these farms were leased to individual farmers who used them for mixed agriculture. After 1994 the farms were sold by the state, in some cases to the same people who had used the land before. When passing the private land, one can see a mountain ridge on the horizon. Behind the ridge lies the next research site, Sediba (Setswana for 'fountain'). The village belongs to the oldest rural communities in the Thaba Nchu area and was established in the early 1880s. Sediba (Appendix 11.2), together with five other settlements, was part of the Sediba 'native reserve' whose socio-political history is illustrated in detail in the following chapters.

Today, Sediba actually consists of two villages of different sizes. The smaller one is called 'Sediba Scheme' – or *skema* as local residents say –

---

needs, i.e. decent housing and water and electricity supply (Lodge 2003:54; Schmidt 2004:283–86).

[9] This might change in the future, however. Shortly before I left the research area, people in the smaller villages could also apply for RDP houses. In 2012, when I made a short visit to the villages again, no subsidised house had been built or was about to be built.

the history of which goes back to an irrigation scheme that was built in the 1930s (Chapter 4.4.2). The larger part of Sediba has two sections subsumed under the name 'Sediba Trust', whose inhabitants were not allowed to use the irrigation scheme. Being the second-largest research site, Sediba had about 194 households and 539 inhabitants in 2011. Sediba is the only research site that has a small clinic, which is open from Monday to Friday and led by a nurse. Minor physical symptoms can be treated here. Moreover, people have access to medication and contraceptives and can have an HIV test. While Rooifontein inhabitants go to larger clinics in Thaba Nchu, the residents of Talla and Middeldeel have to wait for a mobile clinic which comes about once a month, or else go to the other clinics in the neighbouring villages. In severe cases, people call an ambulance.

Sediba has a Thusong Service Centre which was created in the course of a government programme initiated in 1999 for the improvement of the communication between government departments and rural communities. Such service centres exist in rural communities throughout the country. The main aim of the centre is to 'bring government information and services closer to the people to promote access to opportunities as a basis for improved livelihoods' (Government Communication and Information System (GCIS) n.d.). Officially, several government bodies – namely the Department of Home Affairs, South African Social Security Agency, Department of Agriculture and Department of Labour – have regular office hours, and people can apply for identity documents, birth and death certificates, social benefits and unemployment insurance funds, and get information on farming matters. In the first years after its construction the office still functioned quite well, but when I lived in Sediba in 2011 the service centre did not do justice to its name. It was only used for paying the monthly social benefits and as meeting place for smaller gatherings with government authorities. People were at odds over whether the situation was due to the absence of government officers in Sediba, or to rural residents no longer using the services. According to my assistant this was mainly because administrative formalities were often accomplished faster in the Thaba Nchu offices, hence people would go there when they had an urgent issue. Besides the service centre, Sediba has a hall where community meetings are held in irregular intervals.

Excursion 2 – Water and Sanitation: In theory, the inhabitants of all four research sites have access to water free of charge.[10] Within the residential areas, public taps are situated at several street corners where mainly women and children fetch water with 20-litre containers and wheelbarrows. Moreover, 38% of the households in the ethnographic survey (n=45)[11] have a tap on their plot. A few exceptional cases (7%) had a tap inside the house. In practice, however, many villages struggle constantly with a reliable water supply. In Sediba for instance, for several months it was a rare event when the public taps yielded water at all. On the better days, water was at least available during night; on worse days, one would get up at three o'clock in the morning only to find that not even a single drop was available. There are various possible reasons for supply problems, and village residents or officials often did not know what had caused the scarcity. In some cases, the problem was only temporary, e.g. when a pipeline was broken. In other cases, e.g. in Sediba, the water pressure seemed to be too low in general, and consequently the large water tanks in the village were only filled for a few hours during night, when other villages upstream did not use any water. When no water is available the Mangaung metropolitan municipality had to be called, who in turn would assign Bloem Water to erect plastic tanks in the village. However, it normally took several days or even weeks before the 'water-supply-machinery' was under way. In the worst cases, residents had to approach the few households that had a hand pump on their plot and purchase water from them for about R2 per 20-litre canister. Like the water supply, sanitation is in some cases problematic. Most households have a pit latrine on the plot (89%) and some have even a water toilet in the house (7%). Latrines were not of good quality in all the villages, and I heard about several cases where toilets had collapsed as they had not been built on solid foundations.

When leaving Sediba to the north, the gravel road passes by some ruins that belonged to the farm Boston. Where today only ruins are found, the

---

[10] In the past, a chip-card system existed with which people had to pay for their water supply.

[11] All statements in percentages are based on the data of the ethnographic survey, and thus refer to 45 households.

Bophuthatswana government once had an office of the Department of Agriculture. This was also the place where livestock that had been confiscated for illegal grazing was impounded (Chapter 6.4.3). Not far from the farm Boston is the Sediti High school, which is attended by most of the teenagers from the surrounding villages. Sediti has gained regional notoriety as it is one of the schools with the weakest performance in the Free State province, and in 2012 had a matric pass rate of less than 40 % (van Wyk 2013). Whereas children have to walk or drive to the secondary school, primary schools and crèches exist in all four research sites.

Turning right behind the high school, one can drive directly to Talla, the third research site (Appendix 11.1). Talla, a name probably derived from the Setswana word *tala,* meaning 'uncultivated', was also part of the Sediba reserve, and is thus among the oldest villages in the area. It consists of a larger section which was planned during 'betterment', and an adjoining smaller section that developed in the last two decades. Talla is a bit smaller than Sediba and in 2011 had about 150 households and 462 inhabitants. Talla has no community hall, and consequently all meetings are held in the headman's garden or the village crèche.

> **Excursion 3 – Electricity and Heating**: The electricity supply is generally stable. Only when severe storms damage power lines do longer blackouts – lasting for one or two days – occur. Most households have electricity (91%). The only exceptions are those in the informal extensions and households that are not able to afford to have their houses connected to the electricity network (at a cost of approximately R250). In contrast to water, electricity is not free, but must be purchased in the form of pre-paid vouchers.[12] Households that have access to electricity normally possess electrical devices, such as a radio (93%), TV (93%) and fridge (90%). As the winters in the Thaba Nchu grassland are very cold, heating is essential. An expensive option is to use electrical heaters, which are only used by some of the wealthy households; yet they do not rely on these sources alone. A cheaper option is to use paraffin cookers, which can be used for cooking food and water and at the same time warm up the cold rooms in winter. They are used by both wealthier and poorer households. The cheapest option is to collect dry cow dung, which is burnt in old

[12] Kilowatt-hour prices ranged between 0.78 Cents and R1.50 in 2013/2014. Poor households receive 50 kWh each month free of charge (Davis et al. 2008).

> iron bins. The use of wood is less important as the grassland biome offers insufficient forest resources. Yet some poorer households go the mountain ridges now and then, collect wood, and carry it home. Wood can also be purchased when the *Eucalyptus* spp. trees that were planted in each village during the 'betterment' are cut.

To reach the last research site, Middeldeel (Appendix 11.3), one has to drive back in the direction of Sediti high school and then take a turn-off to the right. The gravel road runs southwards through the south-eastern part of the horseshoe-shaped Sediba reserve and then crosses the mountain ridge again. Directly behind the ridge the rangelands of Middeldeel begin. Immediately one recognises a major difference compared to the other villages. In 2011, Middeldeel was the only research site that had fenced-off grazing camps, which were established in the course of the sheep and wool project initiated in the village in the mid-2000s. Like Rooifontein, Middeldeel came into being in the 1960s, when planned settlements were erected on land that had previously been privately used (Chapter 5.4.1). Thus, Rooifontein and Middeldeel are considerably younger than Sediba and Talla. Of my research sites, Middeldeel was the smallest, with 93 households and about 351 people. The village has a shearing shelter, which was also constructed in the course of the sheep and wool project, and which is also used for the community meetings. The assemblies in Middeldeel are conducted each week on a fixed day, and it sometimes happens that people even meet twice a week if something urgent has to be discussed. Thus, communication in Middeldeel is stronger than in other communities. If one stays on the road that passes Middeldeel's rangelands one finally reaches the houses of the Bultfontein townships. Several rural households that refused to resettle during 'betterment' were forcibly removed to these townships (Chapter 5.4.3). After this brief explorative tour, I turn to the historical analysis, beginning in the nineteenth century, when rural Thaba Nchu was still dominated by agro-pastoral production.

3. CONTINUITIES AND CHANGES: LAND USE IN PRE-BETTERMENT TIMES

In this chapter, I provide a detailed analysis of the Thaba Nchu social-ecological system prior to the enforcement of the 'betterment' schemes. The analysis starts with a brief overview of the history of Thaba Nchu in the nineteenth century, from which time the area underwent successive phases of change. At first, these transformations were initiated by political conflicts at the local level. Later, in the early twentieth century, the changes were increasingly driven by legislation made on the national scale (Chapter 3.2). Despite these changes, rural households maintained the agricultural production methods and institutions that coupled the social system with natural resources. In Chapter 3.3 I will illustrate the social and economic organisation of the communities by focussing on the example of the Sediba 'native reserve' (Figure 1). The chapter closes with a section addressing the question of how people coped with environmental shocks in the past.

3.1 Settlement of Thaba Nchu and Disputes in the 19th Century

In 1833, the Barolong boo Seleka and their chief, Moroka, settled together with other Batswana groups in the Thaba Nchu region, which at that time formally belonged to the territory of the Basotho king Moshoeshoe. In the first decades after their arrival at Thaba Nchu the Barolong lived in a large, centralised settlement (Trollope 1878:275), which was a common pattern of territorial organisation in Batswana groups (Schapera and Comaroff 1984:35). The Barolong were agro-pastoralists: they kept cattle and sheep at outposts in the hinterland where young men took care of the animals. Additionally, they cultivated fields with sorghum, maize, wheat, beans, pumpkin, melon and sugar cane. The fields were not necessarily in the immediate vicinity of the central settlement, but close enough to allow the cultivators to keep them in view (Schüler 1965:325).

When the English writer Anthony Trollope travelled to South Africa in 1877, he wrote of Thaba Nchu that it

> 'belongs to one of the most interesting places in South Africa. Thaba 'Ncho is a native town in which live about 6,000 persons of the Baralong tribe under a Chief of their own and in accordance with their own laws. [...] and in the vicinity of Thaba 'Ncho there are other Baralong towns or villages, – within the distance

> of a few miles, – containing from five hundred to a thousand inhabitants each. They possess altogether a territory extending about 35 miles across each way, and within this area fifteen thousand Natives are located.' (Trollope 1878:275–76)

> 'But it seems to me that so large a number of people, living on such an extent of land, which of course is not closely cultivated, may be subject at any time to famine.' (Trollope 1878:284)

Trollope's assumption was seemingly correct. The residents of Thaba Nchu never cultivated enough land to provide food for the whole population, and always depended on wheat imports from Basutoland (Wales 1985:217). Apart from the trading of wheat, the relation between Barolong and Basotho was rather tense, and Thaba Nchu was repeatedly the target of cattle raids. Only after the second Sotho-Boer war (1865-68), when the Basotho territory east of Thaba Nchu was annexed to the Orange Free State Republic, was the tense situation defused (Wales 1985:212–33).

Changing population and settlement patterns in the 1870s marked a significant transformation process in the social-ecological system. First, massive emigration movements, especially of political opponents of Chief Moroka, led to a general decrease in population numbers in the Thaba Nchu area. Second, more and more households left the centralised settlement of Thaba Nchu town and dispersed to rural outposts (Wales 1985:234–45). Clearly, the removal of the threat of Basotho attacks enabled Thaba Nchu people for the first time to erect smaller settlements at a considerable distance from the central town. The reasons for these urban-rural migration movements were mainly of an economic and political nature. Crop failures and a two-year drought made it necessary for the population to disperse and bring larger areas under cultivation. Besides, some people tried to escape the rather arbitrary political style of Chief Moroka (ibid).

Besides changing population and settlement patterns, the second half of the nineteenth century was also characterised by economic changes. A central turn in the region's history came with the discovery of diamonds in the area of today's Kimberley in 1869 and the opening of the gold mines at the Witwatersrand in the 1880s. The quickly developing extractive industry brought about profound economic transformations for both white and black people in Southern Africa. The diamond fields in Kimberley in particular caused changes in the Orange Free State and Basutoland (McCann

1999:160). The emergence of new markets increased the demand for agricultural products. Basutoland for instance exported between 200,000 and 400,000 bags of cereals and wool with a total worth of £75,000 in 1879 (Keegan 1987:11). Not only white farmers of the Orange Free State but also black farmers of Thaba Nchu benefitted from this high export rate, as they gained income from transporting goods across the republic (Keegan 1987:11; Murray 1992:12). Moreover, the diamond fields and gold mines offered new job opportunities in addition to the seasonal farm labour on white-owned land (Wales 1985:248–49).

The death of Chief Moroka in April 1880 sparked a dispute over legitimate succession. Both Tshipinare, an adopted son of Moroka, and Samuel Lehulere Moroka, the youngest son of Moroka's first wife, claimed to be the rightful successors. Both men had a significant proportion of the population behind them. I will not go into detail about the question of inheritance as this has been done by others (see for instance Wales 1985:264–71; Murray 1992:21–25).[13] However, the clashes deserve some attention, as awareness of them is important for a better understanding of the conflicts that emerged in the twentieth century. In June 1880, President Brand of the Orange Free State recognised Tshipinare as Moroka's legitimate successor. Tshipinare was pro-Boer and signed an alliance treaty with the Free State (Watson 1980:364). Samuel and his followers, who were not willing to recognise the new chief, were expelled from Thaba Nchu (Murray 1992:25). Some of Samuel's supporters, many of them not Barolong (Wales 1985:266), remained in the area and settled at a place that later became part of the Sediba reserve. One village elder recollects the origin of the settlements in the Sediba area (Figure 12) in the following way:

> AT: 'Sediba, it's just one village but differs in nations.'
>
> CN: 'Different groups?'
>
> AT: 'Ja, different groups, because in Pathla are Digoja, [...] in Lehurutse are Bahurutshe, [...] in Digojeng are Mosotho under headman Mogalatladi. It was one nation, one tribe of Chief Moroka.'
>
> CN: 'Who came first here? The Digoja?'

---

[13] The question of who was in fact the legitimate successor could not be answered with certainty by either Murray (1992) or Wales (1985).

> AT: 'Okay, let me tell you: Digoja, Bahurutshe, the Ndebele […] they were all one tribe of Chief Moroka. But when they depart from Thaba Nchu to this place, their chief is Samuel. He was [later] arrested by the government after he killed his brother.'
>
> CN: 'Mmh, Tshipinare.'
>
> AT: 'Ja, they send him to exile … So, the government of the Barolong they took his [Samuel's] nation and put them here to Sediba. They said 'this is your place, you stay here', yes, under the headmen Mogalatladi.'
>
> CN: 'So, the people here are connected with Samuel?'
>
> AT: 'Yes, yes, yes Samuel Moroka!' (interview with Andrew Thepe, Sediba, 18 August 2011)

As the quote indicates, the conflict resulted in an attack, and the murder of Tshipinare by Samuel and his supporters in 1884. Before his death, however, Tshipinare had initiated a land survey of the area in 1882 and distributed land titles to his followers. One can only speculate about his motives for doing so, but apparently Tshipinare wanted to both strengthen the administrative power of his political subordinates (i.e. headmen) and secure the land to prevent it from alienation (Murray 1992:28). After Tshipinare's death, Orange Free State President Brand send his troops to Thaba Nchu and declared the annexation of the territory (316,450 ha) in 1884. Robert Tawana, the son of Tshipinare, succeeded his father 'to the now essentially meaningless position of leader of the Thaba Nchu community' (Watson 1980:371). Samuel Lehulere and his main supporters went into exile to Basutoland (ibid. 32) and later, in the late 1880s, to Bechuanaland (Wales 1985:292, 299).

After the annexation of Thaba Nchu, the land titles of the 152 farms were called into question by the Orange Free State government (Murray 1992:33–37). Only 95 farms were granted to black farmers, first and foremost to those who were followers and relatives of the deceased Tshipinare. Land had been valued at 10 shillings per 0.86 ha, and landowners had to redeem their titles at a rate of 1% of the valuation. Additionally, they had to pay the dues for the official surveying of the land. For most landowners, the high costs could only be paid by selling large numbers of livestock or by raising a mortgage (ibid. 40-45). The remaining private land was allocated to white farmers, the Wesleyan and Anglican missions, and the Orange Free State government. Moreover, two 'native reserves' were set aside for black people. The smaller of the two was the Thaba Nchu reserve (6,889 ha), consisting of the town itself; the larger, with about 15,150 ha, was the Sediba reserve (ibid: 37). Within the latter were situated six villages – Sediba, Talla, Rooibult, Bofulo, Morago and Kgalala – each consisting of multiple smaller, scattered settlements. Three decades later both reserves were defined as 'scheduled areas' in the course of the Native Land Act of 1913 (see below). From the late 1930s onwards, the South African Native Trust (SANT) successively purchased land in the Thaba Nchu district and piece by piece consolidated the two reserves.

In summary, during the second half of the nineteenth century the Thaba Nchu social-ecological system underwent crucial transformations. Within a decade, settlement patterns became decentralised, which brought about both socio-political and, especially, economic changes, since larger areas were opened for cultivation. Still, the Thaba Nchu people remained dependent on wheat imports to buffer food shortages. Moreover, the annexation of the territory to the Orange Free State and the establishment of two reserves marked the starting point of interventions by the white government.

3.2 Segregation Legislations until the 1940s

After the founding of the South African Union in 1910, the new government aimed to standardise the various approaches to 'native' administration that had emerged in the pre-existing colonies, i.e. the Cape Colony, Natal, Orange River Colony (ORC)[14] and Transvaal. The Department of Native Affairs was established, and was responsible for all issues related to the administration of black South Africans. In the first decades after its origin, the Union legislated several acts that enforced the segregation of blacks and whites in terms of territory, administration and jurisdiction. In the following, I will briefly present the most important acts and their relevance for Thaba Nchu. Then, I will focus on the 'native reserve' boards which formed the local administrative body for Thaba Nchu and Sediba

[14] After the Second Boer War (1899-1902), the Orange Free State was annexed by the British Empire in 1902, and was thereafter called the Orange River Colony.

reserve. Both boards had been established during the ORC but continued to exist in the South African Union.

3.2.1 Territorial Segregation: The Land Acts of 1913 and 1936

The Natives Land Act of 1913[15] was the 'initial legislative enactment embodying the principle of territorial segregation and separation of land rights' (Union of South Africa 1955:44). The act prohibited the squatting of black people on white-owned farms and the widespread system of shared cropping, which is explained in more detail below. Furthermore, the act aimed at the establishment of so-called 'scheduled areas' to be set aside for black South Africans. While black people were not allowed to buy or hire land outside the 'scheduled areas', whites were not allowed to purchase or lease land within these areas. Provision was made for the appointment of a commission, the Beaumont Commission, which submitted its report in 1916. There the authors identified possible 'scheduled areas' as well as 'additional areas' that were to be added in the long-term (Union of South Africa 1916a). Both the Sediba and Thaba Nchu reserves were declared 'scheduled areas' and the black-owned freehold farms 'additional areas' (ibid. 37-40). Although the titles of the landowners were not directly threatened, the act confronted them with a problem, as it impeded the sale of their farms for two reasons. First, land owners were, according to the Free State law, only allowed to transfer land to black South Africans if they were blood relatives. However, many of the latter could not afford to pay the price. Second, land owners could not sell their farms to whites as the land was provisioned for black occupation in the future (Murray 1992:98). Thus, the farms were basically non-marketable.

In the post-apartheid land reform programme, the Natives Land Act of 1913 marks a critical point, as only land dispossessions that occurred after the passing of the act are considered for restitution. However, in a recent examination of the act's impact, Beinart and Delius (2014) argue that the law itself had only limited effects on the ground. Processes commonly ascribed to the 1913 act, such as the destruction of sharecropping, were in fact caused by other later modifications and legislations (ibid. 668). Moreover, an emphasis upon the 1913 legislation ignores the fact that many dispossessions had taken place before the implementation of the legislation

---

[15] Union of South Africa, Act No. 27 of 1913: Natives Land Act.

and that many additional dispossessions happened much later, in the mid-twentieth century, and were not directly linked to the act.

In 1936 the Natives Trust and Land Act[16] was the second legislation that aimed not only to promote territorial segregation but also to increase state control over communal land. The legislation established the SANT, a fiducial body controlled by the Department of Native Affairs. The Trust was responsible for the purchase and administration of so-called 'released areas', which were to be added to the 'scheduled areas'. For the purchase of the additional land, the South African government set aside £1 million per year for a decade (Davenport 1979: 337). In Thaba Nchu, the newly defined 'released areas' were considerably smaller than the previously targeted 'additional areas' (Murray 1992:129).

3.2.2 Building the Pillars of Indirect Rule

In the pre-Union colonies, the recognition of chiefs and headmen was not handled uniformly (Union of South Africa 1955:66). This changed with the passage of the Native Administration Act in 1927.[17] The act provided the major legal cornerstone of indirect rule, i.e. 'the creation of a separate but subordinate state structure for natives' (Mamdani 1996:62). The governor-general, the highest-ranking member of the colonial administration and a representative of the British monarchy, was declared the 'supreme chief' of all 'natives' in the South African Union, and endowed with despotic powers. *Inter alia* he was entitled to rule 'by decree', i.e. by announcement in the government gazette; to appoint native commissioners as well as additional and assistant native commissioners; to appoint and depose chiefs and headmen, with whom he could vest powers of civil and criminal jurisdiction (Mamdani 1996:71). By inaugurating separate 'native courts' judged by chiefs and headmen according to 'customary law', the jurisdiction of the 'native areas' was segregated from that of the rest of the country (Jacobs 2003a:150).

When the Native Administration Act was introduced in Thaba Nchu, the chieftainship was in 'effective abeyance' (Murray 1992:147). Robert Tawana, who had succeeded Tshipinare, died in 1912. Tawana's son, Albert Setlogelo Moroka, was recognised as his legitimate successor by the South

[16] Union of South Africa, Act No. 18 of 1936: Native Trust and Land Act.
[17] Union of South Africa, Act No. 38 of 1927: Native Administration Act.

African Union government. However, as he was still a minor, he was represented by his uncle, John Pethogane Moroka. A couple of years later Albert Setlogelo Moroka was convicted of stock theft, and the then magistrate decided – without the approval of the South African Union – that his imprisonment would bar him from holding the position as chief. John Pethogane Moroka continued as acting chief. In 1930, he was appointed as chief in the Thaba Nchu reserve with a stipend of £24 per year.[18] However, his jurisdiction did not include the Sediba reserve (Murray 1992:172). Only in 1939 was he additionally appointed as chief in the Sediba reserve and endowed with civil jurisdiction to 'hear and determine claims arising out of Native Law and Custom brought before him by Natives against Natives'.[19] The local headmen in the villages, too, received an annual allowance, which was also raised when their services were 'satisfactory'.[20] In contrast to the inherited position of the chief, Thaba Nchu headmen were elected by the men of the respective community. However, the Native Affairs Department (NAD) had the final say in the appointment and dismissal of headmen.

3.2.3 'Native Reserve' Boards in Sediba and Thaba Nchu

Under Ordinance No. 6 of 1907 of the Orange River Colony, two separate reserve boards were established for the Thaba Nchu reserve (in 1907) and the Sediba reserve (in 1909) as local administration.[21] Both bodies persisted after the constitution of the Union of South Africa and were only discontinued with the Bantu Authorities Act in 1951 (Chapter 5.2). The 'native reserve' boards held regular meetings at which residential certificates for newcomers were allocated or refused, new regulations announced and current problems in the reserve and its locations discussed. The board had authority over some aspects of land use, which are illustrated below. The administrative powers of the reserve boards were nevertheless limited

---

[18] SAB NTS 308 13/54, Thaba Nchu Chieftainship, Report of the committee appointed to enquire into the question of the succession to the chieftainship of the Barolong tribe, 14 March 1941.

[19] SAB URU 1793 1315, Appointment of Chief J. P. Moroka of the Thaba Nchu reserve as chief of the Sediba reserve, 19 April 1939.

[20] SAB NTS 308 10/54, Secretary of Native Affairs, Pretoria, to the Magistrate, Thaba Nchu, 17 May 1926.

[21] Orange River Colony, Ordinance No. 6 of 1907: Native Reserves Management Ordinance.

as the NAD had the final say in all important decisions (Murray 1992:120). The boards consisted of a white chairman – who was initially the local magistrate and later the Native Commissioner – a white vice-chairman, and between five and seven nominated, paid members from the local communities. In the first years of its existence the chairmanship of the boards was rather steady, with most chairmen being appointed for at least five years. However, from the mid-1930s onwards, chairmen were replaced more frequently.

Despite their similar administrative structures, the Thaba Nchu and Sediba reserves differed considerably with regard to socio-economic, ecological and political issues. The semi-urban Thaba Nchu reserve had a comparatively elaborated infrastructure with several so-called eating houses and trading sites, as well as specific places for rubbish disposal and public latrines. The Sediba reserve, in contrast, was rather rural. In 1925, the local magistrate stated that 'there is not a shop on the whole Reserve and the occupation of the natives is purely agricultural and pastoral.'[22] The political distinction of the Thaba Nchu reserve was reflected in the fact that several board members were representatives of the educated, landowning Barolong elite. One of them was Dr James Sebe Moroka.[23] The wealthiest black person in the district, James Moroka had earned a medical degree at the University of Edinburgh, Scotland, and ran his own practice in Thaba Nchu (Murray 1992:103). Between 1949 and 1952 Moroka was president of the African National Congress (ANC). Other influential leaders were, for instance, the Wesleyan reverend Joel D. Goronyane – who was one of the first members of the reserve board – and Walton Zacharias Fenyang, who was the son of Tshipinare's eldest daughter and a central figure in the political establishment of Thaba Nchu (Murray 1992:107). Being strongly represented in the board of the Thaba Nchu reserve, the Barolong elite promoted *inter alia* the interests of the landowning class. As I will show in Chapter 4, they played a significant role in the introduction of the 'betterment' schemes.

With the establishment of the reserve boards, the ORC passed a set of regulations which were subsequently modified, repealed and supplemented.

[22] SAB NTS 8152 22/341, Magistrate T. Kilpin, Thaba Nchu, to the Secretary of Justice, 17 July 1925.

[23] Murray (1992) gives an excellent account on the Barolong elite, their political power and their interests with regard to the land question.

Additional regulations were drafted by the boards and – after being approved by the NAD – gazetted. The initial regulations set up in 1909 for the Sediba reserve referred predominantly to the payment of taxes and the governing of the resources. Particularly important were the following rules[24]:

(1) Each male resident between 16 and 60 was expected to pay a location tax of 2 shillings and sixpence.[25]

(2) Whenever under the Scab Ordinance of 1903 and its amendments sheep or goats had to be dipped, the dipping was carried out and directed by the chairman of the board or his deputy.

(3) The cutting or damaging of any tree or bush growing in the reserve was prohibited.

(4) The owner of livestock illegally grazing on rangelands set aside as winter grazing by the chair or his deputy was expected to pay a fine of one shilling per head of large stock and threepence per head of small stock. The chair and his deputy could detain the animals until the fee was paid.

(5) It was prohibited to erect a hut or house in any part of the reserve except at sites pointed out by the chairman of the board or his deputy.

(6) No cultivation was allowed in any part of the reserve except at places pointed out by the board chairman or his deputy.

(7) Burweed (*Xanthium spinosum*) and other noxious weeds growing in the reserve had to be destroyed by the residents in accordance with orders given by the chairman of the board or his deputy.

Any person that disobeyed the orders or refused to pay the respective fees could be fined £2 or sentenced to imprisonment with hard labour for a pe-

---

[24] SAB NTS 8144 5/341 part 1, Government Gazette No. 237 of 1909, 25 May 1909.

[25] At first the location tax in Sediba was obligatory for men between 18 and 60, however, in 1912 the age limit was reduced to 16; SAB NTS 8142 2/341 part 1; Minutes of Meetings of the Sediba Native Reserve Board, 28 June 1912.

riod of two months. The only exception to this was with regard to the regulation on noxious weeds, for which the punishment for non-compliance was a fine of 10 shillings or fourteen days of imprisonment.[26]

From the government's point of view, the tax payment behaviour of the reserve residents in particular left a lot to be desired. Although the people in Thaba Nchu did not revolt against the taxation, as for instance was the case in the Bambatha rebellion (1906) in Natal (Redding 2000), they expressed their displeasure with it. In a letter sent to the Thaba Nchu reserve board in 1913, 14 petitioners complained that 16-year-olds 'who are school boys and [...] shepherds, and have no lands' are taxed 'in common with men, who have lands, stock etc., and who are under employment.'[27] Exemption from taxes was only approved in individual cases when people were too sick or too old to find work and had nothing to pay with.[28] The increasing tax burden forced men to seek seasonal work outside the reserve. Still, people were often in arrears with their taxes and the chairman had to extend the due date several times. Failure to pay did not immediately lead to a conviction under the reserve regulations. Instead, Native Affairs officials in Pretoria exerted pressure by other means. For example, shortly after the introduction of the Native Urban Areas Act of 1923[29] which required black men to carry passes while staying in towns, reserve residents who were in arrears with their taxes were not given the documents. Thus, they were in fact prevented from earning the necessary money. The NAD only relented after board chairman and magistrate, B. Lefebure, wrote a letter to Pretoria asking for passes to also be granted to non-payers.[30]

Major amendments to the initial regulations followed in 1916 for both the Thaba Nchu and Sediba reserves, giving the chairman of the reserve board more decision power in terms of land management:

---

[26] SAB NTS 8144 5/341 part 1, Government Gazette No. 237 of 1909, 25 May 1909.

[27] SAB NTS 8144 5/341 part 1, Petition to the chairman and the members of the Thaba Nchu native reserve board, 17 January 1913.

[28] SAB NTS 8142 2/341 part1, Minutes of Meetings of the Sediba Native Reserve Board, 11 September 1912.

[29] Union of South Africa, Act No. 20 of 1923: Native Urban Areas Act.

[30] SAB NTS 8142 2/341 part 1, Minutes of Meetings of the Sediba Native Reserve Board, 9 January 1924 and 17 April 1924.

> 'It shall be lawful for the chairman, with the concurrence of the Board, to order the removal of any homestead in the Reserve to some other suitable place upon proof being afforded to him that such homestead is so situated as seriously to interfere with the interests or convenience of other persons residing in the neighbourhood [...]'
>
> 'Whenever it is shown to the satisfaction of the chairman that any arable land is so situated as seriously to interfere with the depasturing of stock in the neighbourhood, the chairman may prohibit the further cultivation thereof, and other land shall be allotted to the person deprived [...]. But in all such cases the allottee shall be entitled to reap any crops that may be on the land before he is required to surrender his rights thereto.'[31]

However, it was explicitly stated in the regulations that the allocation of residential sites and fields was the responsibility of the headmen. Although they needed the approval of the chairman and the reserve board, I found no source that indicated any intervention by the board in terms of land allocation. Any taxpayer was entitled to 'sufficient arable land if available for the purpose of his household'.[32] Absent labour migrants could, with the approval of the chairman, lend out their field to another person for a maximum of three years. After that period, the field was deemed available for re-allocation. The amendments of 1916 reflect that new households were increasingly migrating from white-owned farms in the Free State to the Thaba Nchu region, mainly as a result of the Natives Land Act of 1913. This required a stricter regulation of people's access to resources, particularly arable land.

3.3 Social Organisation and Land Use in the Sediba Reserve

The horseshoe-shaped Sediba reserve was located in the northern part of the Thaba Nchu district (Figure 1). Sediba village, one of six communities located in the reserve, consisted of four scattered settlements, namely Mpharane, Digojeng, Lehurutshe and Phatlha (Figure 12). Each of these was inhabited by one or more patrilineal family groups. Such a group consisted of families whose male members were descendants of a common

---

[31] SAB NTS 8144 5/341 part 1, Government Gazette No. 1049 of 1916, 9 September 1916.

[32] SAB NTS 8144 5/341 part 1, Government Gazette No. 1049, 9 September 1916.

grandfather or great-grandfather (Schapera and Comaroff 1984:40) to whom people would refer when asked about their origin:

> 'Being one family, it's how they used to group themselves. [...] "We are coming from Phatlha!", "Oh from Phatlha, who is your grandfather?", "Monchusi!", It's how we know each other.' (interview with Lydia Pobe, Sediba, 4 September 2011)

According to Schapera (1984), family groups sometimes included other relatives who were not part of the patriline, for instance affines, uterine nephews or even unrelated dependants. The family group was an important unit of cooperation and the associated households rendered mutual assistance with the building of huts, cultivation and the gift or loan of food, livestock and other goods. Moreover, the group as a whole was involved in marriage negotiations, the organisation of feasts and the distribution of estates, and took care of the widows of deceased members. Under the leadership of the family elder, the men settled intra-group conflicts and supported any member that was involved in disputes with other family groups (Schapera and Comaroff 1984:40).

The household was the smallest 'social unit' (Schapera and Comaroff 1984:39). Polygamous marriages were common in the past, though the practice of polygyny was in fact confined to the rich and powerful (Comaroff and Comaroff 1991:132). Wesleyan missionary John Petrus Bertram testified to the Commission on Native Laws and Customs in 1883 that

> 'a man was considered amongst his people a great man for having a number of wives. It had the advantage of improving his status. He enlarged his house, increased his family, possessed greater wealth, and gave him wider family connections.' (Commission on Native Laws and Customs 1970 [1883]:309)

Only one senior informant interviewed in 2011 remembered a polygamous household. This indicates that by the 1930s polygyny was already uncommon. The biographies of the Barolong elite suggest that this was not only the case in rural households (Murray 1992:184–201). I assume that this was mainly an outcome of the Batswana groups' comparatively early encounters with missionaries in general, and the strong influence of the Wesleyan church on the Thaba Nchu people in particular (Lye and Murray 1980:65). The typical household consisted not only of a nuclear family, but also often other relatives as well, for instance a married brother and his family or an unmarried sister and her children (Schapera and Comaroff 1984:39).

3.3.1 Agro-Pastoral Production and a Narrowing Resource Base

The family groups settled predominantly in mountainous areas, which were not useful for cultivation and which enabled people to keep an eye on rangelands and fields. Each of the settlements had its own piece of arable land, which in turn was subdivided into smaller parts allocated to individual households. In contrast, rangelands were used on a communal basis. Aerial photographs of the pre-'betterment' landscape in the Sediba reserve show that a great proportion of land was ploughed (Chapter 5.4.4; also Naumann 2014:50). However, fields also served as grazing area. After the harvest, cattle were driven onto the dry stubble in the fields so that they could eat the leftovers of the crops. Particularly during the winter months this was a valuable source of extra fodder. Usually the owner of a field let his own animals graze first, before he opened his field to others.[33] A positive effect of grazing on arable land was that cattle naturally fertilised the fields (Beinart 2008:344). In addition to the arable lands, a small proportion of a village's grazing was set aside for the winter months. Usually, this land was used between May and June.

In the first half of the twentieth century, the household economies in the Sediba reserve were substantially based on agro-pastoral production. Cattle provided milk, leather, dung, draught power and a means of transport. Moreover, for Batswana people cattle had a high symbolic and social value, since a man's status was determined by the size of his herd (Lye and Murray 1980:114). The specific valuations of cattle are described in detail by others (for instance Comaroff and Comaroff 1990; Ferguson 1992). In Chapter 8.5.1, I will summarise some of the contemporary meanings of livestock. Besides cattle, Thaba Nchu people also farmed with sheep, which provided milk and wool. Meat of both sheep and cattle was only used on special occasions, e.g. funerals or ancestral feasts, and in times of drought when harvest failures forced people to depend more on animal products. Livestock could be acquired and disposed of by various means, but most importantly through inheritance, the payment of bridewealth

[33] SAB NTS 8142 2/341 part 1, Minutes of Meetings of the Sediba Native Reserve Board, 27 November 1912.

(Setswana: *bogadi*)[34] or the lending of cattle (Setswana: *mahisa*) which is described in more detail below. Furthermore, cattle were transacted as gifts or compensation and as payment for service or in barter exchange (Schüler 1965:287–94).

To a greater extent than livestock products, cultivated crops, vegetables and edible wild plants constituted the villagers' main nutritional basis (Schüler 1965:324). Fields were planted with sorghum (Setswana: *mabele*) and maize (Setswana: *mmidi*) as summer crops, and with wheat (Setswana: *korong*) and beans (Setswana: *nawa*) in winter. Moreover, most households cultivated gardens with cabbages, pumpkins, watermelons, peas, potatoes and other vegetables on their residential plots. Oral histories and archival sources indicate that sorghum was the staple diet in the first half the twentieth century (Naumann 2014:45). In the late 1950s, the crop still provided 40% of the summer harvest in the Sediba reserve.[35] The situation was different in other regions. For instance, in the eastern Orange Free State, Basutoland, Pondoland and Kuruman sorghum had been replaced by maize as the major food grain in the 1930s and 1940s (Beinart 1982:99; McCann 2001:263; Jacobs 2003a:129). This aspect is remarkable, since the replacement of sorghum is interpreted by some authors as an outcome of a lack in labour force due to increasing labour migration and school education for children (for instance Beinart 1982). The cultivation of sorghum is more labour-intensive as the crop needs to be protected against birds (ibid. 99). Moreover, it has lower yields than maize.[36] However, sorghum is more drought-resistant, which makes it a suitable crop for dryland cultivation in semi-arid environments (McCann 2001:255). Thus, it suggests that despite the shortage of labour force – which was certainly also true for the Thaba Nchu area – households chose to produce sorghum in order to maintain their food security (Naumann 2014:45–46).

---

[34] According to the report of the Native Economic Commission in 1932 between 3 and 8 cattle were transferred as *bogadi* (Union of South Africa 1932:180).

[35] SAB BAO 5761 H 128/1666/197 part 1, Herwinningsverslag Sediba Gebied, Thaba Nchu, 22 May 1958, p. 3.

[36] In a 'betterment' planning report of 1958 the grain production was estimated at 3 bags of maize, 2.5 bags of sorghum and 4 bags of wheat per morgen; see SAB BAO 5761 H128/1666/197 part 1, Herwinningsverslag Sediba Gebied, Thaba Nchu, 22 May 1958, p.4.

Although livelihoods were substantially based on agriculture, the reserve population was not able to produce enough to fulfil all subsistence requirements. Simkins (1981) estimated that the Sediba farmers produced between 50% and 75% of the necessary food in 1927 (ibid. 270, figure 2). As indicated above, the early reserve regulations provided each household with sufficient land. After the passing of the Natives Land Act of 1913, however, the increasing immigration of evicted families resulted in a shortage of arable and grazing land. However, this was not the only negative outcome of the act. Archival sources and oral histories indicate that reserve inhabitants rented grazing land on neighbouring white-owned farms.[37] This practice was prohibited by the new act, which put further pressure on the narrowing resource base.

3.3.2 Growing Need for Cash Incomes

The shortage of land and increasing need for cash to pay taxes were the main reasons why a growing number of men searched for work outside the 'native reserves'. Moreover, young men sought to defy the authority of their fathers and strengthen their financial independence. Job opportunities were provided at neighbouring white farms, as well as in the diamond and gold mines in Kimberley and at the Witwatersrand respectively. Recruiting agents for the mines actively advertised jobs in Thaba Nchu and Sediba.[38] However, Batswana groups in general did not incline towards work in the mining sector and responded rather reluctantly to the recruitment efforts (see also Comaroff and Comaroff 1997:206; Jacobs 2003a:134). Oral accounts of Sediba residents indicate that male residents of the reserve favoured jobs in Bloemfontein over work in the mines:

> 'Our people they didn't like the mines. Most of our people were working in the railway [...] in Bloemfontein. People who like the mines are the Basothos [...] Our people, I think, they were not so much interested in the mines because [...] when your husband goes he must go for a year ... I think they thought it was too strenuous.' (Lydia Pobe in Naumann 2014:47)

---

[37] SAB NTS 8142 2/341 part 1, Minutes of Meetings of the Sediba Native Reserve Board, 16 October 1916. Interview with Andrew Thepe, Sediba, 18 August 2011.
[38] See for instance SAB NTS 8142 2/341 part 1, Minutes of Meetings of the Sediba Native Reserve Board, 8 September 1914.

Working for South African Railways in Bloemfontein had a major advantage: instead of leaving their homes for several months or years, men could commute on a weekly basis or even more often. From the 1950s at latest a daily bus service ran between the rural locations and Thaba Nchu, from where men could then take another bus to Bloemfontein.[39] Commuting enabled the workers to maintain translocal relations and continue land use activities in the rural areas. Of four former railway workers I interviewed in 2011, all had been engaged in pastoral or even agricultural production during the twentieth century. Commuting also had advantages for the home household, as wage labourers that commute tend to bring a greater part of their incomes home than long-term migrants (De Wet 1995:186). Exact numbers of people employed in the railway sector are difficult to obtain, particularly for the first part of the twentieth century. In the late-1950s, almost 200 men from the Sediba reserve alone were employed by South African Railways in Bloemfontein (Murray 1992:181). By then, about 1,232 households resided in the reserve.[40] Assuming one railway worker per household, this would mean approximately 16% of households received an income from the railway sector.

3.3.3 Cooperation in a Stratified Rural Society

In 1939, the 7,050 residents of the Sediba reserve together owned in total 3,366 cattle, 5,355 sheep, 60 goats, 3 mules, 469 horses and 197 donkeys.[41] Projected onto a household of five persons[42] this would imply an average of 2.4 cattle and 3.8 sheep, and less than one horse or donkey per household. Considering that about six oxen are necessary for a ploughing unit

---

[39] SAB BAO 5761 H 128/1666/197 part 1, Ad-hoc verslag Sediba Besproeiingskema: Thaba Nchu, 28 February 1956, p. 11.

[40] SAB BAO 5761 H128/1666/197 part 1, Herwinningsverslag: Sediba Gebied, Thaba Nchu, 22 May 1958.

[41] SAB NTS 7516 678/327, Locations reclamation report: Thaba Nchu and Sediba Native Reserves and land purchased by the Trust in the Thaba Nchu District, p. 14-15.

[42] I have no data on the average household size in the 1930s. In a memo on soil erosion work in Thaba Nchu, the Director of Native Agriculture, T.G.W. Reinecke, calculated average herd sizes for households with five persons (SAB NTS 9502 138/400/5 part 1, Memo of the Director of Native Agriculture, 3 April 1937, p. 13). Simkins (1981) too, based his calculation of agricultural productivity in the reserves on families of five persons.

(Lewis 1984:18), the cattle numbers would seem to make cultivation difficult. These figures have to be interpreted with caution, as livestock were unequally distributed between the villages and households. A memo of the Director of Native Agriculture, T.G.W. Reincke, indicates that the total number of cattle in Sediba village was more than six times that in Talla. However, both villages had about the same population.[43] Apparently, families with larger cattle herds concentrated in some settlements.

Oral histories indicate that people in Sediba village were rather wealthy. Some elders remembered that their fathers had had herds of about 30 cattle and more. These wealthy men lent out so-called *mahisa* cattle to other households who were in need of livestock. For the lender, this practice facilitated herding and reduced the risk of cattle loss due to diseases as the herd was dispersed in several kraals (Schapera and Comaroff 1984:28). The holder of such *mahisa* cattle rendered services to the livestock owner and could in turn use the animals for milk and as draught power, and sometimes received a heifer (Schüler 1965:288; Schapera and Comaroff 1984:28). As cattle lending required the existence of a substantial herd it was in fact only possible for comparatively rich men. Through the lending of *mahisa* cattle, they could establish patron-client relationships and thus transform their 'wealth in cattle into wealth in people' (Comaroff and Comaroff 1990:205) – or in Bourdieu's terms, they could convert their economic capital into social capital. Oral histories indicate that such *mahisa* relationships were still practised by some families in the first half of the twentieth century.

The example of *mahisa* cattle illustrates that some people were more successful in accumulating capital than others. I have no data on the livestock distribution among households for that time; hence, it is difficult to estimate the extent of stratification[44] in livestock holdings. Only the 'betterment' planning reports of the late 1950s provide data on livestock numbers per households in the Sediba area. However, these reports were produced

---

[43] SAB NTS 9502 138/400/5 part 1, Memo of the Director of Native Agriculture, 3 April 1937, p. 2.

[44] Stratification manifests itself in inequalities in terms of wealth, power and prestige (Peoples & Bailey 2015:295; for a comprehensive discussion of social stratification see Cancian 1976). I focus here mainly on economic stratification that is based on access to and ownership of economic capital, i.e. the unequal distribution of wealth.

after several livestock culls had been enforced. During these culls, the herds of the wealthy livestock owners in particular had been diminished. Thus, the reports do not provide a clear picture of existing inequalities in terms of cattle herds in the first half of the twentieth century. Nevertheless, patterns of socio-economic stratification are illustrated by other archival documents.

Table 2 Cross-tabulation of 'ownership of oxen' and 'ownership of cultivators' in the Sediba irrigation scheme in 1937[45]

Farmers who own	no oxen	1-5 oxen	6-10 oxen	more than 10 oxen	Total
no cultivator	25 (50%)	23 (46%)	2 (4%)	0 (0%)	50
cultivator	0 (0%)	3 (23.1%)	6 (46.1%)	4 (30.8%)	13
total	25	26	8	4	63

In the 1930s an irrigation scheme was established in Sediba village. The history of the scheme is reconstructed in Chapter 4.4.2. Here, I want to focus on a document that the Native Commissioner compiled after the irrigated plots had been allocated.[46] 63 men were registered as users who cultivated a plot in the scheme either individually or in a group of two to three people. The document included a list of the number of oxen and agricultural implements (e.g. ploughs, harrows, planters, cultivators) each farmer owned. In total, the land users possessed almost 193 oxen, 6 donkeys, 39 ploughs, 23 harrows, 14 cultivators, 8 planters and 1 double plough. However, these animals and implements were not equally distributed. For instance, almost 40% of the farmers had no oxen at their disposal (Naumann 2014:47). A correlation of the number of oxen and the number

[45] Data source: SAB NTS 7981 259/337 part 1, Letter of the Native Commissioner, Thaba Nchu, to the Chief Native Commissioner, Pretoria, with an attached list of implements and livestock owned by plot holders of the Sediba irrigation scheme, 7 May 1937.

[46] SAB NTS 7981 259/337 part 1, Letter of the Native Commissioner, Thaba Nchu, to the Chief Native Commissioner, Pretoria, with an attached list of implements and livestock owned by plot holders of the Sediba irrigation scheme, 7 May 1937

of ploughs owned per farmer is statistically significant ($r = 0.443^{**}$, $p < 0.01$). A similar result emerges when looking at the distribution of cultivators (Table 2) and other agricultural implements. Hence, the agricultural means of production were concentrated in the hands of a few irrigation scheme users.

Despite economic inequalities, informants portrayed a rather harmonious picture of past life in the rural areas. They emphasised that their living together was based on mutual cooperation and reciprocal arrangements:

> LP: 'So, it was only hand in hand. You hire this one to look after the cows, this one after the sheep. But we didn't say people were slaves or they were working for [someone]. It was just [a] compromise that people must live together.'
>
> CN: 'But there were also some poorer families?'
>
> LP: 'Ja, ja.'
>
> CN: 'So, they were hired or working for the others?'
>
> LP: 'They were hired but when they [the rich] plough, if I got nothing they take [me] also to go and plough for them […] when the time is coming to reap, you have your […] plot to go and pick up something to eat, you see. People were never selfish; all people were never selfish. They used to fill *pitsa* (Setswana: pot) for people who have nothing. Your children will never sleep without food that time. Because […] if you are hungry you just go to your next door, your aunt: "Auntie I am hungry", "All right, [I] give you something to eat".' (interview with Lydia Pobe, Sediba 4 September 2012)

Several aspects of Lydia's account are remarkable. First, the quote illustrates that support was given to poor households without any obligation of work (see also Jacobs 2003a:145). However, the words 'you just go next door to your aunt' suggest that this generalised reciprocity was rather limited to the personal family group. Second, it indicates that cooperation was essential for agricultural production. Both the richer and the poor and landless households depended on these patron-client relationships. Whereas the former needed labour power, the latter could make a living for a couple of months by providing their labour force during the harvest. At the same time, households who used fields but had no agricultural implements gained access to ploughs and oxen though these reciprocal arrangements.

Third, Lydia's account indicates that these patron-client relationships were increasingly monetised as workers were hired. This is also reflected in the changing character of ploughing and harvesting groups.

An institutionalised form of cooperation is called *matsema* (plural: *letsema*) which was organised during harvesting and ploughing seasons. A landowner in need of help would announce in the village: 'tomorrow I have *letsema'* (Sets: *kamoso ke na le letsema*). The next day up to 50 people would gather together at the landowner's house and then go to the field to help with the reaping. The organising household was expected to provide food and beer:

> 'I cook pap, porridge and [...] some cabbage or beans [...] I make something to make people feel happy. Then I take it with another woman to the *letsema* [...] And they eat there and are happy, happy [she ululates]. We are going to eat; eat and drink and sing the whole day.' (Kedibone Mokgothu in Naumann 2014:46)

Letsema enabled households to plough or harvest a field within a short time period even when they had insufficient labour force within their own family group. The cooperation was based on balanced reciprocity, i.e. the organising household was also expected to help others. Oral histories indicate that these non-monetary reciprocal arrangements began to transform in the mid-twentieth century. A woman that was born in the late 1940s remembered that her father was a successful farmer who owned many cattle and agricultural implements. He usually ploughed other people's fields, but in return expected a part of their harvest. Thus, he had *de facto* 'many fields' (interview with Paulina Mokgothu, Middeldeel, 10 September 2011). On the other hand, the woman's father paid people from Lesotho with cash for their assistance in harvesting his own fields. As he did not rely on reciprocal arrangements, he was not obligated to offer his help 'for free' to other villagers. This example illustrates the bargaining power of wealthier households. When they utilised their production capital strategically, they could easily increase their agricultural profits and thereby their economic capital. Since they had greater access to resources, richer households were able to 'deviate from norms and create new ones' (Ensminger and Knight 1997:5). Indeed, the practice of *letsema* began to decrease and was finally replaced by monetary arrangements. It is difficult to pinpoint when exactly the *letsema* institution ceased to exist; however, as everyone born in the

1930s and 1940s remembered the formation of such working groups, I assume that cash did not completely replace reciprocity until the 1950s (Naumann 2014:46). Similar developments were also noticed in other regions, such as Kuruman (Jacobs 2003a:146–47).

3.3.4 Sharecropping and Labour Tenancy on Commercial Farms

Cooperative arrangements were not only found within the reserve boundaries. Before the land act of 1913, the practice of sharecropping was widespread in the Southern Highveld (van Onselen 2005) and the Orange Free State. Since the 1870s, British and Boer settlers had farmed east of Thaba Nchu and invested capital in the use of the fertile cropland. Many of the new settlers established sharecropping contracts with black tenants (Keegan 1987:12). In these arrangements, tenants ploughed land on white-owned farms. Instead of rent, they gave a portion of their harvest to landowners. Usually tenants had access to additional land where they could keep livestock (Beinart and Delius 2014:671). The number of tenant families allowed per farm was restricted to five nuclear families, though this was not strictly controlled (Keegan 1987:58). Commercial farmers had several motives for such arrangements. First, they needed additional labour force for crop production. Second, many of them lacked agricultural capital, particularly oxen and implements, which the tenants had. Third, white farmers had no experience or skills with regard to cultivation (Keegan 1987:51). Knight (1903) who travelled across the Orange Free State at the turn of the twentieth century, described the advantages of sharecropping as follows:

> 'The poorer settler cannot afford to purchase many head of cattle, but with natives on his land who plough for him with their own waggons, he is enabled to bring much more of his land under cultivation than would otherwise be possible. Moreover, as the reaping can be carried through with great rapidity under this system, the risk of loss through heavy rains in the harvesting season is much diminished.' (Knight 1903:153)

For the tenants, such sharecropping arrangements could be particularly profitable on farms where the owner was absent. In these cases, black farmers could produce rather independently without being too controlled by the landlord. When tenants successfully produced large surpluses that provided their households with food and cash, they could make profit out of such sharecropping relationships. However, not all black farmers were well equipped, and not all landlords were strongly reliant on the means of production of sharecroppers. Particularly in the latter cases, tenants were required to render additional labour service to landlords. In fact, very diverse tenancy arrangements emerged (Keegan 1987:95). Knight (1903) described the rules of such contracts as follows:

> 'Not more than five native families are allowed to squat on one farm without special permission from the resident magistrate. They are allowed to build huts and kraals, to graze stock, and to have the use of lands for ploughing free of charge. In return they undertake to provide one or more young natives for herding, milking, and grooming, who live at the homestead and receive food and £3 or an in-calf heifer per year; one or more girls to attend at the house daily and do the rougher household work; able-bodied natives who assist without payment in doing all ordinary work on the farm; and men for certain extra work, such as dam-making, brick-making, and riding transport, which is done at agreed prices. When natives are working they are provided with food.' (Knight 1903:139-140)

It is important to stress that on almost every farm in the Thaba Nchu district – whether on white-owned or black-owned land – sharecropping was common at the beginning twentieth century (Union of South Africa 1916b: appendix XI, p. 15). The dependence of Boer and British farmers on black tenants was a nuisance to the South African administration. In order to ensure their control over land and the accumulation of resources, officials intervened with measures that protected and promoted white agriculture and suppressed the black rural economy. This was done by providing beneficial loans and facilitating market access for whites (Keegan 1987:96–120). Finally, with the passage of the Natives Land Act of 1913, sharecropping contracts between white and black farmers were prohibited. In practice, however, such arrangements persisted for many years, not least because many white farmers had no choice than to rely on the means of production of black South Africans (Murray 1992:86–91; Beinart and Delius 2014). The act itself did not directly lead to the immediate abandoning of sharecropping. Rather, the mechanisation of white agriculture that was constantly promoted by the South African state eventually made sharecropping unnecessary for British and Boer farmers (Keegan 1987:191).

3.3.5 Changing Roles and Attitudes in the Rural Areas

The division of labour according to age and especially gender in Batswana households has been described by several scholars (for instance Schapera and Comaroff 1984; Comaroff and Comaroff 1990; Jacobs 2003a). Women were responsible for all household chores, including fetching water, dung and wood, cooking food and preparing beer. Moreover, they were responsible for all tasks connected with cultivation, except for ploughing with oxen. In contrast to that, men did all the tasks related to livestock, such as herding, milking, hunting and ploughing. Sometimes they were also involved in planting, cleaning, weeding and reaping (Schapera and Comaroff 1984:27). In the first half of the twentieth century these patterns of gendered labour division became blurred in rural Thaba Nchu. The following three vignettes illustrate the childhoods of two women, both born in the 1930s, and one man, born in the late 1940s, in Sediba village:

> Kedibone Mogalatladi was born in 1930 in Digojeng, as one of eight children. Her father, a member of the Cana Lutheran Church (Figure 3) which was built between 1897 and 1906 on a hill between Lehurutshe and Digojeng, sent her and her siblings to the associated missionary school. Since it was the only school in Sediba 'location', most of the children living in this area attended classes at Cana, even if their parents belonged to other churches. Kedibone's father had about 30 cattle, as well as sheep and horses. In addition, the family ploughed in the dryland fields and cultivated a small vegetable garden inside their plot. As Kedibone had to assist her parents in household and agricultural chores, she and her siblings took turns in attending school. When the lessons were over, the children had to go home and help in the fields or garden. The herding of animals was done by both boys and girls. When it was Kedibone's turn to herd, she spent the whole day in the rangelands. Mostly she was herding sheep. However, she also milked cows and even rode horses. (interview with Kedibone Mogalatladi, Sediba, 9 August 2011; see also Naumann 2014:48)

> Dichutso Mosotho was born in 1935 in Mpharane. She herded the family's cattle and sheep until she was about ten years old. In addition, her family kept some pigs in the yard. In 1946, Dichutso started to attend Cana missionary school, where she completed Standard five, the highest possible grade. When asked

about life in Mpharane, Dichutso replied '*go le monate*' (Setswana for 'it was nice'). In December, the school was closed and the children had a holiday. They were allowed to stay up late, and they prayed and practiced songs for the Christmas service. December was also the time for harvesting the wheat in the household's fields. This was done with a *letsema*. After the harvest, the parents would count all the bags, and set aside seeds for the following year and the grain they needed for self-supply. In Excelsior, the grain for the household was ground and the surplus harvest sold to a cooperative. The family used the money to buy those things they could not produce, e.g. soap, sugar, paraffin and clothes. (interview Dichutso Mosotho, Sediba, 8 August 2011)

Andries Kgware was born in 1948 in Lehurutshe, as one of six children. The nuclear family lived together in one household. However, a sister of his stayed in the homestead of a relative who was elderly and needed care. Andries' family had cattle and sheep as well as fields in the dryland area. In addition, his parents ran a small shop in the village where they sold, among others things, maize meal, potatoes, bread, sugar and tea. Whereas the work in the shop was mainly the responsibility of the mother, the father was responsible for the livestock and also herded the animals. Cultivation was done by both parents with the help of other family members. In contrast to Kedibone and Dichutso, Andries and his siblings were not much involved in herding. In his childhood in the 1950s many parents deemed it better that their children learned to read and write instead of working as herders. Some people combined their flocks and divided the herding work, others paid shepherds to take care of their animals. (interviews with Andries Kgware, Sediba, 17 August 2011, 16 September 2011)

The three examples illustrate that the first half of the twentieth century was characterised by social transformations. Girls and women were increasingly involved in tasks connected with livestock, such as herding, milking and in some cases even ploughing. This relaxation of gender restrictions with regard to animal husbandry was a result of the absence of male labour force. As young men were commuting or migrating, they could not take care of their duties at home. Similar trends were also observed in other regions, for instance in Kuruman in the Cape Province (Jacobs 2003a:144), in Phokeng in Western Transvaal (Bozzoli 1991:45) or in Sekhukhuneland in Northern Transvaal (Delius 1996:28).

Figure 3 Cana Lutheran Church in Sediba (photograph: C. Naumann)

In addition, parents had an interest in sending their sons and daughters to school. Data from the ethnographic census (Figure 4) show that of all the people born before 1951 only 8% never went to school. Older informants stated that they still had to combine school and their household tasks, and thus attended school only from time to time or started when they were ten or eleven years old. In contrast, the younger generation could focus on schooling only. Some parents even sent their children to secondary schools in Thaba Nchu or Bloemfontein. In the ethnographic sample, one man in his 60s held a degree in law, and another man went to college and later worked in the gold mines in Odendaalsrus. However, secondary education was only affordable for families who had enough economic capital to pay school fees and forgo the necessity of their children's labour force. By sending their children to school, these wealthy households converted their economic capital into cultural capital. The latter became more important in a rural society increasingly integrated into the wage labour market. Access to education meant access to more profitable employment sectors, which in turn could increase a household's income.

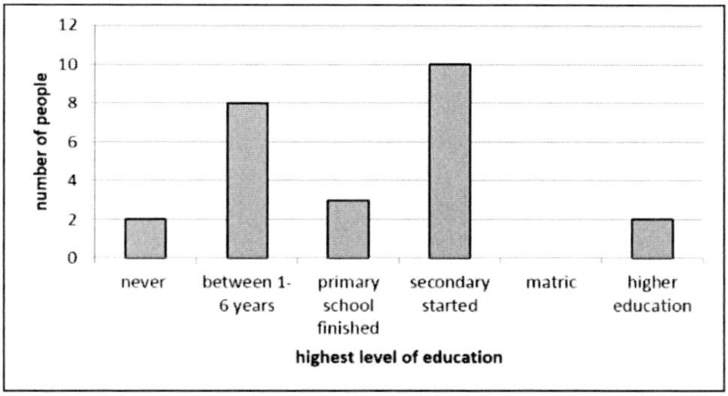

Figure 4 Highest level of education of people born before 1951 (n=25) [47]

As a consequence of wage labour and education, certain agricultural tasks could no longer be undertaken by the households on their own. As Dichutso's childhood account reveals, the processing of grains was done in cooperatives; or with the help of neighbouring farmers:

> '… Natives of the Reserves are dependant [sic] on European farmers owning machines for the thrashing [sic] of wheat. This dependancy [sic] invariably results in the natives having to wait until after such farmers have thrashed [sic] their own wheat as well as that of neighbouring farms, so that the machines are not available until February and sometimes March at the earliest. This season [rainy season 1943/44] on account of the generally large wheat crop coupled with abnormal weather conditions the position is greatly aggravated and quite a fair percentage of stacked native wheat is rotting while waiting for machines to arrive.'[48]

However, the outsourcing of typical household tasks, such as the threshing of grain, was not only a result of a lack in labour force. In general, there was a turn towards advanced agricultural machinery. As in other regions (see for instance Jacobs 2003a:158–61; Beinart 2008:332-366), government officials promoted the intensification and mechanisation of the agri-

[47] Primary school consisted of seven school years (Berzborn 2006:154).
[48] SAB NTS 8141 1/341 part 2, Letter of the Assistant Native Commissioner, Thaba Nchu, to the Secretary of Native Affairs, Pretoria, 25 February 1944.

cultural production in Thaba Nchu by employing agricultural demonstrators drawn from the local population. These demonstrators, who attended training courses in agricultural schools, showed others how to use certain cultivation methods, or demonstrated the operation of agricultural implements on plots, and sometimes also went to schools to teach children gardening. In 1927, an agricultural demonstrator was supplied with implements to cultivate in both the Thaba Nchu and Sediba reserves. In the villages, the services of the demonstrator were accepted only hesitantly. Nevertheless, Sediba residents were impressed by the good yields from these demonstration plots during a drought in the early 1930s.[49] Presumably the demonstrator had been successful because he had not only used advanced machinery (e.g. planters) but also better fertiliser (see also Jacobs 2003a:160). The demonstrations on these exercise plots motivated households to invest more money in agricultural production. Kedibone Mogalatladi remembered that after the presentation by the demonstrator some people bought planters (interview with Kedibone Mogalatladi, Sediba, 9 August 2011). However, the high costs connected with agricultural intensification were only affordable for the wealthy families.[50]

3.4 Coping with Environmental Crisis

During the first half of the twentieth century, the semi-arid environment of Thaba Nchu experienced several ecological crises, often caused by droughts (Setswana: *leuba;* also 'famine') or locusts (Setswana: *tsie*). The destructive effects of the latter on grasses and crops were vividly described by Leonard Flemming (1927) who in 1903 settled on a farm south of the Thaba Nchu district:

> 'Across the eighty-mile-long range of hills in the distance was a pale yellow, in parts brown, cloud. It extended from left to right of the full eighty miles, and was about one hundred miles from the farm when I first saw it. It was a locust swarm [...] Mile by mile, minute by minute, the landscape is gradually blurred out.

---

[49] SAB NTS 8142 2/341 part 1, Minutes of the Meeting of the Sediba Native Reserve Board, 17 March 1932.

[50] I have no data on the prices of planters for this time. However, other implements that became more important were double-furrow ploughs, which cost about £24 in 1937; see for instance SAB NTS 9502 138/400/5 part 1, Memo of the Director of Native Agriculture, 5 April 1937.

> Hill after hill, tree after tree, is hidden, and all the while the big brown cloud mounts higher into the sky and spreads itself over half a world [...] The locusts settle everywhere in millions [...] On the ground nothing but locusts, and so thickly do they pack that not a square inch of earth or grass is visible [...] In ten minutes your two thousand acres of green grass, your hundreds of acres of waving wheat, are razed to the ground. Not one green blade is visible, not one green leaf, not a flower, not a stalk. All eaten right down *into* the soil by millions and millions of locusts.' (Flemming 1927:76–81, original emphasis)

The eradication of locust swarms – predominantly brown locust (*Locustana pardalina*) and red locust (*Nomadacris septemfasciata*) – was a central issue of the agricultural agenda of the government. In 1907, the Department of Agriculture of the ORC passed an ordinance in which guidelines for the control of locusts and *voetgangers* or hoppers – i.e. young locusts that are not yet able to fly – were formulated.[51] The department suggested the eradication of hoppers by spraying an arsenic solution on the plants eaten by the insects.[52] Moreover, when the locusts had laid eggs the soil was to be ploughed in order to prevent hatching, which occurred seven to ten days after good rains.[53]

Black farmers often hesitated to apply the solution and even opposed these campaigns as they feared poisoning their livestock (Beinart 2008:325). Indeed, it did happen that some animals died of poisoning (Price and Brown 2000:42). In the Sediba reserve too, there seemed to be reluctance to use an arsenic solution. Government officials complained that residents failed to assist in the chemical eradication of locusts.[54] Actually, people applied a different strategy that enabled them to destroy locusts and make use of the insects at the same time. As was common in other parts of the Highveld (van Onselen 2005:104), locusts were collected for food. Cooked or 'dried

[51] Orange River Colony, Department of Agriculture: Destruction of Locust Ordinance, July 1907, p. 3.
[52] Ibid. 9-11.
[53] SAB NTS 8142 2/341 part 1, Minutes of the Sediba Native Reserve Board, 30 May and 15 August 1923.
[54] SAB NTS 8142 2/341 part 2, Minutes of the Sediba Native Reserve Board, 23 November 1934 and 25 January 1935.

like *biltong*'[55] (interview Kedibone Mogalatladi, Sediba, 9 August 2011), the insects provided a protein-rich dietary supplement.

In contrast to Sediba land users, most commercial farmers in the Free State supported the governmental approach of arsenic poisoning. However, some commercial sheep farmers in the Karoo opposed the poisoning campaigns (Beinart 2008:325-327). They realised in the mid-1930s that locusts could have positive effects for the specific environment in the Karoo. The insects destroyed the unpalatable, annual *steekgras* (*Aristida* spp.) but did not eat other plants valuable for sheep grazing, such as the Karoo bush, *Medicago sativa*, agave, and prickly pear (ibid.). The spraying of arsenic was eventually discontinued in South Africa in the 1930s. Henceforth, the control of locusts was achieved by using arsenic-laced bait, which could be applied more precisely (Price and Brown 2000:42).

The locust example illustrates the complexity of social-ecological dynamics. The impact of the locusts varied from ecosystem to ecosystem; thus, the issue was actually not effectively addressed with a uniform nationwide strategy. Moreover, actors' approaches within the very same social-ecological system differed significantly. Whereas the state pursued a strategy that sought eradication by all means, local farmers in Thaba Nchu tried to make use of the locusts without risking the health of their livestock.

Besides locusts, perennial droughts had severe effects on the Thaba Nchu social-ecological system. In *The call of the veld*, Flemming (1927) described the harsh, semi-arid climate in the Free State and its threat to vegetation and livestock:

> 'And then a month went by without any rain – and then another. It was the beginning of one of the worst droughts the country has ever gone through.[56] [...] For days and weeks and months – drought, and only drought [...] If you walk on the veld, it crackles under foot like so much dry straw; but in most places it is not long enough to crackle, it is more like short quills embedded in cement [...] Week by week the crops change colour. The fresh bright green has gone completely [...] And it gets to the yellow white stage – a sapless, shrivelled yellow – and you know then

---

[55] Biltong is a culinary specialty in Southern Africa consisting of cut, dried and spiced strips of beef or game.

[56] Unfortunately, Flemming (1927) did not specify to what year or period he is referring to.

> that all the rain in the world will not help your crop now [...] Cattle – mere skin and bone – crawl hopelessly about, nibbling at the dry, brittle tufts of veld; and all the water that they have is probably miles away, around which the cattle lie in the blazing sun for hours, too tired or too weak to move. Some are too weak to get up. These you kill.' (Flemming 1927:170–73)

When asked whether they remembered such drastic droughts – personally or from accounts of elders – my informants mentioned the early 1930s. At that time, Southern Africa was struck by 'the most devastating drought of the twentieth century' (van Onselen 2005:179). In August 1931, after two below-average rainy seasons (Figure 5), it became obvious that the situation in the Thaba Nchu district was critical. Members of the Sediba 'native reserve' board reported that some of the reserve inhabitants were starving, and asked the government for assistance.[57] Despite moderate rains between September and December 1931, the food shortage was not mitigated. In March 1932, the government announced that it would provide the reserve residents with 900 bags of maize as famine relief.[58] However, in 1933 the situation even got worse when the 'red dust' (Sets. *lerole le le hibidu*) swept across the dry earth 'for a long time' (interview with Kedibone Mokgothu, Middeldeel, 11 August 2011). In those years, 'there was too much hunger' (interview with Kedibone Mogalatladi, Sediba, 9 August 2011). In some cases, famine forced people to eat dry, ground grasses just to appease their hunger. Government officials were aware of the critical situation in the Thaba Nchu reserves, as the report of the circle engineer of the Irrigation Department in Bloemfontein reveals:

> 'The recent severe drought throughout the Free State has caused heavy losses in most agricultural products and stocks, and nowhere has the effect been so bad as in the Native Reserves [...] Most mealie and kaffir corn crops were either dead or dying, and those which might pull through could only be expected to produce the poorest of yields. Thus, Natives who have experienced at least three bad years in succession are in dire straits without food, money or seed even; it means rationing large numbers of

---

[57] SAB NTS 8142 2/341 part 2, Minutes of Meetings of the Sediba Native Reserve Board, 28 August 1931.

[58] SAB NTS 8142 2/341 part 2, Minutes of Meetings of the Sediba Native Reserve Board, 17 March 1932.

> them and/or providing work and affording some the chance of earning a bare existence.'[59]

The severe impacts on agricultural production were exacerbated by the Great Depression, which also affected other parts of Africa. The economic crisis brought about massive unemployment particularly in the mining industries and cash crop production (Marx 2004:241). Due to a high gold price South Africa's economy soon recovered, with an annual growth rate of 5 % between 1933 and 1939 (Bernstein 1996:6). However, the economic crisis reduced the possibilities of remuneration in the early 1930s. This had severe consequences for the rural households. In order to supply their families with food, farmers of the Sediba reserve needed cash. A few jobs were made available when the dam and irrigation scheme were built in Sediba between 1933 and 1935 (Chapter 4.4.2). In desperation, 180 reserve inhabitants offered to work for food only. This was welcomed by government officials as they wanted to keep the construction costs for the irrigation scheme as low as possible.[60] Thus, people had to pay for drought relief by offering their labour force. Finding work was one coping strategy during severe droughts; emigration was another. I assume that at least some households searched for better conditions elsewhere and left the Thaba Nchu region. Although I have no precise data on emigration as a result of the 1930s drought, archival sources indicate that several families had left for Bechuanaland during earlier droughts.[61]

---

[59] SAB NTS 7981 259/337 part 1, Report of the circle engineer of the Irrigation Department, Bloemfontein, to the Director of Irrigation, Pretoria, on the possibilities for irrigation in the Thaba Nchu native reserves, 16 March 1933, p.1.
[60] SAB NTS 7981 259/337 part 1, Letter of the Director of Irrigation Department, Pretoria, to the Director of Native Agriculture, Pretoria, 23 March 1933; Letter of the SNA, Pretoria, to the Director of Irrigation, Pretoria, 16 August 1933.
[61] SAB NTS 8142 2/341 part 1, Minutes of Meetings of the Sediba Native Reserve Board, 17 November 1916.

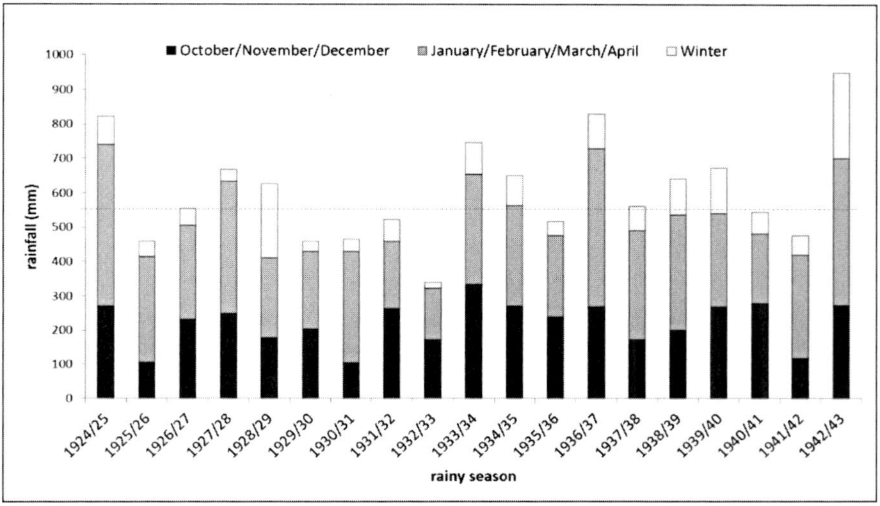

Figure 5 Total seasonal and annual rainfall in Thaba Nchu, 1924-43[62]

The drought ended with heavy precipitation in the 1933/1934 rainy season. Oral accounts illustrate the aftermath as a miracle:

> 'After the red dust, it rained so much [...] People found tomatoes, carrots, cabbage and spinach [...] They were finding them without anyone having planted them.' (interview with Kedibone Mogalatladi, Sediba, 9 August 2011, translated by Tsametse Pobe).

> 'Our grandmothers said that after that [the drought] there came a lot of rain [...] After the rain, a few months [later], when you'd go there into the field you found [...] everything you want to eat. [...] It was the miracle of the Garden of Eden.' (interview with Setadimo, Middeldeel, 11 August 2011)

---

[62] The grey, horizontal line indicates the long-term average annual rainfall of the region (560 mm). I adopted the presentation of rainfall data by rainy seasons rather than by year from Jacobs (2003a:8). The columns should be read as follows: 1924/25 shows the rainfall from October 1924 to September 1925. Source of rain data: SAB NTS 7526 717/327 part 2, Memo of the Assistant of Native Agriculture, Pretoria, to the Controller of Native Settlements, Pretoria, 17 January 1944, Annexure A.

Nevertheless, the serious effects of the drought were still visible in the following years. Many animals had died, and those that had survived the drought were in a very bad condition. Furthermore, people had used every single grain as food and lacked seeds for the cultivation of their fields. Thus, the reserve population remained dependent on assistance by the government. In June 1934, the NAD provided the Thaba Nchu and Sediba reserve with 100 bags of various wheat species for the purpose of sowing. The recipients of the seed donation were carefully recorded as they were expected to return the same amount of wheat after the harvest. Moreover, the seed was chemically treated with copper carbonate 'to ensure that the people will plant and not consume it'.[63] In the Sediba reserve, bags were not distributed immediately since the people were expected to provide 125 extra oxen for the construction of the irrigation dam in Sediba location.[64] This was an extortionate expectation, as not only were the people in need of seeds, they also had no oxen in good shape.[65] Eventually, the reserve inhabitants had no other choice but to accept the government's requirements.

The 1930s crisis was not restricted to the Thaba Nchu region or communal areas. Farmers nationwide lost large numbers of animals (Beinart 1996:152, figure 8.1). The number of pauperised white South Africans increased massively after the economic depression,[66] a trend which substantially contributed to the rising radicalisation of Afrikaner nationalism (Marx 2006:173).

Despite all the difficulties during the early 1930s, one cannot conclude that the Thaba Nchu social-ecological system was extraordinarily vulnerable. In fact, the system reorganised and recovered comparatively quickly, as suggested by several indicators. In 1936, inhabitants of the Sediba reserve refused to provide their oxen for dam construction work in Kgalala village because they needed the animals for ploughing purposes. People were no

---

[63] SAB NTS 8154 32/341, Letter of the SNA, Pretoria, to the NC, Thaba Nchu, 1 June 1934.

[64] SAB NTS 8142 2/341 part 2, Minutes of Meetings of the Sediba Native Reserve Board, 14 June 1934.

[65] SAB NTS 8142 2/341 part 2, Minutes of Meetings of the Sediba Native Reserve Board, 10 April 1934.

[66] The Carnegie Commission of 1932 classified 300,000 out of 2 million white South Africans as 'very poor' (Bernstein 1996:7).

longer willing to work for food, but expected cash payment;[67] a sign that famine was averted. Moreover, the restriction of livestock numbers was a major point of discussion at the reserve board meetings, which reveals that herd sizes had increased again.[68]

The system's quick reorganisation after a severe drought was not unique to the early 1930s, but could also be observed in the early 1940s. Although not as extreme as in the decade before, the 1940s started with two seasons of below-average rainfall (Figure 5). After a subsequent year with almost 950 mm of rainfall in 1942/1943, however, the vegetation soon improved again. An agricultural officer in Thaba Nchu concluded:

> 'The effect of such a year's rain as may be expected is almost phenomenal; even the *worst eroded areas* where the camps have been closed to grazing for the last three years gives [sic] one the appearance of being *under grazed.* [...] In a 300 morgen [257 ha] camp near Blydskap [a government farm southwest of Thaba Nchu town] 300 cattle were put in five weeks ago and [it] still appears as if no grazing had taken place there this season.'[69] (emphasis added)

Strikingly, during the first half of the twentieth century, the social-ecological system was still able to recover quickly. Thus, the historical data indicate that the system was resilient, which is in sharp contrast with the narrative invented by government officials at that time discussed in Chapter 4.

3.5 Preliminary Summary

I have shown in this chapter that the Thaba Nchu social-ecological system was strongly coupled in pre-'betterment' times. Settlement patterns were adapted to ecological conditions, i.e. family groups lived together in mountainous areas that were not suitable for cultivation. Within the patrilineal family group, households mutually supported each other. Institutions of

---

[67] SAB NTS 8142 2/341 part 2, Minutes of Meetings of the Sediba Native Reserve Board, (day unknown) September, 1936.

[68] SAB NTS 8142 2/341 part 2, Minutes of Meetings of the Sediba Native Reserve Board, 28 October 1938.

[69] SAB NTS 7526 717/327 part 2, Memo of the Assistant of Native Agriculture, Pretoria, to the Controller of Native Settlements, Pretoria, 17 January 1944.

support and cooperation provided the structure for economic organisation in the settlements. Working parties based upon balanced reciprocity enabled households to harvest their individual fields within a short period of time, even when they did not have enough labour force. Through the lending of *mahisa* cattle, wealthier households established patron-client relationships with poorer ones. In this way, the former expanded their social network, while the latter mainly received access to livestock products and services. Data suggest that the pre-'betterment' social-ecological system was vulnerable to extreme environmental shocks, which resulted in livestock loss, harvest failure and starvation. However, when environmental conditions improved, the system soon recovered and returned to its former state.

The social system, particularly social institutions, underwent significant transformations during the first half of the twentieth century: for instance, gendered labour division was relaxed and *letsema* was increasingly replaced by monetary arrangements. Institutional changes were initiated by various forces from within and outside the social-ecological system. Bargaining between actors with differential capital was an internal factor that resulted in institutional transformation. Households endowed with greater economic capital aimed at the accumulation and diversification of their capital. On the one hand, they invested in formal education, which improved their access to the labour market. On the other hand, wealthy individual farmers deviated from established norms – namely, participation in reciprocal working groups – and expected money or a proportion of the harvest as remuneration for their assistance. Thereby, they accumulated economic capital.

The internal factors driving institutional change within the system cannot be understood without considering external forces. All households in rural Thaba Nchu had to reorganise their livelihoods in a system that was increasingly tied to the money economy, as well as regional and national labour markets. Rural households could not escape the taxes that had to be paid, and thus had to engage in wage labour. Wage employment was also a means for young men to defy the authority of their parents. Moreover, salaries could be invested in agricultural implements to buffer the household against harvest failure. Hence, the engagement in wage labour was not only an adaptation to changing political conditions, but was also a strategy to increase a household's resilience. Since rural residents were comparatively closely situated to labour markets in Bloemfontein, they could maintain close ties to their homes while continuing agricultural production.

At the end of the nineteenth century, after the annexation of the Barolong territory, the state increasingly intervened at the local level. One important step was the creation of formal regulations for land use practices, which has been partially described here. Although some bureaucratic institutions had been created early – for instance, the Scab Ordinance of 1903, or the rule that only five sharecropping families were allowed on private farmland – the most critical formalisation of institutions considered here were the reserve regulations passed in 1907 and 1909 in the Thaba Nchu and Sediba reserves, respectively. By giving the board chairman the right to define appropriate places for the erection of houses or cultivation of land, a foundation was laid for the definition of clear-cut land use categories in later regulations. However, access to land for cultivation or a homestead was still obtained through the local headman. Additionally, major acts, such as the Natives Land Act of 1913, did not have immediately negative effects, as others have pointed out (Beinart and Delius 2014). This was the case, for one reason, because these early regulations contrasted with the social practices found at the local level. For instance, the 1913 land act could not take effect as long as white farmers still depended on black people's ploughs and cultivation skills. The full repercussions of these top-down-implemented institutions only developed successively over time, as the following chapter will show.

4. THE INVENTION OF A DEGRADATION NARRATIVE, 1920S–1950S

Whereas the previous chapter elaborated upon social-ecological dynamics on the local level, I want to focus now on the South African state as an actor that is creating new institutions and discuss how these actions have been politically legitimised. The first part of the twentieth century is characterised by official debates on land degradation, and particularly soil erosion, and the way in which these issues should be addressed both politically and technically. In the following section, I will first provide a brief overview of discussions on resource degradation and the emerging conservationist thinking in colonial South Africa. Whereas early colonial botanists ascribed the deterioration of rangelands in the Cape colony mainly to white settler agriculture, in the later years officials were increasingly concerned about the situation in black farming areas. During the 1920s and 1930s, government commissions and departments emphasised the ecological crisis in these regions, but, as I will show, did so based on questionable observations and generalisations. Yet the narrative of environmental collapse was used as a way of legitimizing state intervention and resulted in various policies, bureaucratic institutions and the constitution of a large state apparatus in order to control land use on the local scale. After introducing the legislative basis for intervention and the bureaucracy, I focus on the actual state interventions on the local level in Thaba Nchu. By using the example of the Grobler irrigation scheme, which was established in Sediba, I will start with a detailed account of the planning and technical implementation of government schemes and the problems encountered. The main three foci of conservation efforts in Thaba Nchu, i.e. soil, rangelands and livestock, will then be addressed, and the increasing coercion associated with state intervention elucidated. The chapter concludes with a final evaluation of the conservation measures.

4.1 Emerging Debates of Degradation and Soil Erosion

With his seminal book *The Rise of Conservation in South Africa* William Beinart (2008) provides a vivid narrative of the origin of 'conservationist thinking' (ibid. xvi) in the Cape Colony. He illustrates that environmental change and degradation had already been addressed in the eighteenth century. Botany was one scientific field that was significantly influenced by observations made in the Cape. Debates arose over the *veld,* a South African concept referring to vegetation and soil but also including much broader issues such as nature and landscape (ibid. 64). The Swedish botanist Anders Sparrman and his compatriot Carl Peter Thunberg were two of the early scientific travellers who visited the Cape in the 1770s. Particularly the former considered white settler agriculture as exploitative since high animal numbers had deteriorative effects on soil and vegetation. In contrast to that, Sparrman argued, the land use practices of Khoikhoi groups were better adapted as they moved more frequently with their dwellings and kraals (Grove 1989:165; Beinart 2008:66–67). Many subsequent scientists referred to Sparrman's comments when describing the complex of overgrazing and vegetation change. Botanists such as the German Ludwig Pappe, who became the first colonial botanist in the Cape in 1858, advanced these explanatory models and linked the exhaustion of rangelands and the disappearance of specific plant species to pastoral farming. By the 1860s, elements of a 'degradationist critique' (Beinart 2008: 77) had been formulated.

The complex of drought was another topic that dominated South African environmental debates in first decades of the nineteenth century. Particularly in the early 1820s, during severe droughts narratives of declining rainfall, desiccation and denudation became common amongst farmers, scientist and missionaries (Grove 1989; Jacobs 2003a:61–62; Beinart 2008:77–88). One of the latter was the Scottish missionary Robert Moffat who arrived at Kuruman in 1820. In his writings, he blamed the Batswana people amongst whom he worked for both 'moral and environmental disorder' (Grove 1989:166), and he linked the expulsion from Paradise with the existing drought caused by the 'sinful' behaviour of the Batswana. His pejorative descriptions of indigenous knowledge and land use practices reveal that the earlier thinking with regard to indigenous farming systems had changed (ibid. 165–6). However, the negative impact of white settler agriculture on natural resources remained, along with other topics,[70] an official concern during the nineteenth century (Beinart 1989:148).

---

[70] Officials were also concerned about the hunting of game as wild animal populations were considerably decimated in the late nineteenth century (Beinart 1989:149). Several scholars have dealt in detail with the environmental history of hunting and the protection of game in reserves (for instance Carruthers 1989; MacKenzie 1997; Beinart and Hughes 2007).

John Croumbie Brown, who became the colonial botanist in 1862, saw both white and black farming methods as potentially destructive and as the cause of desiccation, floods, soil erosion and rangeland deterioration (Grove 1989:178–79). Brown addressed the complex relation between desiccation and *veld* fires which according to him were the main cause of degradation processes. The burning of *veld* was a common management practice applied by many land users in order to clear the land of dry grasses and invader plants; to produce ash, which was an excellent fertiliser for the soil; and to encourage plant growth. Brown argued for the abandonment of *veld* fires, except for the burning of belts around houses or forests, and promoted the resting of rangelands and the afforestation of the Cape. Furthermore, official concern arose in the sheep farming districts over the issues of soil erosion and vegetation change, and particularly that of the increase of unpalatable plant species and poisonous herbs (Beinart 2008:99–127). To sum up, major environmental problems in pastoral areas in the Cape had already been addressed in the nineteenth century. This does not mean, though, that the lines of argument were not contested and redefined later. Despite the debates amongst academics and practitioners, the colonial government in the Cape did not invest large amounts of money in conservation efforts at the turn of the twentieth century. Environmental legislation was aimed first and foremost at the curtailment of resource exploitation, but not at a proactive rehabilitation.

At the turn of the century, officials of the Department of Native Affairs in the Cape increasingly emphasised environmental issues and regulated the control of weeds and pests as well as the treatment and prevention of animal diseases (Beinart 1989:148–49). These aspects were also discussed in Thaba Nchu and Sediba, where initial reserve regulations were aimed at the cutting of trees and bushes, control of rangeland use, eradication of *Xanthium spinosum* and other noxious weeds, and the dipping of sheep to prevent scab disease. While those early regulations focussed on particular resources or pests, the conservationist approach became more holistic – and more invasive – in the following decades. Thaba Nchu was one of several districts declared as prime sites of soil erosion, which resulted in the implementation of numerous conservation measures.[71] Before I elaborate

[71] Other districts that were the focus of early soil conservation works were Glen Grey and Whittlesea in Ciskei; Nquntu and Bergville in Natal; and Tzaneen in Transvaal, and particularly the Herschel district in the Cape (Beinart 2008:349).

on the interventions in more detail, I will briefly shed light on the official narratives of soil erosion prevalent at that time.

In the first decades of the twentieth century government officials regarded, *inter alia*, overstocking, kraaling animals at night and ploughing against the contour as the main drivers of soil erosion processes. As counteractive measures, government officials suggested the limitation of livestock numbers and kraals, as well as the establishment of fenced camps with dispersed watering points. Moreover, ploughing and the herding of livestock were to be conducted along the contour in order to prevent gully formation. Officials were most anxious about the effects of gully erosion and so-called *dongas*[72], but were also concerned about sheet erosion processes, which are characterised by the removal of fertile topsoil due to runoff of rainwater (Beinart 1984:56p.). Since the late nineteenth century, missionaries, travellers and government officials critically observed the increase of *dongas* in Basutoland (Showers 1989; McCann 1999:141–73). In contrast to the landscape of the Orange Free State, gully erosion was most striking there (McCann 1999:141). To control and prevent this type of erosion government officials suggested that *dongas* were to be filled with stones and barriers of wood, or planted with trees. Sheet erosion processes were, at least in the earlier decades of the twentieth century, perceived as a lesser threat in Basutoland (Showers 1989:269, 278). This was apparently not so in other parts of Southern Africa, where sheet erosion was regarded as the major problem. On cultivated lands, government officials experimented with levelled terraces or ridges of heaped up earth along contour lines (Beinart 1984:57–58).

From the late 1920s onwards, the issue of soil erosion was increasingly promoted on the political agenda of the South African government. There are several reasons why the topic became prominent at that point of time. A serious drought in 1919/20 causing massive livestock losses – about 200,000 head of cattle and 5.4 million head of small stock (Beinart 2008:250) – resulted in the appointment of the Drought Investigation Commission chaired by the Afrikaans-speaking senior agricultural officer Heinrich Sebastian Du Toit. In its final report in 1923, the Commission concluded that the South African drought was not caused by a lack of rain

---

[72] The term *donga* is a Nguni word but is commonly used in Southern Africa to describe the deep gullies resulting from running water (McCann 1999:170, endnote 2). In Setswana, *donga* means *lengope* (Pl. *mangope*).

alone but was also a consequence of bad land management, particularly because of over-stocking, kraaling and transhumance (Delius and Schirmer 2000:720; Beinart 2008:251–57). The findings of the Commission, which focussed mainly on pastoral areas, stimulated a debate on the necessity of a broader governmental intervention with regard to soil conservation. As a consequence, official authorities established a National Soil Erosion Council, called a national conference on that topic and institutionalised soil conservation as an integral part of agricultural extension services (Beinart 1984:60). Besides national debates, officials were also influenced by a growing body of international literature on that topic, and by documented experiences in other countries, particularly the USA, where severe droughts in the 1930s led to the Dust Bowl, causing serious damage to environment and agriculture (Delius and Schirmer 2000:719; Beinart 2008:333–34).[73]

With the growing awareness of soil erosion, official concerns increasingly concentrated on 'native reserves', which by the late 1920s were commonly described as eroded (Beinart 2008:338). Overstocking was declared as one of the major problems, a phenomenon rooted in the cultural importance of cattle for a man's status and as bridewealth payment. Anthropological accounts of the 'cattle complex' (Herskovits 1926) provided the scientific reasoning for official concerns. Moreover, the practice of shifting cultivation was regarded as resource exploitation as arable land was scarce and population numbers rising. In short, as Beinart (2008) puts it, the farming methods of black South Africans 'seemed destined to precipitate ecological collapse' (ibid.338).

Without doubt, there was increasing pressure on natural resources in black rural areas between the mid-nineteenth and early twentieth century. On the one hand, land accessible to peasants was becoming scarce as 'native reserves' were set aside for black occupation and tenants were driven off the

---

[73] In the article *Soil Erosion, Conservatism and Ideas about Development: A Southern African Exploration 1900-1960* Beinart (1984) initially stressed the importance of the USA as a kind of role model for environmental conservation measures introduced in South Africa. Later Beinart (2008) qualified this statement and suggested that environmental degradation in the Great Plains did indeed influence South African debates, but that the conservationist thinking originated mainly from debates that arose in the Cape Colony in the nineteenth century (ibid. 333–34).

white-owned farmland. More and more people were pushed into the reserves due to rigid segregation policies, such as the Natives Land Act of 1913 and the Native Trust and Land Act of 1936. On the other hand, agricultural production considerably intensified during that time. The hoe was replaced by the plough, which required in turn that peasant households keep more livestock so as to have sufficient draught power. The engagement in labour migration enabled households to invest cash in cattle (Beinart 1984:66). In the Transkei, significant increases in sheep and cattle numbers took place during the 1920s (Beinart 2008:339, figure 10.1). In Basutoland too, the negative effects of agricultural intensification – initiated by missionaries – had become obvious at that time (Showers 1989; McCann 1999). Having said this, it would be wrong to assume that rural economies in Southern Africa were transformed everywhere in the same manner and intensity (Beinart 1984:67). Within the same district developments varied considerably, as the case of Thaba Nchu suggests. Rapid urbanisation and a reduction of resources were less problematic in Sediba than in the Thaba Nchu reserve, where grazing for livestock was already limited in the 1930s.

According to Beinart (1984), one reason for the general criticism of black peasantry was based in the government's aim to secure white agricultural production. Using the example of livestock diseases, he argues that their effective control required a broader approach as diseases would not stop at the boundaries between black and white farming areas (ibid. 62). Such reasoning not only made sense with regard to animal sicknesses but also extended to other pests, as the following example illustrates. In 1935, the Thaba Nchu locust officer threatened 'serious actions' against the headmen of the Sediba reserve locations if they would not prompt the citizens to assist in the eradication of locust swarms. Although he argued that the inhabitants should be worried about the destructive effects the pests would have on their own crops, the locust officer was presumably not only concerned about their land. His brother's farm directly adjoined the 'native reserve' and was thus just as threatened with complete destruction as the neighbouring land.[74] The two examples show that even if reserves – and, later, homelands – were treated separately in political terms, ecological dy-

[74] SAB NTS 8142 2/341 part 2, Minutes of Meetings of the Sediba Native Reserve Board, 25 January 1935.

namics were not confined to the area within or outside the reserve boundaries. Policies had to address threats, such as diseases and plagues, more broadly. In other words, the boundaries of a social-ecological system are not fixed or closed, but in fact can shift depending on which aspects are considered.

The official desire for the protection of white agriculture does not provide an exhaustive explanation for the increasing governmental intervention with regard to conservation. Another reason was that government reports on soil erosion in 'native areas' were largely generalising a phenomenon that did exist, but not necessarily everywhere and not to the same extent and intensity (Beinart 1984:63–65). As Showers (1989) shows for Basutoland, official accounts of soil erosion were based on localised observations of sites exposed to high levels human and animal pressure, e.g. land next to roads and villages. Nevertheless, colonial reports of the early twentieth century constructed the narrative of a country that was eroding away (Showers 1989:268–72). In other parts of Africa too, degradation narratives emerged that justified official interventions and the control of local land use practices (for instance Fairhead and Leach 1996; Leach and Mearns 1996). Bearing in mind these incidences, it seems necessary to critically analyse contemporaneous accounts of soil erosion in the Thaba Nchu district.

4.2 Searching for Evidence of Ecological Crisis

A new stimulus that enhanced the idea of pertinent environmental crisis and legitimised government intervention in reserve areas was provided by the report of the Native Economic Commission in 1932. Although not all the report's recommendations were taken on, it had a strong influence on official thinking in the following decades (Delius and Schirmer 2000:720). The Commission, headed by agricultural economist Dr John Edward Holloway, was appointed because politicians, officials and mine-owners were increasingly concerned about the future of the 'native areas'. The mine-owners worried about the supply of substantial labour force from the rural areas. The government members feared that black people would migrate from destitute reserves into townships of urban areas inhabited by white people (ibid.).

The report argued that South Africa was shaped by two economies that were in conflict with each other; i.e. the advanced money economy dominated by whites, and a 'primitive', black subsistence economy. The conflict could only be solved either by 'the extermination or the absorption or the development of the backward race' (Union of South Africa 1932:3). The main problem, according to this argument, was rooted in the reserves, where over-stocking was likely to lead to an ecological collapse within the next two decades. Most severe was the situation in Middeldrift, Herschel and Glen Grey in the Ciskei, but other 'native areas' were also threatened:

> 'Unless precautionary measures are taken against overstocking the condition in the Transkei and the Native areas in the rest of the Union will be to-morrow what that of the Ciskei is to-day. The same causes are at work there, and they will inevitably produce the same effects in the near future – denudation, donga-erosion, deleterious plant succession, destruction of woods, drying up of springs, robbing the soil of its reproductive properties, *in short the creation of desert conditions.*' (ibid. 11; emphasis in the original)

Besides over-stocking, cultivation methods were described as destructive as people allegedly used no crop rotation or fertilisation. Moreover, the Commission complained that farmers used low quality seeds and did not remove the stumps of the harvested crops (ibid.). However, the fact that animals were driven onto those fields to eat the remainders of the harvest and concurrently fertilise the soil was not taken into consideration.

The Native Economic Commission made several suggestions for approaching the 'native economic question', though some of them were not new. Most important from a conservationist perspective were (i) limitation of livestock, especially donkeys and goats, since they were the main threat to the vegetation; (ii) improvement of the livestock breeds; (iii) strengthening of the agricultural education, particularly through 'native demonstrators'; (iv) fencing-off of rangelands; (v) improving the water supply in the rangelands; (vi) application of soil-reclamation measures in heavily eroded areas; (vii) afforestation; (viii) purchase of better seeds; (ix) better access to markets; and, in the long term, (x) the concentration of the settlements. Furthermore, it was pointed out that the purchase of additional land would be necessary to extend the scheduled areas set aside in the Natives Land Act of 1913. The Commission wrote against the 'one man, one lot' principle practised in the Transkei, as that would allow land use only at a very low economic level. Instead, a certain 'scope' was to 'be given for advanced natives' (ibid. 15–22): in other words, progressive farmers should be allowed to cultivate more than one land holding. Needless to say, the report gave little evidence of how this should be done in a situation of severe land scarcity.

As before, debates about ecological deterioration in South Africa were influenced by experiences reported in other countries, and particularly in other British colonies. Commission reports in both Kenya and South Africa repeatedly referred to each other to support their arguments; e.g. the Kenyan Land Commission of 1932/34 cited the South African Drought Investigation Commission and Native Economic Commission which in turn drew on the Kenyan Agricultural Commission of 1929 (Union of South Africa 1932:16; Anderson 2002:135). The international debates were not restricted to government and colonial offices, however, but were also reported in agricultural magazines. For instance, *The Farmer's Weekly* published an article about the 'moral justification' of destocking in the devastated Kenyan reserves.[75] Debates in both southern and eastern Africa cross-fertilised with ideas about appropriate conservation measures. Hence it is not surprising that rehabilitation and development policies in Kenya and South Africa show obvious similarities, such as compulsory culling, enclosure of rangelands and construction of irrigation schemes (for Kenya see Anderson 2002).

The Native Economic Commission also gathered evidence in Thaba Nchu, 'but circumstances did not allow of more than a cursory inspection of part of the Native areas in the District, one adjoining the town and another known as the Seliba Reserve a little way outside' (Union of South Africa 1932:46). An addendum by the lawyer Frank Archibald William Lucas gives an overview of the Thaba Nchu district, its history of land ownership and its socio-economic conditions. Lucas stresses several times that the Barolong owned farms that were severely overstocked and degraded. Apart from that localisation, it often remains vague as to what part of the district – i.e. the private farms, the urban locations in the Thaba Nchu reserve or the rural locations in the Sediba reserve – Lucas' statements are actually referring to. With regard to the ecological status, he came to the conclusion that the '[v]egetation is becoming scanty and the bush is being destroyed

[75] SAB NTS 7335 127/327 part 2, Press-clipping 'Over-stocking problem in Kenya', The Farmer's Weekly, 16 November 1938.

which is causing enormous soil erosion' (ibid. 180). Irrespective of the question of whether a lawyer has the expertise to judge the state of the environment, it is essential to ask to what extent a degradation of the *veld* was actually occurring. When answering this question, one should keep in mind that the early 1930s marked a critical phase in large parts of Africa, as it was a time of a protracted drought period which led to massive livestock losses and harvest failures in southern and eastern African countries (see for instance Murray 1981:14–15; McCann 1999:165; Anderson 2002:76–77; Jacobs 2003a:137–38; van Onselen 2005:173–79; Bollig 2006:161). The reserves in Thaba Nchu district too, were stricken by famine (Chapter 3.4). The Native Economic Commission's assessment of the reserves' status seemed to gain validity in the context of the prevalent ecological and economic conditions. Hence, the call for rigorous interventions by the state became louder.

4.3 Providing the Legal Basis for Government Intervention

After the passing of the Native Trust and Land Act of 1936, all land defined as 'scheduled areas' was vested in the SANT. Moreover, the Trust was responsible for the purchase of the so-called 'released areas' that were to be added to the reserves. Land that was purchased by the SANT was automatically subject to the 'betterment' proclamation of 1939 (see below) and improvements on that land were financed by the Trust's funds. The land act and the statements on its policy formulated afterwards laid the cornerstone for the reclamation of the reserves. A major emphasis was put on stock limitations as a means of rehabilitation of the devastated lands (Yawitch 1982:9). Furthermore, the act made provisions for detailed location surveys with the aim of identifying arable, grazing and residential land, and planning the protection of forest resources and the provision of water (Delius 1996:55). For the Thaba Nchu district a survey-based 'reclamation report' was compiled in 1939 by the Pretoria-based controller of native settlements, the Director of Native Agriculture and an engineer.[76]

Proclamation 31 of 1939 was aimed at the 'Control and Improvement of Livestock in Native Areas' and legitimated draconian interventions with

---

[76] SAB NTS 7516 678/327, Locations Reclamation Report, 14 November 1939.

regard to pastoral farming.[77] It enabled the NAD to (i) declare, after consultation with the residents, any reserve or SANT land as 'betterment' area; (ii) assess its carrying capacity[78]; (iii) count all stock in the area; (iv) enforce livestock culls of the surplus animals conducted by assigned culling officers; (v) fix the date by which all culled animals 'shall be either slaughtered or removed from the betterment area'; and (vi) castrate all bulls, stallion, donkey jacks and sheep or goat rams that appear to be too old or not of a high grade. Thus, the proclamation was not only aimed to limit animal numbers but also focussed on a systematic elimination of poor quality stock.

4.3.1 Introducing the 'Betterment' Proclamation in Thaba Nchu

Soon after its passing, the 'betterment' proclamation was introduced in the Thaba Nchu reserve. The quick implementation was also due to the Barolong elite, who were in favour of the new regulations. For them the rangeland condition in the reserve was of particular interest, despite the fact that they owned land in the district or rented it at a comparatively cheap rate (Murray 1992:164). Being legitimate taxpayers, many wealthy farm owners possessed residential certificates for the Thaba Nchu reserve and hence were entitled to hold livestock on those grounds. Dr James Moroka and other private land owners not only lent *mahisa* cattle to reserve residents [79] but also kept parts of their herds for some time in Thaba Nchu to fatten them up:

> 'Many native farmers [private land owners] who have residential rights in the Reserve, are completely defeating the object of stock limitation by grazing in the course of a year many more stock than they are actually entitled to. The following is an example of what takes place. A native farmer had the right to 20 head [of cattle] in the Reserve. These he fattens up and then either sells them outside the Reserve or exchanges them for 20 other stock

---

[77] Union of South Africa, Proclamation 31 of 1939: Control and Improvement of Livestock in Native Areas, Government Gazette.

[78] According to the proclamation's definition, the 'carrying capacity' was measured in cattle units defined as one head of large stock or five head of small stock. The term 'large stock unit' referred to cattle, horses, asses and mules, while 'small stock unit' encompassed sheep and goats.

[79] SAB NTS 7526 717/327 part 1, Memo of the Director of Native Agriculture, Pretoria, 15 March 1941, p. 1.

> which are moved to the Reserve from his farm [...] Many similar exchanges are made in the course of a year, so that instead of actually sustaining 20 definite animals on the Reserve he fattens anything up to 100 head [of cattle] in the area.'[80]

Land owners had an interest in reducing grazing pressure on the reserve rangelands, as they used them as additional grazing lands. On the other hand, access to private farms enabled the landowning elite to circumvent the 'betterment' proclamation. In other words, their economic capital – in the form of land titles – made them less vulnerable to sanctions by state officials. Thus, the elite's approval of the 'betterment' proclamation was a strategic decision from their point of view; after all, cooperation with the state was necessary to maintain their power. Archival sources indicate that 'betterment' was accepted by more groups than just the Thaba Nchu elite, however. The Chief Native Commissioner reported in September 1938 that various chiefs and councils had unanimously agreed to the implementation of the proclamation – then still a draft – for the limitation of stock in their areas.[81] Other case studies have also shown that not everyone disapproved of the idea of 'betterment' entirely, such as the chief of the Shixini area in the Transkei (McAllister 1989:348) as well as some people in Kuruman (Jacobs 2003a:188–89).

In contrast, the members and residents of the Sediba reserve mostly opposed the limitation of livestock numbers. When the 'betterment' proclamation was to be applied to the reserves and SANT land in the Thaba Nchu district,

> 'the residents of the Seliba and Carolusrus[82] Reserves [...] were opposed to such a step. [...] The Assistant Native Commissioner is of the opinion that it will be unwise to force the Proclamation upon the Seliba and Carolusrus Reserves. I am in agreement with this view for I feel that with the application of the Proclamation to the neighbouring Thaba Nchu Reserve the residents of the

---

[80] SAB NTS 7526 717/327 part 2, Assistant Native Commissioner, Thaba Nchu, to the Secretary of Native Affairs, Pretoria, 15 January 1943.

[81] SAB NTS 7335 127/327 part2, Chief Native Commissioner of the northern areas, Pretoria, to the Secretary of Native Affairs, Pretoria, 29 September 1938.

[82] North of the Sediba reserve was a small community of coloureds who were descendants of Carolus Baatje. In 1940, the households were resettled to Thaba Phatshwa east of Thaba Nchu (Murray 1992:259–61).

> other Reserves will in time become educated up to the obvious advantages to be gained and will of their own volition fall into line.'[83]

However, the special treatment of the Sediba reserve soon ceased, as neither government officials nor the Barolong leaders were willing to accept opposition to 'betterment'. In early October 1940, a general assembly of the Barolong under Chief John Sepinare Moroka was held at Sediba location. Of the 1,510 taxpayers resident in the reserve only 270 (about 18 %) attended the assembly, which indicates the people's disapproval of the government and Barolong officials. At the meeting, it was resolved to introduce the Moroka Tribal Levy[84] in order to pay off the SANT for the purchase of land for black occupation. Furthermore, 200 attendees voted in favour for the declaration of the Sediba reserve and the adjunct farms as 'betterment' area. Despite the fact that *de facto* only 15% of the total number of taxpayers agreed on the stock limitation proclamation, the Assistant Native Commissioner regarded the voting as a complete success.[85] Only one month later Government Notice No. 1732 declared the Sediba reserve and its adjoining farms a 'betterment' area.[86] The example illustrates that the notion of 'consulting the residents' that officially legitimised state intervention was merely rhetorical, as approval was only obtained from a minority of the residents. However, it also is doubtful whether those people that agreed to the 'betterment' were in fact fully aware of the actual procedures it entailed. Hendricks (1990) points out that when such consultations of residents took place, the implications of 'betterment' were not comprehensively discussed (ibid. 97).

4.3.2 Modifications of the Proclamation

The 'betterment' proclamation was only rarely implemented before the Second World War. After its outbreak, the shortage of materials and government staff – about one third of the NAD employees were at war – hampered the large-scale execution of livestock culls (Hendricks 1989:316). In

---

[83] SAB NTS 7526 717/327 part 1, Chief Native Commissioner of the northern areas, Pretoria, to the Secretary of Native Affairs, Pretoria, 8 July 1939.
[84] In 1949 the levy was replaced by a grazing fee (Murray 1992:174).
[85] SAB NTS 7526 717/327 part 1, Assistant Native Commissioner, Thaba Nchu, to the Secretary of Native Affairs, Pretoria, 3 October 1940.
[86] SAB NTS 7526 717/327 part 1, Government Notice No. 1732, 1 November 1940.

this regard, the Thaba Nchu district was one of the exceptions as livestock culls continued to be regularly conducted, even during the war period. The rehabilitation approach employed by the NAD changed in 1945 with a statement of policy entitled 'A New Era of Reclamation' made by the Secretary for Native Affairs at a special session of the Ciskeian General Council. The 'New Era' document[87] claimed a 'need for a bigger effort', as in most reserve areas 'the downward trend has not yet been stopped'. Smit called for a 'settlement of the land in the most advantageous manner' and the demarcation of zones of residential, arable and grazing land, among other things. The latter had to be divided into camps to allow the rotation of the herds and the resting of certain parts of the rangelands. Arable lands were to be redistributed and all land users would receive an equal share of land. Moreover, 'rural villages' were to be developed where 'surplus people' would have to settle. The planning was to be supervised by so-called 'planning committees' consisting of an administrative officer of the NAD, an agricultural officer, an engineer and a soil chemist, as well as a surveyor, a draughtsman and a clerical staff member.[88] De Wet (1995) points out that a new way of thinking was expressed in this document, evident in the assertion that there was not enough land to enable every household in the reserve to derive a livelihood from agriculture, and that a viable subsistence agriculture could only be achieved by separating full-time farmers and full-time wage labourers (De Wet 1995:43). This idea was later taken on by the Tomlinson Commission and elaborated in more detail (Chapter 5.3).

With Proclamation No. 116 of 1949, named 'Limitation, Control and Improvement of Livestock and of Pastoral and Agricultural Resources in Native Areas', government authorities gained more power of intervention than with the original 'betterment' proclamation (Hendricks 1990:117; De Wet 1995:43). From then on, in addition to control over livestock, which affected only the grazing areas, commissioners had control over all types of land. The demarcation of land into specific zones was only one aspect; much more far-reaching was that the Native Commissioner could set up rules for the use of rangelands and the manner of their cultivation, such as

[87] D.L. Smit, 'New Era of Reclamation': Statement of policy made by Mr. D.L. Smit, Secretary for Native Affairs at a Special Session of the Ciskeian General Council at Kingwilliamstown in the 8th January 1945 (Government Printer, Pretoria, 1948); in SAB NTS 8899 140/362 (1) part 2.
[88] Ibid.

cropping, tillage, crop rotation and fertilisation. Moreover, the proclamation empowered the government to terminate any right of use or occupation of land in case that land was in need of reclamation work or if it was being used for purpose other than that defined by the Native Commissioner.[89]

Although this was an extension of the Native Commissioner's powers, one should keep in mind that the government official had already similar rights as a chairman of the reserve boards in the Thaba Nchu district. For instance, theoretically he was entitled to remove households that erected their homesteads in an unsuitable area. However, I found no instances where this was actually enforced by the Native Commissioner. Moreover, until the discontinuation of the reserve board in 1953 (Chapter 5.2), the allocation of land was effected through the headmen of the respective village and the approval of the board members and the chair. The only exception was the Grobler irrigation scheme, which is described below in detail. Here, the Native Commissioner had more powers in terms of allocation and cancellation of leasehold rights. Yet one thing did change after the passing of the new proclamation, in that the Department of Native Affairs increasingly tried to intervene with regard to cultivation practices (see below).

To sum up, a conservationist thinking had already evolved in the Cape Colony in the nineteenth century. The main threats to the soils and vegetation of the rangelands, i.e. over-stocking and kraaling, had been identified at an early stage. While initially concerns for the most part addressed the effects of European settler agriculture, placing the blame on black people's farming methods was common amongst government officials at the turn of the century. The debate accelerated with the publishing of the Native Economic Commission's report in 1932, which painted a rather bleak and pessimistic scenario of the situation in the 'native reserves'. Intervention in those areas seemed urgently necessary in order to prevent an ecological collapse. Shortly after the report, measures against soil erosion were implemented in the districts that were said to be eroded (Chapter 4.4.3). But the alleged crux of soil erosion, i.e. high numbers of inferior livestock, could not be solved by constructing contour lines and filling up *dongas* with stones. Strict control of animal husbandry, so the argument went, was necessary to prevent a further deterioration of the rangelands. It did not

[89] Union of South Africa, Proclamation 116 of 1949: Limitation, Control and Improvement of Livestock and of Pastoral and Agricultural Resources in Native Areas, Government Gazette.

take long until the first proclamation was formulated, which was aimed at the direct control of the major 'evil' of peasant agriculture, and yet after the Second World War it was claimed that more regulation was needed to achieve a real reclamation of SANT land. This shift 'from conservation to control' (Delius and Schirmer 2000:727) became even more evident from the 1950s to 1970s (Chapter 5). During that period, the initial policy changed its focus from reserve rehabilitation to stabilisation and resettlement (Hendricks 1989:307). Beinart (2008) argues that this change of mind was based on a 'post-war optimism', the idea that a state could actively guide an entire society, combined with a renewed willingness to pay for the costs of such programmes (ibid. 365).

4.4 From National Debates to Local Intervention, 1930s–1950s

Being one of the first districts chosen for soil conservation works, Thaba Nchu was in official accounts commonly described as overgrazed and degraded. Comprehensive state intervention with regard to land use started in the 1930s. At first, the government only approached issues of soil erosion, but after a few years conservation measures became more wide-ranging and aimed at the converging fields of erosion control, *veld* and water management, and livestock husbandry. The NAD created separate files for each subject; therefore, thematically subdividing the historical analysis of environmental regulation seems justified. Having said that, it must be stressed that of course the management approaches to soil, vegetation, water and livestock were interconnected, and thus affected one another. Before I provide a detailed description of the rehabilitation efforts in the Thaba Nchu district, I will briefly introduce the main actors involved in conservation work at the local level.

4.4.1 Local-Level Bureaucracy in the Thaba Nchu District

The implementation of national acts, proclamations and policies and the supervision of conservation efforts in the district required a dense network of staff whose work had to be coordinated. On the local level, the hierarchies were led by native commissioners and their assistants, who were based in Thaba Nchu and were responsible for the administration of the issues of black people living in a specific district. They had to report regularly to the NAD (De Wet 1995:43–44). As outlined above, the proclamations of 1939 and 1949 endowed the commissioners with manifold powers with regard to land use.

The numbers of the agricultural personnel fluctuated as people were employed only temporarily or were transferred to other areas. In 1944, the staff included one agricultural officer in Thaba Nchu; an engineer; four overseers, who were based at different stations across the district; and four foremen who stayed in Thaba Nchu and were responsible for specific issues such as dams or erosion. Moreover, several employees were drawn from the local population: two tractor drivers; a bricklayer and a general worker; four local agricultural demonstrators, and seven rangers.[90] The latter two groups were at the lower end of the administrative hierarchy and lived at locations in the reserves. Rangers were in charge of the fencing as well as *veld* and wood resources; they monitored the compliance of the rotational grazing system and controlled the wool classification during shearing season.[91]

The agricultural officer's duties contained not only agricultural issues in general, but also the supervision of soil erosion works and, from the 1940s onwards, the enforcement of livestock culling in particular. Moreover, it was part of his job to write monthly reports to the Native Commissioner about weather, cultivation, *veld* and animal conditions, inoculations, and the holding of lectures, as well as miscellaneous extension and development work such as the distribution of seeds.[92] The training offered by the agricultural staff addressed various topics ranging from the usage of advanced technologies to the shearing of sheep. From time to time it seems that the agricultural officers drew in their reports a less bleak picture of the conditions in the Thaba Nchu area than did those officials from the outside who came only now and then for a short visit. There are two possible explanations for this. On the one hand, local officers did not want to appear in a bad light and therefore gave their reports a positive spin. On the other hand, those officers had a much better knowledge of the region and the actual conditions there. In contrast to the Thaba Nchu-based officer, the agricultural overseers were responsible for day-to-day supervision. One of

---

[90] SAB NTS 8142 2/341 part 1, Minutes of Meetings of the Sediba Native Reserve Board, 19 July 1927; SAB NTS 7394 335/327 part 2, Monthly agricultural report: Personnel, September 1944.

[91] SAB NTS 7526 717/327 part 2, Senior Agricultural Officer and Agricultural Officer, Thaba Nchu, to the Assistant Native Commissioner, Thaba Nchu, 12 August 1944.

[92] Monthly agricultural reports between 1936 and 1945 can be found in SAB NTS 7394 335/327 part 1, SAB NTS 7394 335/327 part 2.

them lived on a portion of the Rakhoi farm next to the Sediba reserve (Figure 1). Overseers and foremen were often drawn from the lower classes of white society and many of them enjoyed exercising power over black people (Delius and Schirmer 2000:722).

Native affairs commissioners often changed after a couple of years. As said above, this was also true for agricultural personnel. Commissioners and other staff differed not only with regard to sympathy towards local populations, but also in terms of ideas and their focus for action. In some cases, this resulted in sudden – and unpredictable – changes of policies. Sometimes when government staff were too ruthless or engaged in illegal activities, residents of the Thaba Nchu area approached the next higher level of the hierarchy in order to make complaints. In some cases, residents were even successful, and the problematic staff were transferred to another region.[93] Generally, however, people had to bear the uncertainties caused by personnel fluctuations, which resulted in locals' reduced trust in state officials.

4.4.2 'It's When the Trouble Started': The Grobler Irrigation Scheme

The Grobler scheme was the first irrigation project on SANT land in the district; two more schemes were built about 20 years later in Feloana and Woodridge. Interestingly, the soils of all three schemes were in the late 1980s described as unsuitable for irrigation by soil scientists of the UFS (Chapter 6.3.2). Ever since its construction the Grobler scheme had served as a government showcase, despite the fact that its history was never unproblematic. During a community meeting with the Department of Agriculture and Rural Development in September 2011, government officials stressed the uniqueness of the irrigation scheme and asserted that it had been an important agricultural development project in the past. They completely disregarded, perhaps unwittingly, the fact that the scheme had never been a genuine success. By focussing on the Grobler scheme, I will provide an illustrative example of rehabilitation debates and politics and the mistreatment of local populations, as well as one of technical, social, financial and ecological bad planning and implementation.

---

[93] In 1936, for instance, the Assistant Native Commissioner was replaced after the headmen of the Sediba Reserve complained about him (SAB NTS 8142 2/341 part 2, Minutes of the Meeting of the Sediba Native Reserve Board, 19 June 1936).

Construction of the Scheme

In the Thaba Nchu district the rainfall during the rainy seasons of 1929/30, 1930/31 and 1931/32 and 1932/33 was considerably lower than average, and the ecological system was in a critical state (Chapter 3.4). Besides handing out maize as a means of short-term drought relief, the NAD officials regarded long-term measures as necessary to prevent such crises in the future. Therefore, the circle engineer of the Irrigation Department in Bloemfontein, E. Fincham, was sent to the Thaba Nchu district in 1933. He was instructed to identify three possible sites for the construction of large-scale dams (Setswana: *letamo*) that would serve not only for livestock purposes but also for irrigation projects.[94] One suitable dam site was about 24 km northeast of Thaba Nchu town, in the immediate vicinity of the Sediba location. 'Here, a small spruit [small river] runs through a short rocky (dolerite) gorge at the entrance of which it is proposed to build for storage. A fair basin exists which, were the catchment area bigger, could be made to hold a good body of water'.[95] Two other possible sites were identified north and south of the Kgalala location; however, only the dam at Sediba was completed. There is no evidence in the archival sources that the local population was ever asked whether they wanted or needed a dam or irrigation project. It is also not recorded whether the chief was consulted; in any case, the chief gained civil jurisdiction over Sediba only in 1939 (Chapter 3.2.2). In one of my first conversations with elders from Sediba, a woman said, with regard to the dam construction, 'it's when the trouble started';[96] thereby implying that it marked the begin of governmental interventions that eventually culminated in the resettlement of hundreds of people in Sediba.

The site was situated at the southeastern boarder of the Sediba reserve. It was estimated that the designated dam would encroach to the extent of about 25 ha on the adjoining farm, 'Milton' – a portion of farm Rakhoi No. 5 – owned by Percival Robert Miller. Fincham suggested that the neigh-

---

[94] SAB NTS 7981 259/337 part 1, Report of the circle engineer of the Irrigation Department, Bloemfontein, on irrigation possibilities in the Thaba Nchu district, 16 March 1933.

[95] Ibid. 2.

[96] Lydia Pobe during a group interview with village elders, Sediba, 03 November 2010.

bour 'should be approached diplomatically' for a so-called right of servitude for storage. The circle engineer reckoned that Miller would be happy to gain access to a dam for his livestock basically for free.[97] Yet the landowner was not willing to offer part of his land without compensation. In addition to free usage of water, he demanded a piece of Sediba reserve land of a size equal to the area used for the dam on his farm. He was eager to obtain certain dryland fields, i.e. fields that were not irrigated, but this was strongly opposed by the village residents. In the end, Miller received a financial compensation of £150.[98]

Officials estimated that the costs for the dam construction would amount £2,600. This was a rather cheap calculation which could only be achieved by paying low wages of eight pence per day to the local labourers. Working for a period of twelve months, the employees would each have earned about £12. Compared to a daily wage of 30 pence – which the Native Economic Commission regarded as a reasonable income for a black person (Union of South Africa 1932:149) – the budgeted salary for the dam workers was exploitative. However, officials looked for even cheaper options and the idea arose to pay 'the natives' with food only, which reduced the prospective construction costs by £200.[99] Sediba inhabitants who were still suffering from starvation had no choice but to accept this meagre payment. In August 1933, about 180 persons agreed to work for food alone (Chapter 3.4).[100] At that time, it was common practice to pay food rations to black people employed for conservation work, as examples from Transvaal[101] or Ovamboland in South-West Africa show (Hayes 1998).

An officer of the Irrigation Department who had previously worked at the Gopani dam in Transvaal was selected as foreman for the construction of

[97] SAB NTS 7981 259/337 part 1, Report of the circle engineer of the Irrigation Department, Bloemfontein, on irrigation possibilities in the Thaba Nchu district, 16 March 1933, p. 7.

[98] SAB NTS 7981 259/337 part 1, Native Commissioner, Thaba Nchu, to the Secretary of Native Affairs, Pretoria, 12 September 1933.

[99] SAB NTS 7981 259/337 part 1, Calculation of the costs for the proposed dam at Sediba, author unknown, 8 July 1933.

[100] SAB NTS 7981 259/337 part 1, Secretary of Native Affairs, Pretoria, to the Director of Irrigation, Pretoria, 16 August 1933.

[101] A letter by the Secretary of Native Affairs suggests that in Zeerust in Transvaal, workers were only paid with maize; SAB NTS 7981 259/337 part 1, Secretary of Native Affairs, Pretoria, to Director of Irrigation, Pretoria, 23 August 1933.

the Sediba dam. It was envisaged that the work would begin in September 1933; however, the effects of the drought thwarted the official planning. The villagers' oxen, which were to be used for the work, were in a very bad condition. The director of the Irrigation Department, who paid a short visit to the Thaba Nchu district, estimated that the animals would need at least four weeks of additional feeding before they could be used for heavy construction work. In the meantime, labourers should commence with work that could be done without oxen, such as clearing the site, putting up sheds and digging a cut-off trench.[102] Subsequently, however, the lack of animal power remained problematic. In April 1934, when the heavy work at the dam started, the chairman and members of the Sediba reserve board argued about 125 oxen that had to be provided by the local population – apparently including inhabitants of the other locations – for the dam construction. As a consequence of the drought oxen were a scarce resource; those still alive were in use for ploughing.[103] Eventually, the NAD provided oxen for the dam construction; soon after the work was finished they were shifted to the Herschel district.[104]

In November 1934, about one year after the foreman had arrived in Sediba, the dam was completed. Three days before Christmas Eve a meeting was held at the Sediba dam, which was attended by Native Commissioner Meaker, Assistant Native Commissioner Wilson, agricultural officer Wolmarans, and the interpreter, I.T. Mokgothi, as well as the headman of the location, the local member of the reserve board and about forty residents. During that meeting, the Native Commissioner suggested renaming the dam the 'Grobler Irrigation Scheme', after the then Minister of Native Affairs, Peter Gert Wesler Grobler.[105] One important issue that was raised several times during the meeting was the question of who should be entitled to a plot in the irrigation scheme, which in fact still had to be constructed. Officials stressed that it was not their intention to give plots to inhabitants of other locations, and that the first people to receive a piece of

---

[102] SAB NTS 7981 259/337 part 1, Director of Irrigation, Bloemfontein, to Secretary of Native Affairs, Pretoria, 8 September 1933.
[103] SAB NTS 8142 2/341 part 2, Minutes of Meetings Sediba Native Reserve Board, 10 April 1934.
[104] SAB NTS 8142 2/341 part 2, Minutes of Meetings Sediba Native Reserve Board, 21 September 1934 and 23 November 1934.
[105] SAB NTS 7981 259/337 part 1, Minutes of Meeting held at the Sediba Dam, 21 December 1934.

ground would be those that had no access to arable land in Sediba at all. Furthermore, it was announced that the distribution of plots would be undertaken by the agricultural officer with the help of the headman and the people.[106] Not addressed during the meeting was the question of what was to be done with the people's dryland fields that were situated at the prospective irrigation site. Archival sources indicate that it was planned to offer those people another dryland field as compensation.[107]

Figure 6 Furrow of the Grobler irrigation scheme in 2010 (photograph: C. Naumann)

In January 1935, the planning of the furrows (Figure 6) and the layout of the plots commenced. The construction of furrows, feeders and fences was again done by local men employed 'at the lowest possible rates' under the

---

[106] SAB NTS 7981 259/337 part 1, Minutes of Meeting held at the Sediba Dam, 21 December 1934.

[107] SAB NTS 7981 259/337 part 1, Secretary of Native Affairs, Pretoria, to the Native Commissioner, Thaba Nchu, 31 January 1935.

supervision of the local agricultural officer.[108] During this initial construction work the first problems occurred. The materials available were insufficient – e.g. timber – or inadequate. The concrete that had been used to build parts of the furrows was too rough and thus not water-resistant: those parts of the furrows collapsed shortly after their construction. Other parts were filled up with soil after rain fell.[109] In short, the 'everything as cheap as possible' attitude of the NAD – with regard to labour power and materials – began to take its toll even before the irrigation system was completed.

Running the Scheme

The construction problems not only resulted in higher costs but also in delayed completion. In January 1936, the irrigation scheme was eventually finalised. By then, the scheme comprised 102 ha subdivided in 49 plots ranging from 0.8 ha to 4 ha each; at a later date, the allotments were further subdivided and their sizes standardised. After its completion the Deputy Director of Native Agriculture remarked that four of the plots were too stony and *de facto* useless, and these were therefore not immediately allocated to land users.[110] Aerial photographs from the 1950s indicate that three of these stony fields were later used for cultivation; the fourth was planted with trees by the local agricultural officer.[111] An initial irrigation trial performed in March 1936 was not successful as the furrow could not carry the 4 cusecs (about 113 litres per second) for which it was designed. The reasons were that some of the banks were not high enough and the furrow was

---

[108] SAB NTS 7981 259/337 part 1, Secretary of Native Affairs, Pretoria, to the Native Commissioner, Thaba Nchu, 2 March 1935; SAB NTS 7981 259/337 part 1, Acting Director of Irrigation, Pretoria, to the Circle Engineer of the Free State, Bloemfontein, 5 March 1935.

[109] SAB NTS 7981 259/337 part 1, Assistant Engineer, place unknown, to the Director of Native Agriculture, Pretoria, 2 May 1935; SAB NTS 7981 259/337 part 1, Extracts from the monthly reports of the agricultural officer, Thaba Nchu, 19 June 1935.

[110] SAB NTS 7981 259/337 part 1, Memo of the Deputy Director of Native Agriculture, Pretoria, May 1937.

[111] Ibid.

choked by weeds and grasses.[112] Archival documents suggest that the latter issue was permanent problem at the Grobler irrigation scheme.

In 1937, 31 plots within the scheme were allocated to a total of 63 people; nine of them were farming alone, the others shared plots in groups of two to three people.[113] However, the users were *de facto* acting in a grey area. As no regulation for the scheme and its use had been approved, the farmers' rights to the land were not protected. When in July 1938 a draft regulation for the Grobler irrigation scheme was published in the government gazette,[114] the first step was to cancel all existing usufruct rights. The reallocation of the land was done through the Native Commissioner in consultation with the Sediba reserve board. In accordance with the regulations, the maximum size of rentable plots was 1.7 ha, for which an annual fee of 2 shillings and sixpence per 0.4 ha had to be paid. The fee was collected by the reserve board, which was expected to use the money for the maintenance and repair of the scheme. Furthermore, the board was responsible for ensuring that the plots were 'properly levelled, ploughed, fertilised and crops rotated in accordance with good farming practice'. The allocation, transfer and cancellation of land rights were recorded in a register kept by the Native Commissioner. He could cancel the rights of a user for various reasons, for instance (i) when the user did not cultivate the ground for a period of three successive years; (ii) when the user was in arrears with the fees for more than two years; or (iii) when the plot was used for purposes other than those authorised. A plot holder's duties were to supply his labour force free of charge when levelling or any other work needed to be done, to direct water only onto the plot and during the determined times of irrigation, and to use the allotment for cultivation only. Furthermore, land users were not allowed to erect a fence or to plant trees on their plots without the prior permission of the Native Commissioner.[115] Interestingly, no fur-

---

[112] SAB NTS 7981 259/337 part 1, Assistant Engineer, place unknown, to the Director of Native Agriculture, Pretoria, 14 April 1936.
[113] SAB NTS 7981 259/337 part 1, Native Commissioner, Thaba Nchu, to the Chief Native Commissioner, Pretoria, with an attached list and sketch map, 7 June 1937.
[114] SAB NTS 7982 259/377/2 part1, Draft Proclamation: Regulations for Grobler Irrigation Scheme, District of Thaba Nchu, Government Gazette 14 April 1938.
[115] Ibid.

ther details on the land use practices – e.g. the planting of certain crop species, the ploughing techniques used, and the usage of fertiliser – were determined in the regulations.

Government officials from Pretoria encouraged the collective rather than individual use of the plots as done in the dryland fields. Farmers should form associations,

> 'whose objects will be collective and communal crop production and maintenance of the irrigation canals and furrows [...] Each plotholder will be expected to contribute his quota of man and animal power, implements and tools, etc. or to compensate his association in whatever manner is decided on for his shortcomings, if any. He will share equally in either winter or summer grain crops, which his association may produce.'[116]

Director of Native Agriculture, T.G.W. Reineke, stressed in memos that collective working groups should be supported with governmental ploughing units, i.e. oxen and ploughs that could be leased at a lower rate. In contrast, individuals that did not want to work in teams were denied access to such ploughing units.[117] This discrimination was primarily aimed at the better-off farmers, who could afford to work alone. It was consistent with the NAD policy with regard to the culling of livestock (Chapter 4.4.5). Despite the governmental efforts, collective farming did not work well in the Grobler irrigation scheme. In June 1938, the Assistant Native Commissioner suggested that people should not be forced to cooperate. He argued that residents are 'against communal working on a large scale. It would lead to no end of trouble'.[118] Although his statement is in contrast to the oral histories, which stress the importance of cooperative institutions, it is likely that in the case of the irrigation scheme teamwork did in fact fail. First, with the formation of collective cropping groups, the state created new commons on the irrigation scheme. As pointed out in Chapter 3, dryland fields were held individually. Thus, the communal use of arable land

---

[116] SAB NTS 9502 138/400/5, Report on soil erosion works in Thaba Nchu of the Director of Native Agriculture, Pretoria, 3 April 1937.
[117] SAB NTS 7981 259/337 part1, Comment of the Director of Native Agriculture *ex post* written on a report of the Deputy of Native Agriculture on the Grobler Irrigation Scheme, 2 June 1937.
[118] SAB NTS 7981 259/337 part 2, Assistant Native Commissioner, Thaba Nchu, to the Chief Native Commissioner, Pretoria, 9 June 1938.

differed from the socially embedded institutions of land tenure. Second, the top-down approach of land allocation ignored existing social networks and kinship patterns. Maps of the scheme indicate that on several plots, people from different family groups were expected to cooperate.[119] Third, I assume that the repeated cancellation and re-allocation of usufruct and leasehold rights – at first the dryland fields and later the irrigated plots – led to insecurity and distrust amongst the farmers, which reduced their willingness to cooperate.

Despite the teamwork difficulties, the agricultural officer's monthly reports sent between 1936 and 1941 draw a rather positive picture. He repeatedly stressed the growing interest in the scheme among the locals, and the successful yields that irrigation farmers had achieved, particularly in the dry winter of 1940.[120] Surprisingly, a senior agricultural officer of Thaba Nchu, who visited Sediba in 1942, had a contrary opinion, and reported that the scheme had become run down.[121] According to him, the problems were manifold and not restricted to a lack of collective action: (i) people applied conservative farming methods; (ii) water for irrigation was wasted; (iii) and the soil of the plots was degrading. The main cause of erosion was the inappropriate levelling of the fields.[122] Proper cultivation of an irrigation scheme required advanced equipment in order to prevent the plots from washing away. However, even government officials realised that local land users did not have the necessary agricultural capital for proper levelling.[123] Yet the people lacked not only capital but also experience and knowledge of irrigated cultivation. The last of which was true of both inhabitants and the agricultural demonstrator who was based there.[124]

---

[119] SAB NTS 7981 259/337 part 1, Native Commissioner, Thaba Nchu, to the Chief Native Commissioner, Pretoria, with an attached sketch map of the irrigation scheme, 7 May 1937.
[120] SAB NTS 7982 259/337/1, Monthly reports of the agricultural officer between February 1936 and January 1942.
[121] SAB NTS 7981 259/337 part 2, Senior Agricultural Officer, Thaba Nchu, to the Secretary of Native Affairs, Pretoria, 4 June 1942.
[122] Ibid.
[123] SAB NTS 7981 259/337 part 2, Assistant Native Commissioner, Thaba Nchu, to the Secretary of Native Affairs, Pretoria, 8 June 1942.
[124] SAB NTS 7981 259/337 part 2, Director of Native Agriculture, Pretoria, to the Controller of Native Settlements, Pretoria, undated 1942.

In 1944, the furrow and feeder system of the scheme was renewed and the fields were re-levelled and extended by about 17 ha. In connection with the extension work, plots were again redistributed.[125] In the same year, responsibility for the scheme was handed over from the Sediba native reserve board to the SANT. Reserve inhabitants were thereby deprived of the little decision-making power they had with regard to the administration of the irrigation scheme.[126] However, the change of the authority did not result in a better management. In 1948, the senior agricultural office was outraged after a visit of the irrigation scheme:

> 'I was so disappointed that I could not express my view in words [...] the maize in this scheme is a shame compared to drylands adjoining the scheme. Cultivation is almost non-existant [sic] and the lands are no longer level. [...] I also found that plotholders are in arrear with their rents and the whole scheme is best described in one word as a *flop*.'[127] [emphasis added by author]

To sum up, the Grobler irrigation scheme was meant to be a prestigious development project benefitting the inhabitants of the Sediba reserve. However, as the case study has shown, its construction, maintenance and utilisation were characterised by various problems and conflicts from the very beginning. In the following, I will pay more attention to conservation practices in the Thaba Nchu district in general. I want to thereby illustrate that the scheme was not an exceptional case of bad planning but rather representative of the more general rehabilitation efforts of the NAD.

4.4.3 Combating Soil Erosion in Rangelands and Fields

In December 1917, soil erosion is mentioned for the first time in the minutes of meeting of the Seliba reserve board. Officials observed with concern the formation of *dongas* in several locations. Problematic from the chairman's point of view was the attitude of the local inhabitants. They

---

[125] SAB NTS 7981 259/337 part 2, Sediba Irrigation Scheme: summary of proposals as a result of discussion between Director of Native Agriculture and Senior Engineer, 5 January 1944.

[126] SAB NTS 7982 259/337/2 part 2, Secretary of Native Affairs, Pretoria, to the Assistant Native Commissioner, Thaba Nchu, 3 March 1944.

[127] SAB NTS 7981 259/337 part 3, Senior Agricultural Officer, Thaba Nchu, to the Chief Native Commissioner of the western areas, Potchefstroom, 9 March 1948.

were said to 'neglect' the subject of soil erosion, with dangerous consequences for the land. To prevent the spreading and deepening of *dongas* headmen were advised to block the gullies with stones at both ends and in the middle.[128] Small *dongas* were to be treated first to avert further deterioration.[129] Despite this early attention, the subject of soil erosion was rarely mentioned in either the Thaba Nchu or Seliba reserves for the next couple of years. Official environmental concern focused mainly on grazing pressure, droughts, locusts and unwanted weeds, but not explicitly on soil erosion.

Combating soil erosion in the Thaba Nchu district became a major focus for NAD officials in the mid-1930s. Comprehensive conservation work was initiated in 1934; at first in the Thaba Nchu reserve and shortly afterwards also in the Sediba reserve. It is difficult to estimate in retrospect how severely the areas were actually affected by soil erosion processes. However, archival sources suggest that *dongas* and sheet erosion particularly occurred close to areas with a high human and animal concentration, and in mountainous terrain. In 1936, government officials found in all locations of the Sediba reserve sites that were characterised as eroded and in need of conservation work. In total about 814 ha were affected – approximately 5% of the total area of the Sediba reserve. However, the extent of soil erosion varied strongly and ranged from 'very little erosion' (in Talla, for instance), to 'very badly eroded'. The latter category applied to 171 ha of the grazing land in Rooibult location, and thus covered about 1% of the total reserve area. Hand-drawn maps of the sites indicate that erosion occurred mostly close to roadsides – a situation very similar to that in Lesotho (Showers 1989:274–75). The only exception was Morago, where a *donga* developed at the bottom of a mountain ridge. Here, there were two settlements, which evidently put high pressure on the soil.[130]

The early soil conservation work was exclusively conducted on grazing land and concentrated on the fencing off of eroded areas, control of *dongas*,

---

[128] SAB NTS 8142 2/341 part 1, Minutes of the Meeting of the Sediba Native Reserve Board, 14 December 1917.

[129] SAB NTS 8142 2/341 part 1, Minutes of the Meeting of the Sediba Native Reserve Board, 15 February 1918.

[130] SAB NTS 9509 138/400/19 vol. 2, Native Commissioner, Thaba Nchu, to the Chief Native Commissioner of the northern areas, Pretoria, 21 October 1936, see also annexes.

construction of contour banks, and the re-establishment of vegetation on bare patches, i.e. spots in the *veld* where no vegetation was growing. *Donga* control was mainly achieved by filling the gullies with stones; a practice that was only suitable in the case of light erosion. Deeper *dongas* were blocked with dams and planted with tree species such as *Robinia pseudo-acacia*, grey poplar and willow.[131] Contour banks were usually earth walls about 60 cm high that were heaped up manually or by means of ploughs across the slope of the grazing land.[132] After their construction these earth ridges were planted with grasses in order to make them more robust against water runoff. Very good results had been achieved with *Pennistum clandestinum* (kikuyu grass), *Cynodon dactylon* (kweekgras or couch grass), *Digitaria eriantha* (woolly finger grass), *Chloris gayana* (Rhodes grass) and *Paspalum dilitatum*. The first two species were also successfully planted in Lesotho (Showers 1989:282). According to government officials in Pretoria, the local *Cynodon dactylon* was most suitable for protecting contour banks. *Pennisetum clandestinum,* whose sods are directly planted in the soil (van Oudtshoorn 2004:259), needed more moisture to grow successfully.[133] Experiments by the agricultural officer in Thaba Nchu, however, indicated that *Pennistum clandestinum* adapted itself better than *Cynodon dactylon* or *Digitaria eriantha*.[134]

In contrast to the rangelands, where local inhabitants apparently accepted governmental intervention to a certain degree, fields were a more sensitive topic. It was only in the 1940s that arable land and ploughing practices became a focus of attention for soil conservation work. The comparatively late intervention by government officials with regard to cultivated areas was mainly the result of a 'local antipathy to disturbances of ploughed lands', as the assistant engineer expressed it in 1939. According to him, arable lands of the district laid out on slopes of less than 4% showed signs of soil erosion. However, erosion was 'nowhere sufficiently well marked

---

[131] SAB NTS 9502 138/400/5, Report of the Director of Native Agriculture, Pretoria, 3 April 1937.
[132] SAB NTS 9502 138/400/5, Report of the Soil Erosion Officer, Thaba Nchu, December 1934.
[133] SAB NTS 9509 138/400/19 vol. 1, Director of Native Agriculture, Pretoria, to the Soil Erosion Officer, Thaba Nchu, 14 December 1934.
[134] SAB NTS 9502 138/400/5, Report of the soil erosion officer, Thaba Nchu, December 1934; SAB NTS 9502 138/400/5, Report of the soil erosion officer, Thaba Nchu, re the growing of grasses, 12 February 1936.

to cause the alarm which alone, to the lay mind, justifies action'.[135] Moreover, intervention was unwanted by both locals and officials as long as fields were cultivated with crops. Consequently, intervention with regard to arable land could not be carried out as rigidly as it could in the case of grazing land.

South African farmers – both black and white – had often ploughed up and down the hills in order to save the energy of the draught animals; particularly when they had to pull heavy machinery. Commonly, the soil was ploughed while the oxen were moving downhill; on their way uphill oxen would just pull the machine without actually ploughing the ground (Beinart 2008:346). Government officials saw this common practice as a main cause of the development of gullies on cultivated land. Thus, the NAD intensively promoted ploughing along the contour (Beinart 2008:346). In Thaba Nchu district too, the agricultural officer was keen to persuade the land users to turn their soil along, rather than across, the contour lines. His efforts were not without effect, and in September 1947 many fields of the Sediba reserve – though not all – were, according to his report, laid out in this pattern.[136]

In the long run, at least from the perspective of government officials, the prevention of the erosion of fields required more intensive measures. During a two-day visit to the Thaba Nchu district in the late 1940s, the principal agricultural officer and an engineer from Pretoria came to the conclusion that the conservation and improvement of the cultivated land,

> '[...] cannot be delayed any longer, as an increased population pressure demand [sic] an increased production of food, which can only be expected from a stabilised soil with increased productivity. Not only has the productivity of the soils been seriously impaired through wrongful cropping systems, but in addition there is loss of fertile surface soil through water and wind erosion with a consequent lowering in production. In some cases, the lands have not been laid out on the contour and no protection works or terraces have been constructed in any of the areas.

[135] SAB NTS 9502 138/400/5, Assistant Engineer, Pretoria, to the Engineer in the Northern Areas, Pretoria, 3 January 1939.

[136] SAB NTS 9509 138/400/19 vol. 3, Report on soil erosion works in Sediba Reserve, September 1947.

> There is approximately 16,000 morgen [13,705 ha] under cultivation which demands attention i[n] some way or other if it is to be hoped that the owners should continue to produce sufficient for themselves and a surplus for landless families. There can be no more urgent service than to ensure […] the stability and increased productivity of the arable areas in this district.'[137]

In order to achieve stable agricultural production, the visitors from Pretoria recommended the construction of terraces on about 10,278 ha of the total area under cultivation in the Thaba Nchu and Sediba reserves. It was estimated that this would require about 4 years of work. Thereafter, terraces were to be constructed on the newly acquired SANT farms. However, these ambitious plans were not put into practice. The construction of contours and terraces within the arable land was connected with heavy – and expensive – work. More than three years later not even a single kilometre of terraces existed in the district.[138]

In 1951, Pretoria officials made a completely different evaluation of the urgency of anti-erosion works on cultivated land. A report envisaged the contouring and terracing of only 3,426 ha of the fields. According to that about 10,278 ha – i.e. exactly the same amount of land as in the 1948 report – only needed grass strips and tree belts as they were situated on gentle slopes. Furthermore, these areas would 'have to be beaconed with precast concrete beacons with their numbers embossed on them. The arable lands will be demarcated at approximately three morgen each'.[139] These rather contradictory shifts in the evaluation of soil erosion and rehabilitation measures were typical. After all, grass strips and tree belts were the cheaper option and easier to implement. However, in the 1950s only a few fields in the Sediba reserve were surrounded by *Eucalyptus* ssp. tree belts, which mainly served as windbreakers. Additional plantations were established in the 1960s.

---

[137] SAB NTS 9509 138/400/19 vol. 3, Report on soil and water conservation in Thaba Nchu, 13 March 1948, p.2.

[138] SAB NTS 9509 138/400/19 vol. 3, Secretary of Native Affairs, Pretoria, to the Chief Native Commissioner, Potchefstroom, 28 November 1951.

[139] SAB NTS 9509 138/400/19 vol. 3, Report of the principal agricultural officer and engineer of the western areas, Potchefstroom, on the proposed works in connection with the reorganisation of soil conservation in the Thaba Nchu district, 12 February 1951, p.2.

4.4.4 Improving the Veld: Grasses, Camps and Water

Early official accounts of the natural vegetation in the Thaba Nchu district remain rather superficial and do not provide comprehensive information about species composition or density of vegetation cover. Sometimes the agricultural officer reported more detailed observations of the *veld* condition and wrote, for instance, that 'the *veld* in the district is on the whole good and consists of the different types of bush, also Rooigras [*Themeda triandra*], species of Woolly Finger [*Digitaria eriantha*] and species of *Panicum'*.[140] Yet government officials were increasingly interested in the species compositions and vegetation dynamics of South African rangelands. New ecological concepts, such as the theory of plant succession formulated in 1899 and elaborated in Frederic Edward Clements' works after 1916 (Archer 2003:115), had influenced the thinking of South African botanists and ecologists (Beinart 2008:369–70). This was also reflected in the reports of the Director of Native Agriculture, who explained vegetation changes in the Thaba Nchu area in 1937 in the following way:

> 'Due to the light stocking of late years [animal numbers were reduced due to the severe drought in the early 1930s, CN] the veld in most sub-locations is well covered with sweet grass. The plant-succession of the veld grasses seems in this area to be either Steek grass or Queek via species of *Eragrostis* to Rooi grass [or] *Chloris Virgatum* to Rooi grass (and some *Panicum Coloratum* as well as species of Finger grass).'[141]

The perennial climax grass *Themeda triandra* whose quality had already been recognised by botanists in the nineteenth century (Beinart 2008:124–25) was regarded as one of the most important climax species and an indication of a good grazing management (Beinart 2008:370). Interestingly, archival sources of the 1950s still describe *Themeda triandra* as the predominant grass in the Sediba reserve.[142] These historical sources allow the

---

[140] SAB NTS 9502 138/400/5, Report of the agricultural officer respective of the growing of grasses at Thaba Nchu, 12 February 1936.

[141] SAB NTS 9502 138/400/5, Report of the Director of Native Agriculture about soil erosion works in Thaba Nchu, 3 April 1937.

[142] See for instance SAB BAO 5761 H128/1666/197 part 1, Ad-hoc report Sediba Besproeiingskema, Thaba Nchu, 28 February 1956, p. 2; SAB BAO 5761 H128/1666/197 part 1, Rehabilitation report: Sediba Reserve, Thaba Nchu, 22 May 1958, p. 3.

conclusion that no drastic shift of the state of the vegetation took place over the long-term; and therewith the data suggest that the vegetation of the area had at least a moderate resilience.

The plant succession model is based on the idea that disturbed rangelands are dominated by low-valued, so-called pioneer species that are better adapted to unfavourable conditions. Areas with better conditions, for instance low-stocked rangelands that rest for certain periods during the year, are dominated by sub-climax or even climax species. When the latter are abundant, the *veld* is considered to be in its best state (van Oudtshoorn 2004:26). Accordingly, an improvement of the *veld* requires a careful utilisation of the natural resource and a sophisticated knowledge of the effects of grazing animals. While the early conservation works in the grazing lands in the Thaba Nchu district focussed on the construction of contour banks, the NAD policy regarding the erosion of rangelands changed in the following years. After Pretoria officials had visited the target districts for the soil conservation works in March 1939 they recommended in their report:

> 'That the construction of contour banks in the grazing areas be reduced to a minimum and that more attention be paid to the regaining of vegetal cover by means of controlled grazing, establishment of suitable grasses, fertilizing where necessary, and that generally more regard should be given to the animal plant relationship.' [143]

The Need for Dams and Boreholes

Uncontrolled grazing was one of the main issues that government officials regarded as a threat to the vegetation. Therefore, the establishment of fenced-off grazing camps was considered an urgent matter in Thaba Nchu. *Veld* improvement, so the common thinking went, could only be achieved by adapting a controlled grazing system by which animals could be excluded from rangelands.[144] In 1939, most locations in the Seliba reserve had at least two grazing camps at their disposal; others were soon to be set

---

[143] SAB NTS 9509 138/400/19 vol. 2, Report on the inspection of soil erosion works, 13 April 1938, p. 11.

[144] SAB NTS 9502 138/400/5 part 2, Secretary of Native Affairs, Pretoria, to the Chief Native Commissioner northern areas, Pretoria, 2 March 1939.

up.[145] However, official enthusiasm and intervention with regard to rangeland management occasionally caused the opposite of the effect that was aimed for, i.e. the improvement of the natural *veld*. Particularly in the case of the grazing camps, unprofessional planning resulted in a deterioration of the rangeland vegetation. Already in 1939, officials worried that some camps lacked an adequate water supply. When these camps were in use, animals were 'driven daily to the water' resulting in 'serious damage' to the veld.[146] The NAD therefore, saw it as 'essential that where possible, adequate water facilities should be provided in grazing paddocks. As far as possible, all camps to be erected in the future should be constructed to have water available'.[147]

In 1948, the principal agricultural officer and an engineer stated in their report on soil and water conservation in Thaba Nchu that the *veld* condition in that district had deteriorated and that the area had a reduced carrying capacity. Although the authors of the 1948 report mentioned the 'heavy pressure of the livestock population', at no point did they claim that the deterioration was a result of overgrazing. Instead, they stressed that the insufficient water supply was one of the main obstructions to an 'effective and efficient *veld* management'. Besides, the deterioration was a result of the fact:

> '[...] that different types of *veld* are within the same grazing camps, that is, mountain areas, rolling country and *vleis* [a shallow, intermittent lake] are within the same grazing zone because water may only be available in the *vlei*. The result is severe selective grazing and a lowering in [sic] the carrying capacity of the *veld*.'[148]

This is not to argue that the deterioration of natural vegetation and soil in the reserves were only the result of a bad planning, but the example of the grazing camps clearly shows that governmental interventions could have

---

[145] SAB NTS 7981 259/337 part 2, Map of the Sediba reserve and adjunct farms (plan no. 829/39), 4 March 1939.
[146] SAB NTS 9502 138/400/5 part 2, Secretary of Native Affairs, Pretoria, to the Chief Native Commissioner northern areas, Pretoria, 2 March 1939.
[147] Ibid.
[148] SAB NTS 9509 138/400/19 vol. 3, Report on soil and water conservation in Thaba Nchu, 13 March 1948, p.2.

negative effects on the ecological system – and that officials began to realise this. This awareness did not, however, result in a re-evaluation of the farming methods of Thaba Nchu land users.

As natural surface water in the form of permanent or intermittent streams is only accessible in a few parts of the Thaba Nchu district, water for both domestic and agricultural use was mainly provided by artificially created dams or underground water sources. Access to the latter was, and still is today, gained by drilling boreholes and installing wind-powered pumps. These technologies probably appeared in South Africa by the 1870s (Archer 2003:120). Before then, communities and farmers depended on fountains, rivers and earth dams (Beinart 2008:158). The introduction of boreholes allowed the irrigation of arable land and access to grazing land that was previously not suitable in the dry season. As with wire for fencing, wind-powered pumps were initially imported from the United States (Archer 2003:121). Developed in the second half of the nineteenth century, the steel wind-powered pumps were particularly suited to the environmental conditions of the Great Plains. As they were self-regulating, had a basic design with standard parts and were easily transportable, they found their way into arid environments all over the world. Until the Second World War about 80 % of the total supply of wind pumps in the South African market were imported from the United States (Archer 2003:122–23).

Drilling for groundwater was difficult and risky, as boreholes sometimes yielded no water, or the water was too brackish (Beinart 2008:171). Moreover, the installation of a water point with borehole, wind pump, iron pipes and reservoir was a 'substantial investment cost relative to other farming inputs such as land, livestock and the labour required' (Archer 2003:124).[149] Thus, white farmers were initially suspicious and only reluctantly accepted the sinking of boreholes and new pumping technologies (Archer 2003:124; Beinart 2008:171). This changed in the first decades of the twentieth century, when drilling was subsidised by the government and numbers of boreholes increased rapidly (Turton et al. 2004:148; Beinart 2008:192). Between 1926 and 1955 the recorded numbers of wind pumps in South Africa increased from 44,000 to 151,000 (Archer 2003:122).

It is not clear from the archival documents when the first boreholes were drilled in the Thaba Nchu district. What can be said with certainty is that

[149] According to Archer (2003), the landed price of a wind pump was about £27 (ibid.124).

in the early 1930s a few boreholes had already been sunk in the Sediba reserve. However, in 1932 the inhabitants of some locations still depended on small dams or water holes. The Sediba reserve board was concerned about that, not least because the enduring drought had reduced the available surface water dramatically. Both humans and animals were drinking from the same water sources, a practice that was considered potentially harmful. Hence, the board decided that additional boreholes should be sunk in Talla, Kgalala, Morago and Rooibult location.[150]

Figure 7 Borehole No. 21080 in Sediba location, November 1942[151]

In 1939, 24 boreholes existed in the Sediba reserve (about 15,150 ha); some in the direct vicinity of a settlement, others closer to the grazing camps. Not all of them were equipped with a wind pump; other technologies included pumps operating by human or animal power (Figure 7). However, no information was given on the efficiency of the boreholes. Besides the

---

[150] SAB NTS 8142 2/341 part 1, Minutes of Meeting of the Sediba Native Reserve Board, 4 August 1932.

[151] Photograph in SAB NTS 10967 T21083. The borehole was temporarily worked with a food pedal as the animal gear had been stolen. The following description is written on the reverse of the photograph: 'Boy of seven [behind the pump] pumping water for his sister [in the foreground]. Semi-skilled labourer amazed at the easy attitude of the small boy [...]. Cattle on the horizon on their way to be watered.'

underground sources, four dams provided water for livestock. Compared to the adjoining ex-commercial farms that were purchased as 'released areas' in the course of the Native Trust and Land Act of 1936, the reserve was less developed in terms of water access. In the late 1930s the 6,980 ha area included 19 boreholes and 20 livestock dams.[152]

Despite the existing boreholes, officials stressed the necessity of additional watering points in the rangelands on SANT farms. The officials recommended to 'expedite the construction of earth dams throughout the area, in order to improve the position where supplies are inadequate and to augment the underground supplies'.[153] The hilly and rolling landscape of the Thaba Nchu area was thought to be optimal for the construction of dams. The officers suggested a total of 25 earth dams of different sizes. Smaller dams should provide water for about two to three months, larger ones should hold an eight to nine months' supply. The principal agricultural officer and the engineer from Pretoria were sure that a better distribution of watering points in the rangeland would reduce trampling and increase the grazing capacity of the *veld*. Furthermore, they assumed that earth dams would have positive effects on the underground water.[154] To lend weight to their recommendations, they emphasised:

> 'At the time when the carrying capacity was first assessed [1940] and the culling of livestock had to be done, the Natives were told that the carrying capacity would be steadily increased as a result of management, increased water supplies, etc. The Department is therefore duty bound to ensure the services required to bring about this increase in carrying capacity is undertaken with vigour. It is obvious that *at the present rate of dam construction the desired progress cannot be made in veld rehabilitation.*'[155] (emphasis added)

From the late 1940s onwards new earth dams were created across the whole Thaba Nchu district. Although the earth dams were not as expensive as boreholes in their maintenance, their construction costs were significantly

---

[152] SAB NTS 7516 678/327, Locations Reclamation Report, 14 November 1939, p. 18.
[153] SAB NTS 9509 138/400/19 vol. 3, Report on soil and water conservation in Thaba Nchu, 13 March 1948, p.3.
[154] Ibid.
[155] Ibid.

higher. In the late 1950s, the total cost of so-called stock dams, i.e. dams that were constructed only for the watering of livestock, amounted to £1,000 up to £2,000.

Provision of Additional Resources: Fodder Plants and Wood

Grazing camps, boreholes and dams were not the only measures taken in relation to the rehabilitation of the *veld*. In both Thaba Nchu and Sediba reserves a couple of hectares were set aside for the establishment of arboreta. The Director of Native Agriculture recommended the planting of drought-resistant fodder trees, such as *Prosopis* sp., as well as certain hardy trees that provide wood, such as *Tamarix* sp., *Gleditsia tricanthos*, *Cedrus deodara*, *Pinus halepensis* and *Eucalyptus* species. Furthermore, several fodder plants were to be cultivated, for instance various species of spineless cactus (*Opuntia* spp.),[156] Cape Vaal-brak (*Atriplex* spp.) and various *Salsola* species.[157]

Drought-resistant fodder plants, some of them imported to the Cape Colony as early as the seventeenth century, were widely planted in arid and semi-arid South African districts as they could be utilised as supplementary fodder, particularly in dry years. Based on the archival sources at hand, I cannot reconstruct with certainty which of the aforementioned fodder plants were actually planted in the Thaba Nchu district. In none of the documents is it explicitly stated. However, one can take a look at the plants that grow in the villages today. In all four villages, spineless cactus and *Agave americana* are found in the present residential areas, and some even in the former ones. Both were, along with *Atriplex nummularia*, also important for livestock production in the Karroo (Beinart 2008:269). As indicated in the introduction to the research sites (Chapter 2.2) all villages also have access to small *Eucalyptus* spp. groves.

---

[156] Spineless cactus was a hybrid of wild prickly pear species (*Opuntia* spp.). The latter were fast-spreading invaders and became a major pest in South Africa at the turn of the century. Farmers and agricultural officers tried to eradicate the prickly pear species, and were eventually successful (for a detailed account of the eradication of prickly pear (see Beinart 2008:266–303).

[157] SAB NTS 9502 138/400/5, Report of the Director of Native Agriculture, Pretoria, about Soil Erosion Works in Thaba Nchu, 3 April 1937.

4.4.5 Increasing Coercion: Livestock Improvement and Culling

As in the Transkei, Ciskei and other 'native areas', governmental interventions in Thaba Nchu focussed not only on rangeland ecology but also on pastoral farming in general. Officials were concerned that the establishment of grazing camps and water points alone was not sufficient to achieve a better *veld* quality. The presence of too many inferior animals was thought to present another major problem for the ecology and economic development of the 'native areas'. The idea was that better breeds would produce more milk, meat and wool and thereby reduce the need for large herds (Beinart 2008:353). Hence, the cattle-breeding policy formulated by the Director of Native Agriculture in 1937 had the aim of improving the 'native' livestock by introducing pedigree bulls, rams and stallions, as well as by the castration of inferior animals. Moreover, through the organising of public auctions, black farmers were to be encouraged to market their animals.[158]

With regard to cattle, government officials initially favoured the Afrikander breed. Starting from 1935, several bulls served the cows in locations of the Thaba Nchu district.[159] Afrikander cattle were preferred as they were good draught animals and produced excellent meat, but female progenies were not good dairy cattle (Beinart 2008:356). Hence, from the 1940s onwards bulls of breeds that provided more milk, such as Red Poll and Brown Swiss, were introduced in order to maintain and increase the milk supply of the reserve households.[160] Besides the improvement of cattle, government officials initially promoted sheep farming. In 1937, the Director of Native Agriculture stated that the 'systematic development of the wool industry in Thaba Nchu district is of the greatest importance to the economic advancement of the residents in the Reserves'.[161] However, following observations in the Ciskei and Transkei the policy was questioned in the late

---

[158] SAB NTS 7335 127/327 part 2, Memo of the Director of Native Agriculture, Pretoria, 6 August 1937.

[159] SAB NTS 9502 138/400/5, Memo of the Director of Native Agriculture, Pretoria, 3 April 1937, p. 12.

[160] SAB NTS 7526 717/327 part 2, Memo of the Director of Native Agriculture, Pretoria, 4 May 1946; SAB NTS 7527 717/327 part 4, Senior Agricultural Officer of the western Areas, Potchefstroom, to the Chief Native Commissioner of the western areas, Potchefstroom, 7 March 1955.

[161] SAB NTS 9502 138/400/5, Memo of the Director of Native Agriculture, Pretoria, 3 April 1937, p. 13.

1930s as sheep were increasingly regarded as the main threat to grazing land (Beinart 2008:355–56). Unsurprisingly, sheep were the major focus of the culling operations that started in 1940 in Thaba Nchu and other South African districts.

The improvement of animals alone could not solve the assumed overgrazing problem. Both government officials and the Barolong elite supposed that the limitation of livestock in the Thaba Nchu area was necessary. During a reserve board meeting in 1929, Dr James Moroka and W.Z Fenyang proposed that the number of animals grazed for free on reserve land was to be limited per taxpayer. One year later an additional regulation was passed that allowed each holder of a residential permit to graze 10 head of large stock and 20 head of small stock free of charge. For all heads of large and small stock that exceeded this number a payment of 2 shillings or 3 pence respectively had to be paid.[162] Ten years later, the Thaba Nchu reserve was the first part of the district that was declared as a 'betterment' area under proclamation 31 of 1939. As illustrated above, Sediba followed a couple of months later.

As a first step, the grazing capacity of the rangelands in the new 'betterment' areas was reassessed. Native affairs officials visited the reserves and determined a rather low carrying capacity of 6.8 ha per large stock unit (LSU) and 1.7 ha per small stock unit (SSU).[163] Such assessments were conducted several times in the 1940s and 1950s. Results often varied significantly depending on the rainfall in the previous years. After the assessed carrying capacity was announced in the government gazette, the culling of surplus livestock commenced. In 1941 alone, the local culling officers – i.e. the senior agricultural officer and the agricultural officer based in Thaba Nchu – culled about 30% of the animals of the Thaba Nchu reserve, 25% of those on the Sediba reserve, and 30% of livestock on the new villages that were established on SANT land, namely Tweefontein,

---

[162] SAB NTS 8144 5/342 part 1, Government Notice no. 228, 7 February 1930.

[163] When the carrying capacity was assessed in Thaba Nchu government officials defined one 'large stock unit' as a full-grown cow or four sheep; see SAB NTS 7526 717/327 part 1, Report of the Director of Native Agriculture, Pretoria, 2 March 1940, p. 3.; SAB NTS 7526 717/327 part 1, memo, unknown author, Pretoria, 27 November 1940.

Rietfontein, Klipfontein and Gladstone.[164] During these first culling operations, the planned procedure was that (1) goats and donkeys were to be eliminated completely; (2) numbers of horses should be reduced to a minimum of one or two per owner; (3) numbers of sheep were to be reduced by 50% or more; (4) cattle were only to be culled if they were aged animals, or if the culling of aged animals and unsuitable breeds had not fulfilled the required reduction of an individual's livestock.[165] Moreover, each pastoral farmer received a stock book as a record of the total number of animals allowed.[166]

Culled livestock was branded with a 'C' and had to be removed from the 'betterment' areas by a predetermined date, either by selling or slaughtering. If the animals were not removed by the respective date they could be declared forfeit to the SANT.[167] Animals were sold at public sales or at governmental-funded auctions held once or twice per year in Thaba Nchu town by Roderick & Co., a Bloemfontein-based auctioneer. In June 1941, shortly after the above-mentioned culling operation, in total 369 head of cattle, 39 horses and 24 sheep were sold for £2,121 and sixpence. About 380 cattle remained unsold as their owners 'wanted unreasonable prices'.[168]According to the Assistant Native Commissioner good prices were achieved for slaughter oxen (about £9) and tollies (about £3 and 10 shillings).[169] In the South African Union prices for livestock remained high and even increased during and after the Second World War; however, people were not very keen to sell their animals (Beinart 2008:354). Generally, a post-World War price increase was also evident in Thaba Nchu (Figure 8). Yet the prices for livestock varied; those for cattle more than those for sheep. In the late 1940s, during a sequence of years with rainfall below the average (De Vielliers and Booysen 1998:3), prices for animals sold by black farmers were low. In 1948, livestock owners complained that cattle

---

164 SAB NTS 7526 717/327 part 1, Senior Agricultural Officer, Pretoria, to the Director of Native Agriculture, Pretoria, 28 February 1941 and 26 May 1941.
165 SAB NTS 7526 717/327 part 1, Memo of the Director of Native Agriculture, Pretoria, 15 March 1941.
166 SAB NTS 7526 717/327 part 2, Senior Agricultural Officer, Thaba Nchu, to the Additional Native Commissioner, Thaba Nchu, undated 1946.
167 SAB NTS 7526 717/327 part 2, Government Notice no. 265, 13 February 1942.
168 SAB NTS 7526 717/327 part 1, Assistant Native Commissioner, Thaba Nchu, to the Secretary of Native Affairs, Pretoria, 6 June 1941.
169 Ibid.

branded with a 'C' never achieved good prices. The senior agricultural officer suggested that this complaint might be justified. As buyers 'knew very well that such stock must be sold within a certain period' they could make agreements in order to keep the price low.[170] However, after better rains in the early 1950s prices increased significantly, particularly for cattle.

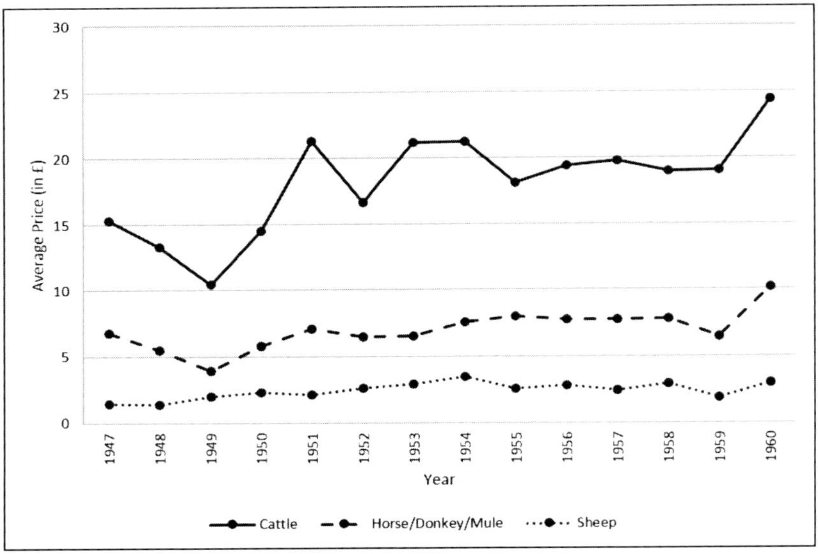

Figure 8 Average prices for 'native' livestock at public and SANT auctions, 1947-1960[171]

[170] SAB NTS 7526 717/327 part 2, Senior Agricultural Officer, Thaba Nchu, to the Chief Native Commissioner of the western areas, Potchefstroom, 25 October 1948.

[171] Data are drawn from auction reports between August 1947 and September 1960 in NTS 7256 290/326. For each of the irregulary conducted public and SANT auctions (n= 228) the agricultural officer reported the total number of (i) cattle; (ii) horses, donkeys and mules; and (iii) sheep sold by black farmers. Moreover, the total price achieved for each animal type was provided. Based on those reports I calculated the mean price per auction per animal type. As no differentiation is made between young and old animals, average prices only serve as a rough indicator of price trends. To allow a better visualisation of the data, I illustrate the mean price per year.

After the first culls government officials realised that the main obstacle to an improvement of the animals was the selling behaviour of the local farmers. Instead of bringing their old animals to the auctions livestock owners tended to sell the calves and foals of the introduced pedigree bulls and stallions in order to keep their animal numbers low. From the farmers' perspective this was reasonable though, as the young animals were useless for ploughing purposes and did not provide milk. The Assistant Native Commissioner came to the conclusion that '[i]t is clear [...] that our efforts to relieve the overstocking problem are in direct conflict with efforts to improve stock'.[172] Therefore, the Commissioner suggested, livestock owners with five or less large stock units might retain the progeny for a period of three years. After that the full-grown animals would eventually replace the older ones.[173]

Although the admission of young cattle led to an increase of the LSU in total, officials were not concerned about the rangelands as they had improved considerably 'in spite of the fact that officially the area is regarded as overstocked'.[174] The good condition of the grazing lands was more due to exceptionally high rainfall – with almost 950 mm from October 1942 until September 1943 – than to the reduction of livestock numbers. Since government officials feared the under-utilisation of the *veld*, the carrying capacity of the reserves and the SANT farms was increased from 6.8 ha per LSU to 4.7 ha per LSU in 1944.[175] Yet culling was continued in the same year, particularly in the Sediba reserve, where livestock numbers still exceeded the determined carrying capacity. In total, about 27% of the Sediba livestock were culled in 1944.[176]

Despite the governmental efforts with regard to livestock improvement and culling, officials considered additional control measures. From their point of view there was still a lot to be done:

---

[172] SAB NTS 7526 717/327 part 2, Assistant Native Commissioner, Thaba Nchu, to the Secretary of Native Affairs, 3 December 1942.
[173] Ibid.
[174] Ibid.
[175] SAB NTS 7526 717/327 part 2, Memo of the Assistant Director of Native Agriculture, Pretoria, 17 January 1944.
[176] SAB NTS 7526 717/327 part 2, Senior Agricultural Officer, Thaba Nchu, to the Assistant Native Commissioner, Thaba Nchu, 27 September 1944.

> 'there has been no adequate control of stock owing to the absence of dipping control regulations or some means of identification of the stock. It is patent that large numbers of stock have been illegally introduced into the district and in almost every case it is impossible to prosecute the offenders due to lack of proof. Exchanges in and out of the betterment areas has also been taking place without authority and as previously reported deliberate stealing of Trust grazing by both Europeans and Natives is a common practice and most difficult to counteract. The position in this district is complicated further by the fact that a large number of small private farms [owned by black people] are situated inside the released areas and in the majority of cases completely surrounded by Trust holdings. These farms are all heavily overstocked and their animals, to eke out a living must trespass into Trust ground.'[177]

The quest for better control led to the introduction of a compulsory branding and marking of all approved animals – not only the livestock due to be culled, as before. In Thaba Nchu and Sediba reserves and on the SANT farms the new branding consisted of an 'S' within a circle, representing the initial of the then chief, Albert Setlogelo Moroka.[178] Furthermore, the appointment of additional ranger staff was regarded as necessary to achieve better control of illegal stock movements.[179] Nevertheless, people tried to avoid the forced selling by not presenting their animals during the assessments and culling operations. Hence, the agricultural officer suggested that all animals without an official brand should be shot and their owners tracked down.[180]

Medium-Term Effects of Livestock Improvement and Culling

Culling operations were often connected with large-scale castration of inferior bulls and rams, which were typically mixed breeds. In 1954, a report

---

[177] SAB NTS 7526 717/327 part 2, Assistant Native Commissioner, Thaba Nchu, to the Secretary of Native Affairs, Pretoria, 17 August 1944.
[178] SAB NTS 7526 717/327 part 2, Assistant Native Commissioner, Thaba Nchu, to the Secretary of Native Affairs, Pretoria, 3 August 1944.
[179] SAB NTS 7526 717/327 part 2, Assistant Native Commissioner, Thaba Nchu, to the Secretary of Native Affairs, Pretoria, 17 August 1944.
[180] SAB NTS 7526 7171/327 part 2, Agricultural Officer, Thaba Nchu, to the Assistant Native Commissioner, Thaba Nchu, 4 April 1949.

of the senior agricultural officer indicated for the various locations in the Thaba Nchu district that cattle herds consisted of between 0.74% and 2.5% bulls; or in other words, they had a bull-to-cow ratio between 1:134 and 1:40. This low ratio[181] had severe effects on the reproduction of herds, which ranged between 1.1 and 5 cows per calf. The bull-to-cow and cow-to-calf ratios were slightly better in the newly established SANT-villages of Moroto, Tiger River, Merino, Paradys, Gladstone, Tweefontein, Woodbridge, Commissiedrift, Klipfontein and Rietfontein (Figure 1), as they had been under strict state control since their origin.

Although the bull percentage was better in the following year, agricultural officers continually remarked on the low reproduction rate in the region.[182] Indeed, the ratio of bulls and cows was not the only problem. A veterinary examination revealed that in addition to two Brown Swiss bulls that served several Trust-villages numerous cows were also found to be infected with epivag, a venereal infection affecting the reproductive systems of cattle. Cases of the disease, which occurs in East and South Africa, greatly increased in the 1930s. In contrast to imported breeds, indigenous African cattle possess a high natural resistance to epivag (Biovision Foundation 2012). Thus, the introduction of foreign breeds had made herds prone to new diseases. The state veterinarian, fearing a further dispersion of epivag into other parts of the Orange Free State, Basutoland, the Cape Province, Natal and the Transkei territories, immediately suggested drastic measures: (i) cattle movement between villages was forbidden; (ii) all 104 SANT bulls serving in the district were to be isolated on the governmental farm Eureka; and (iii) newly-born calves were to be immediately castrated.[183] In later documents on this matter the tone was less agitated. By 1956, the bulls, along with 140 infected cows, were branded, sold and removed from

---

[181] In her study on small-scale cattle farming in post-apartheid Thaba Nchu, Moorosi (1999) found a bull-to-cow ratio of 1:46, which she defined as 'extremely low' (ibid.41). The Department of Agriculture of Kwazulu Natal recommends a bull-to-cow ratio between 1:12 and 1:30 depending on the age of the bull (Gertenbach 2000).

[182] SAB NTS 7527 717/327 part 4, Senior Agricultural Officer, Potchefstroom, to the Chief Native Commissioner of the western areas, Potchefstroom, 7 March 1955.

[183] SAB NTS 7527 717/327 part 4, Memo of the state veterinarian, Bloemfontein, 14 May 1955.

the district, and epivag was no longer considered a significant threat in the district.[184]

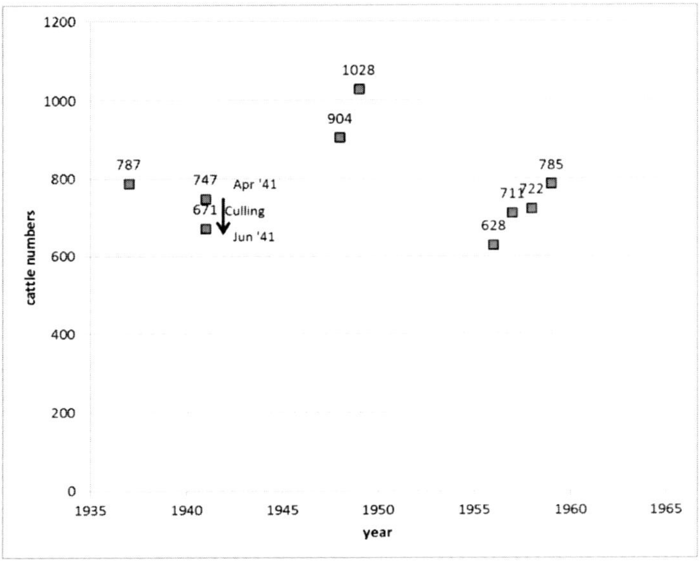

Figure 9 Cattle numbers Sediba village, 1937-1959[185]

Figures 9 and 10 show the cattle and sheep numbers held in Sediba village between 1937 and 1959. For the year 1941 two values are provided for both animal types. The reason for this is that two counts were made in that year; one in April 1941 before a livestock cull and one in June 1941 after the cull. When comparing both diagrams it is striking that animal numbers developed differently over time. In 1941, numbers of sheep were reduced by about 50%, and thereafter herd sizes were kept to a minimum. In contrast to this, cattle herds were reduced by less extreme amounts in 1941, and total numbers even peaked in the late 1940s. On the one hand, the carrying capacity of the Sediba reserve was re-assessed – and eventually increased – in the mid-1940s, which might explain the rise in the absolute

[184] SAB NTS 7527 717/327 part 4, Senior Agricultural Officer, Potchefstroom, to the Chief Native Commissioner of the western areas, Potchefstroom, 29 March 1956.

[185] Data source: SAB NTS 7526 717/327 part 1-4; and SAB 9502 138/400/5.

cattle numbers in Sediba village. On the other hand, culling programmes concentrated on goats, donkeys, horses and sheep. Because of their importance for ploughing and milking, cattle were only 'touched' when no other animals could be culled. The Native Affairs officials had realised that bovine animals were the 'most valuable possession[s]' of the black farmers and therefore 'should in general provide [the] main source of revenue'.[186] Still, multiple culling operations had severe effects on the number of cows and oxen, too. After two decades of culling, the livestock owners of the Sediba reserve (n = 855) owned on average about five LSUs each (Figure 11).[187] However, this value subsumed cattle, horses and sheep, and therefore the average number of draught and dairy animals that each household owned was definitely smaller.

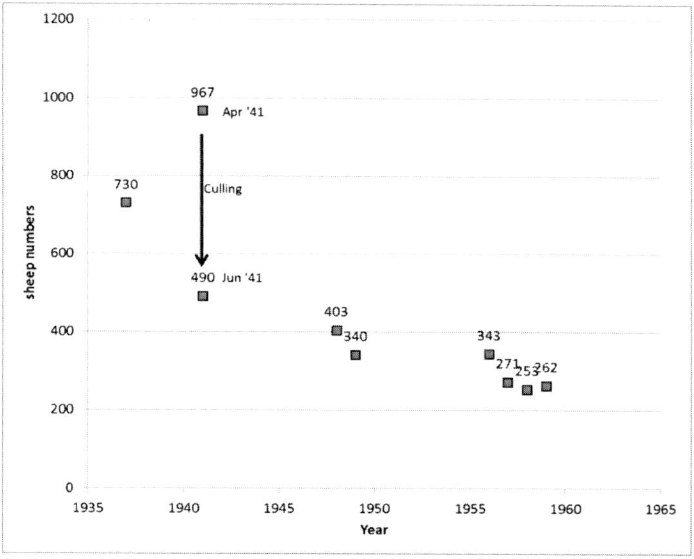

Figure 10 Sheep numbers Sediba village, 1937-1959[188]

---

[186] SAB NTS 7526 717/327 part 2, Memo of the Director of Native Agriculture, Pretoria, 26 August 1947.

[187] SAB NTS 7527 717/327 part 4, Native Affairs Commissioner, Thaba Nchu, to the Chief Native Commissioner, Bloemfontein, 11 December 1959, see attached list.

[188] Data source: SAB NTS 7526 717/327 part 1-4; and SAB 9502 138/400/5.

Livestock culling not only undermined the household economy by causing a massive lack of draught power and milk; it had also significant consequences for the social system, since cattle were a means to create and maintain social relations. Moreover, marriages were legitimated by the transfer of cattle (Sets. *bogadi*). A woman was not considered married, and a child still regarded as a descendant of its mother's father or brother, until *bogadi*-cattle had been paid (Lye and Murray 1980:112). There is very little evidence that allows conclusions to be drawn about the persistence of bridewealth payments in Thaba Nchu. F.A.W. Lucas wrote in his addendum for the Native Economic Commission that an average price was about three to eight cattle; 'substitutes for cattle are sheep, horses, goats, and occasionally a cart' (Union of South Africa 1932:180). According to Lucas' account, the *bogadi*-system was dying out. More recent statements by government officials in archival documents imply that the payment of *bogadi* had lost its importance in the twentieth century. When in August 1947 members of the Sediba reserve board pled to exclude *bogadi*-cattle from the cull, the board chairman argued that the *bogadi* custom was almost never resorted to and that in cases when such payments were made people should use cash rather than cattle.[189] However, there exist reasonable doubts that this assessment was a realistic one. Schüler (1965), who studied the customs of the Barolong in Thaba Nchu in the mid-1960s, stated that bridewealth payments were still one of the most important ways to acquire cattle (ibid. 292).

Culling made the payment of *bogadi* difficult as household herds had been reduced to a handful of animals. When young men intended to marry, they had to spend a couple of years as migrants to earn the bridewealth.[190] Although labour migration was not a new trend, the forced decimation of livestock herds presumably increased the pressure on young men to search for work outside the Thaba Nchu district. This was not necessarily an unwanted situation from the perspective of the young. Beinart (1982) suggests for Pondoland households that by earning at least part of their bridewealth, sons could increasingly defy the authority of their fathers and other patrilineal relatives; and they gained independence with regard to marriage (ibid. 96–97).

---

[189] SAB NTS 8142 2/341 part 3, Minutes of Meeting of the Sediba Native Reserve Board, 12 August 1947.
[190] Informal interview with Mosela Temeku, Sediba, 18 August 2011.

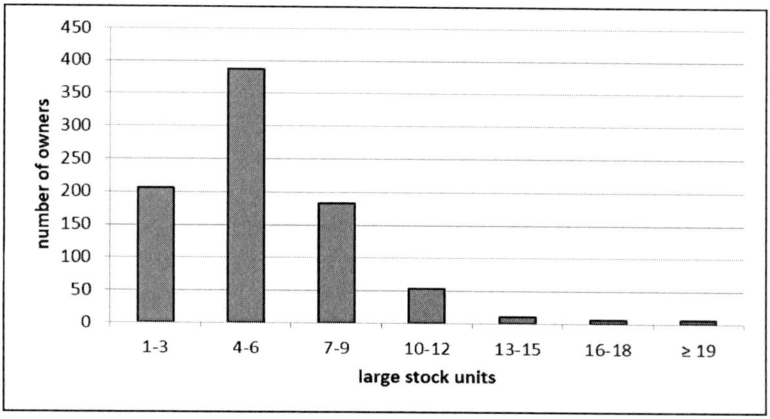

Figure 11 Number of LSUs per livestock owner (n = 855) in the Sediba reserve in 1959[191]

The effects that livestock culling had on patron-client relationships were more severe, as it undermined the social organisation of entire communities. Government officials wanted to equalise livestock ownership; hence, culling impacted the big livestock owners most severely (Beinart 2008:361). Despite the fact that their herds were often of better quality, the animals of those owning large herds were also culled when they were 'above standard'.[192] Thereby government authorities hoped to educate the people to dispose of their 'stock while it still has good market value'.[193] Furthermore, they thought that the higher prices fetched by quality livestock would encourage people to improve the condition of their animals. Last but not least, owners of large herds were a major focus because they usually kept many sheep, which since the late 1930s were regarded as the main cause of overgrazing.[194] There is evidence from the oral accounts that some wealthy livestock owners managed to avoid the reduction of their

---

[191] SAB NTS 7527 717/327 part 4, Native Affairs Commissioner, Thaba Nchu, to the Chief Native Commissioner O.V.S., Bloemfontein, 11 December 1959, attached list.

[192] SAB NTS 7526 717/327 part 2, Culling guide of the senior agricultural officer used in the Sediba Reserve in 1944, Thaba Nchu, 2 December 1944, p.2.

[193] Ibid.

[194] Ibid.

herds by lending out parts to other residents.[195] For the majority of households, however, culling meant the rigorous branding of a 'C' on most of their animals.

According to government officials, the harsh culling policy was necessary to guarantee that more people could obtain rights to livestock without exceeding the carrying capacity of the land. In 1944, the senior agricultural officer was contented to note that in the Sediba reserve about 144 households had acquired rights to livestock because of culling. The situation in the Thaba Nchu reserve was similar. People would approach him daily to 'obtain a stock-book and the necessary permit to introduce one or two head'. [196] However, this report only dressed up the outcomes of culling: firstly, the number of cattle that people owned – one or two head – was too small to gain little more than a bit of milk every day; secondly, I doubt that all those who applied for a stock-book were indeed new pastoralists. In the very same letter, the officer stated that the number of people who kept stock without having a permit was not a 'trifling amount'.[197] Since in such cases animals were often immediately culled people feared their discovery. I assume that many of those reserve residents who illegally owned stock were also those who applied for a permit. It appears questionable whether culling did indeed enable the very poor to keep at least some animals.

By reducing the numbers of animals in individual herds to a minimum, the patron-client relationships were deprived of their material basis, namely cattle. Thus, the effects of the culling were critical for both the wealthy households and the poorer ones that depended on the oxen of the former for ploughing. Parallel to the culling operations, government officials promoted the purchase of tractors and argued that the machines would help people to use the soil in a more productive manner. While Sediba reserve members were unanimously against this idea – mainly because they feared

[195] Interview with Paulina Mokgothu, Middeldeel, 10 September 2011.
[196] SAB NTS 7526 717/327 part 2, Senior Agricultural Officer, Thaba Nchu, to the Assistant Native Commissioner, Thaba Nchu, 27 September 1944, p. 2.
[197] SAB NTS 7526 717/327 part 2, Senior Agricultural Officer, Thaba Nchu, to the Assistant Native Commissioner, Thaba Nchu, 27 September 1944, p. 1.

the costs connected with the repair of tractors – Thaba Nchu reserve members seconded the authority's proposal and even suggested buying more than one tractor.[198]

4.5 Conservation Efforts in Thaba Nchu: A General Evaluation

Although the Thaba Nchu district was one of the first declared as a soil conservation district, the work done there was not as comprehensive as in other areas. Most of the state's money and time were spend in the Herschel district in the Cape. By the late 1930s, about £30,000 had been paid for soil conservation work in Herschel, and by 1940 the NAD had built over 1,800 km of contour banks on arable land and more than 500 small earth dams for livestock. Together with Glen Grey in the Ciskei, the Herschel district served as a model area. New advanced techniques and conservation methods were applied and agricultural officers and engineers of the NAD were sent there in order to observe the results (Beinart 1989:144). The work done in Thaba Nchu until the mid-1950s was on a medium scale and, in contrast to Herschel, was clearly focussed on rangelands (Table 3).

An evaluation of conservation efforts that goes beyond a quantitative analysis should address two fundamental questions:

(i) was severe resource degradation actually a problem in Thaba Nchu?

and, connected to that,

(ii) was the official reasoning for intervention – i.e. to discontinue bad-farming practices – justified?

The first question has to be answered with both a yes and a no. Since in parts of the district signs of soil erosion were clearly visible, it would be wrong to argue that the issue was completely invented by officials. However, as in the case of Lesotho (Showers 1989), it was not a severe threat affecting the whole area. In the early twentieth century soil erosion and the degradation of rangelands were concentrated in certain areas with high human and animal pressure. This indicates that overgrazing, which certainly existed in terms of high livestock numbers, did not have as negative an impact on rangelands as officials thought. In the cultivated area too, soil

[198] SAB NTS 8142 2/341 part 1, Minutes of Meetings of the Sediba Native Reserve Board, 20 October 1947; SAB NTS 8141 1/341 part 1, Minutes of Meetings of the Thaba Nchu Native Reserve Board, 29 October 1947.

erosion was not severe. This explains the reluctant intervention of the NAD officials. The regulation of cultivation practices was meant to improve the agricultural production in order to feed the increasing population. However, as the case of the Grobler irrigation scheme has shown, the forced improvement of production was misplanned and – at least in some cases – led to more severe soil erosion than the so-called 'bad-farming' methods of the local residents.

Table 3 Soil erosion work in the Thaba Nchu district from the beginning of inter-vention until September 1954[199]

Type of conservation work	Work done
Contours (in grazing land)	464 km
Training walls (alongside river courses)	5 km
Terraces	0
Area worked	5,281 ha
Dams	172
Fencing	120 km
Area fenced off	2,800 ha
Trees planted	12,470
Grass planted	82 bags
Aloes	23,028

The answer to the second question must be no. Indeed, archival documents indicate that in some years rangeland condition was bad and crop yield low. However, these cases occurred in years with below-average rainfall. The opposite happened in rain-laden seasons, such as 1948, when agricultural officials recorded a very good maize harvest in the Sediba reserve. Due to late winter rains in 1947, grazing was still abundant in June 1948 and accordingly cattle achieved better prices than in the years before.[200]

[199] Numbers must be interpreted with caution as a systematic recording of erosion works was apparently only undertaken from the mid-1940s. Thus, the figures can only be estimates. SAB NTS 9509 138/400/19 vol. 4, Agricultural Officer, Thaba Nchu, to the Additional Native Commissioner, Thaba Nchu, 8 November 1954.
[200] SAB NTS 9509 138/400/19 vol. 3, Soil erosion report of agricultural foreman, Thaba Nchu, 29 June 1948.

Hence, the farming methods – though not always optimal – were not the main reason for ecological and economic crises. Rainfall had a major effect on vegetation and its variability. The availability of grazing and the success of cultivation varied significantly depending on the amount, timing and distribution of rain. In other words, the rangelands in Thaba Nchu resembled a disequilibrium system (Behnke Jr. et al. 1993).

4.6 Preliminary Summary

In this chapter, I have reconstructed how narratives of degradation became established in official debates in South Africa and how the causes of the ecological crisis were increasingly ascribed to farming methods of black South Africans. The state created an image of the 'native' areas as social-ecological systems with low resilience. The environmental threat and the danger of a system collapse offered justification for intervention. In the Thaba Nchu case – perhaps not so much in the urban area, but certainly in the rural Sediba reserve – the ecological disaster was exaggerated, since a realistic assessment of rangeland degradation and soil erosion would not have legitimated such drastic actions. Apart from an analysis of land use control on the local scale in Thaba Nchu, the archival documents also allow more general conclusions to be drawn about the effects of the environmental governance.

First, historical data show that the NAD's conservation measures and strategies for economic development were strongly shaped by debates prevalent at that time. Explanatory models of environmental processes changed, and accordingly so did the foci of the policies – as shown by the example of sheep farming, which was at first promoted and later curtailed. Often government authorities applied conservation measures that were more suited to white-owned commercial farms, without taking into account local socio-economic conditions. Officially, the approval of residents was sought, but when people disapproved of the plans of Native Affairs, conservation measures were enforced at any cost. Moreover, policies and policy changes were not thoroughly explained to black farmers, but rather put directly into practice.

Second, the measures and institutions implemented were not always effective, and sometimes even triggered environmental degradation. The planning of grazing camps was not elaborated properly, and due to a lack of water points rangelands deteriorated. The example of the Sediba irrigation scheme furthermore illustrates that the NAD normally chose the cheapest options, which were not necessarily the best. Inappropriate and cost-cutting planning that resulted in resource deterioration rather than conservation was not restricted to Thaba Nchu, as Showers' (1989) case study of Lesotho has shown. On the other hand, agricultural staff were not always the most knowledgeable experts in all areas (i.e. in soil erosion, dryland and irrigation farming, *veld* improvement, tree plantations, water supply, livestock improvement and culling). At times, the memos and letters of the Director of Native Agriculture and other senior officials of the NAD reveal a certain resentment toward the work of local personnel. Pretoria officials sent officers and demonstrators for agricultural training to fill the gaps in their knowledge, for instance at the Glen Agricultural College near Bloemfontein. But apart from technical and practical knowledge, the successful regulation and control of land use practices required a person with intercultural competences – skills that not all staff members had developed.[201]

Third, from an institutional perspective, I have shown that bureaucratic state-implemented institutions were often based on insights gained from natural sciences and included allegedly objective measures, as illustrated by the example of the carrying capacity. That this 'objectivity' was relative will be demonstrated in the following chapter. Here, I want to focus on the multipurpose character of institutions, which is best exemplified by looking at the livestock culls. Officially, culling was aimed at a reduction of grazing pressure and an improvement of livestock breeds. However, the latter aim was not achieved, as a farmer's idea of useful animals differed from the official view. Unofficially, the state sought to increase its political control over local farmers by curtailing their access to rangeland resources; moreover, the culling undermined existing institutions, such as the *mahisa* and *bogadi* systems, as a by-product. Whether or not this was intended when the institution was crafted is difficult to say; certainly, however, the complication of *mahisa* relationships and *bogadi* payment was a positive development from the state's point of view. Socially embedded institutions

---

[201] For example, in 1942 Sediba reserve inhabitants complained repeatedly about an agricultural overseer who was said to be stern and unsympathetic and apparently used camps reserved for winter grazing for his own animals. Although the senior agricultural officer came to his defence, the NAD eventually transferred him in order to relax the situation; SAB NTS 8142 2/341 part 2, Minutes of the Sediba Native Reserve Board, 14 July 1942, 17 September 1942 and 25 January 1943.

were not discontinued immediately after the first cull. In the medium term, the culling had severe effects, but these effects were not always the same. Recent ethnographic data show that the payment of *bogadi* cattle still occurs in the area; in contrast, the lending of *mahisa* cattle is no longer practised. However, other arrangements seem to have replaced the *mahisa* system, as I will show in Chapter 8.4.2.

Fourth, a general trend noticeable over two decades of soil conservation is that governmental interventions were of an increasingly controlling and coercive nature. A programme that started with the filling of *dongas* with stones in the 1910s culminated in the strict regulation of the use of each and every hectare of land and head of livestock. A close monitoring of compliance at the location level was certainly not conducted everywhere with the same intensity. As oral histories indicate, people tried to avoid culling by not presenting their animals or by lending them out to other households. I assume that in the case of agricultural land, too, people made informal arrangements beyond governmental control. However, in the following chapter I will illustrate that this breaking of rules was increasingly punished and entire families bullied by government authorities.

5. 'STABILISING' THE THABA NCHU AREA, 1950S AND 1960S

In the previous chapter I showed that 'betterment' in the Thaba Nchu district initially emphasised reclamation, i.e. intensive conservation work to restore and improve natural resources in a comparatively small area (Beinart and Bundy 1980:306). The idea of resource conservation was a basic principle which was reflected in the measures taken against soil erosion, improvement of water supply and limitation of livestock numbers. This approach shifted in the 1950s – not only in Thaba Nchu but also in other regions – towards a programme of 'stabilisation', which meant first and foremost the prevention of further resource degradation, as this was a precondition for an increase of the economic viability of the rural communities. Stabilisation included the planning of larger areas and comprehensive resettlement schemes (Beinart and Bundy 1980:306; Hendricks 1989:307). The tools of stabilisation were objective measures, i.e. maps and surveys by which rural communities were standardised. This standardisation made the population more 'legible' (Scott 1998:3) and thus easier to govern and control.

In this chapter, I will elucidate how the idea of conservation and improvement of the rural areas was dismissed in order to prepare the Thaba Nchu region for the resettlement of thousands of displaced people and gain more control over the social and economic life of families. Strict regulation of the rural areas was on the one hand necessary to perpetuate their function as a 'labour reservoir' (Hendricks 1989:316). On the other hand, more control was needed to simplify 'classic state functions' (Scott 1998:2), such as taxation, and suppress opposition, which erupted in Thaba Nchu as it did in other homelands. Particularly in this regard the newly established Barolong tribal authority played a critical role. Before I discuss the new directions of 'betterment' policies in the 1950s, I will briefly shed light on national economic and political developments that are important for an understanding of the events on the local level. In Chapter 5.4 I illustrate in detail the enforcement of the resettlement in the Thaba Nchu area, in both the rural and urban parts. Thereafter, I will address a point that was only marginally considered in the previous chapter, namely the people's response to 'betterment'. In particular I will show here the forms of protests that occurred and discuss who belonged to the group of 'strikers', as the government called them, and who did not. The chapter closes with a preliminary summary that draws some conclusions on the effects of 'betterment' and on the Thaba Nchu social-ecological system and its resilience.

5.1 Industrialisation, Unemployment and the Control of Labour Migration

When the National Party won the elections in 1948 the phase of 'Grand Apartheid' was initiated, i.e. the comprehensive and large-scale segregation of the various South African population groups. In the years before, the South African Union had experienced major economic transformations. The increasing industrialisation and mechanisation in the manufacturing industries had resulted in a high demand for additional labour force. Since wages in the mining and agricultural sector were considerably lower, many people left their former workplaces in order to seek jobs in the manufacturing industries. The Union Party under General Smuts did not impede the influx of black South Africans into towns, with the effect that mine owners and commercial farmers faced severe labour shortages. Moreover, the white working class feared the undermining of their jobs by low-paid black South Africans. Therefore, the National Party gained a substantial part of their votes from commercial farmers and members of the white working class (Unterhalter 1987:10–11; Hendricks 1989:308–09).

One important aim of the new government was to control the movement of black South Africans in order to prevent them from leaving their former workplaces, i.e. white-owned farms, and entering the cities. Labour migration and population influx into urban areas was regulated by introducing a restrictive system of 'reference books' which replaced the previous passes black people were expected to carry. Reference books were now compulsory for all black South Africans over 16, including women.[202] These documents included a photograph of the holder and information such as their place of origin, employment history and tax payment records, as well as details of police encounters (Davenport 1978:270). Moreover, in 1952 the Urban Areas Act of 1945 was amended by rules that allowed black people to stay in towns only under the condition that (i) they had been born and

[202] With exception of the Orange Free State, where already in the late nineteenth century black women living in urban areas were expected to carry residential passes, women were excluded from the pass laws until the Abolition of Passes and Coordination of Documents Act in 1952 (Wells 1983:56; Lodge 1984:140).

lived there continuously, (ii) were long-term employees or had lawfully stayed for 15 years in the area, (iii) were dependents of people who fulfilled both previous requirements, or (iv) had permission to stay, for instance because of a short-term contract. Other black South Africans were only allowed to stay for a maximum of 72 hours in the urban area (Unterhalter 1987:12). Later in 1968, the control of labour force was reinforced by introducing tribal labour bureaus where people had to get a permission in order to search for work in towns (ibid.20).

From the early 1950s onwards, the National Party emphasised the separate development of the 'native areas' which was to be achieved by the creation of so-called 'homelands' or 'Bantustans'. The homeland policy coincided with a phase of widespread opposition by black South Africans in both urban and rural areas and an investment boom with significant effects on the South African economy. Foreign investments and an influx of technology in the 1960s led to a huge expansion of the manufacturing sector as well as of commercial agriculture. While the former sector faced again an increased need for additional labour power, the mechanisation of the latter resulted in a rise in agricultural unemployment (Yawitch 1982:44; Unterhalter 1987:15). Between 1960 and 1977, the number of unemployed farm workers increased from 589,000 (24%) to 1.14 million (37.9%). In total, the unemployment figures increased from 1.24 million (18.3%) in 1960 to 2.3 million (22.4%) in 1977 (Simkins 1978:34, 36, 41). At the same time, the homelands experienced a dramatic population influx of (i) unemployed people who were removed from the towns, (ii) former labour tenants and ex-farm workers who had lost their jobs in the agricultural sector, and (iii) people who had been living on so-called 'black spots', i.e. freehold land owned by black South Africans in areas designated for whites (Yawitch 1982:44; Unterhalter 1987:93). It is against this background that one must understand the 'betterment' policies of the 1960s and the reasons why the focus shifted from the implementation of small-scale conservation measures to the large-scale creation of grid-patterned villages that would provide enough space for newcomer households.

5.2 Paving the Way for Self-Governance

In Chapter 3.2.2 I showed that the basis for indirect rule had been laid with the Native Administration Act of 1927, which established 'native courts' based on 'customary law', thereby creating a separate legal system for black South Africans. Although the act extended the power of the chiefs with regard to jurisdiction, the major aspects of the administration of 'native reserves' were still vested in the NAD. This changed in the early 1950s, when the South African government paved the way for the creation of self-governing homelands and thereby initiated a shift from the 'era of segregation' to the 'era of apartheid' (Mamdani 1996:100–01). The Bantu Authorities Act of 1951[203] made provision for the establishment of three administrative bodies in the rural areas by introducing tribal, regional and territorial authorities (Murray 1992:156; Jacobs 2003a:176). Tribal authorities constituted local councils that, in the words of Nancy Jacobs (2003a), were 'the basic building block of Indirect Rule' (ibid.176). They were established in order to

> 'render assistance and guidance to its [the tribal authority's] chief [and] advise and assist the Government and any territorial and regional authority [...] in connection with matters relating to the material, moral and social well-being of natives resident in that area, including the development and improvement of any land within that area.'[204]

Tribal authorities consisted of several members who were paid by the government and appointed by the chief. Two or more tribal authorities constituted a regional authority, whose chairman was selected from the chiefs, headmen or councillors of the respective tribal authorities. Regional authorities administrated education, public works, health care and agriculture – in particular the control of stock disease, improvement of farming and afforestation. In some cases, such as Thaba Nchu, tribal authorities were granted the powers of a regional authority. The highest level of the indirect rule hierarchy constituted the territorial authorities, which consisted of two or more regional authorities.[205]

One of the main tasks of chiefs and tribal authorities was to convince local populations to accept rehabilitation measures, such as livestock culling, fencing-off of rangelands, division and reallocation of arable lands and resettlement in centralised villages (Murray 1992:156). I have shown in

---

[203] Union of South Africa, Act No 68 of 1951: Bantu Authorities Act.
[204] Union of South Africa, Act No 68 of 1951: Bantu Authorities Act, section 4.
[205] Ibid. section 5; Union of South Africa, Proclamation No. 136 of 1953: Conferment of Powers, Duties, Functions and Obligations upon the Barolong Tribal Authority, Thaba Nchu; in SAB NTS 8154 37/341.

Chapter 4.3.1 that in the Thaba Nchu area 'betterment' was already officially approved in 1939. However, in other South African regions the implementation of the schemes took a long time in coming as communities refused to give their approval.[206] To achieve the consent of the local population it was therefore critical to obtain the cooperation of the chiefs (Delius and Schirmer 2000:729). By introducing a system of local government, the chiefs were released from their accountability to the *kgotla*, the assembly of all men. Chiefs and their councillors were not obliged to consult with local residents and thus could be used to enforce unpopular decisions and laws (Yawitch 1982:20; Lodge 1984:266; Delius and Schirmer 2000:729).

Although in theory, a rejection of the Bantu Authorities Act was possible, many Batswana groups accepted the legislation (Schapera and Comaroff 1984:67). On the one hand, chiefs were in a difficult situation, as they could be replaced when they did not accept the legislation (Yawitch 1982:20). On the other hand, the increasing salaries for Bantustan officials and the access to development resources encouraged chiefs to accept it (Murray 1992:156). On 19 January 1952, Chief Albert Setlogelo Moroka[207] and 700 Barolong agreed in a 'tribal meeting' to the inauguration of a Bantu tribal authority in terms of Act no. 68 of 1951.[208] One year later, on 1 May 1953,

---

[206] There existed legal differences with regard to 'betterment' implementation. Land that was bought by the South African Native Trust from the late 1930s onwards could be subjected to 'betterment' without the consent of the local population. However, the 'native reserves' that were defined in the Natives Land Act of 1913 were not owned by the SANT. Here, rehabilitation schemes could only be implemented after consultation and approval by local residents. It was particularly in these old reserve areas that Native Affairs officials struggled with the enforcement of 'betterment' (Delius and Schirmer 2000:729).

[207] Despite the fact that the former magistrate had decided to disinherit Albert Setlogelo Moroka due to a criminal offence he had committed (Chapter 3.2.2), the latter was appointed as chief in 1941. After the death of John Pethogane Moroka in 1940, the question arose whether his son or Albert Setlogelo would be his lawful successor. A committee appointed to resolve the issue argued that the magistrate's decision in 1922 had not been approved and that John Pethogane Moroka should have never been appointed as chief. Moreover, they argued that Albert Setlogelo was still a young man when he was convicted for stock theft; see SAB NTS 308 13/54, Thaba Nchu Chieftainship, Report of the Committee appointed to enquire into the question of the succession to the Chieftainship of the Barolong tribe, 14 March 1941.

[208] SAB NTS 8154 37/341, Tribal resolution, Thaba Nchu, 19 January 1952.

the Barolong tribal authority was established. With the creation of the body both the Thaba Nchu and Sediba 'native reserve' boards were discontinued and all their assets, liabilities and properties transferred to the Barolong tribal authority.[209] The new authority consisted of ten councillors, i.e. nine ordinary members and one chair (Schüler 1965:95). Many members of the old landowning Barolong elite were incorporated into these new administrative structures, for instance by being employed as councillors. Later these people played central roles in the emerging Republic of Bophuthatswana. Hence, the landowning elite transformed from the 1950s onwards into a 'modern bureaucratic, professional and commercial elite' (Murray 1992:156-158, 185).

A further important step towards the self-government of the 'native areas' was made with the Promotion of Bantu Self-Government Act No. 46 of 1959.[210] One pillar of the act was the recognition of 'national' groups, namely Xhosa, Zulu, Basotho, Batswana, Pedi, Venda, Tsonga and Swazi, who were assigned to specific homelands. The territory of these homelands corresponded approximately with the 'scheduled' and 'released areas' identified in the land acts of 1913 and 1936 respectively and consisted of little more than 13% of the total land in South Africa. The homeland for all Setswana-speaking groups included six larger and 20 smaller blocks of land; Thaba Nchu being one of the latter. In 1961, the Tswana territorial authority was constituted, being only an advisory council since Pretoria retained political and financial control. In 1968, the Tswana territorial authority was given more powers and its own bureaucracy, which also included an agricultural department led by white officials (Unterhalter 1987:16; Erasmus and Krige 1998:10; Jacobs 2003a:195).

To sum up, in the 1950s and 1960s the way was paved for self-governance with the introduction of 'native authorities'. By creating a system of separate local governments, the principle of indirect rule, the foundations of which were laid with the Native Administration Act in 1927, was expanded. At first glance, it may seem that the black South Africans in these areas were granted more decision-making power. However, 'the indirect

[209] Union of South Africa, Proclamation No. 136 of 1953: Confernment (sic) of Powers, Duties, Functions and Obligations upon the Barolong Tribal Authority, Thaba Nchu, 1953, in SAB NTS 8154 37/341.

[210] Union of South Africa, Act No. 46 of 1959: Promotion of Bantu Self-government Act.

rule state was not a weak state' (Mamdani 2012:8) and the influence of government officials remained strong. Chiefs and their local councils were used to enforce the unpopular 'betterment' schemes and to suppress any form of local resistance, as I will illustrate in Chapter 5.5.

5.3 Defining 'Economic Farming Units': The Report of the Tomlinson Commission

At the beginning of the 1950s the NAD had realised that the rehabilitation efforts outlined in the 'New Era of Reclamation' policy (Chapter 4.3.2) were largely ineffective. Since planning committees had to conduct detailed surveys for each 'betterment' area first, the work proceeded only slowly. NAD officials deemed a realignment of the policy necessary in order to achieve stabilisation of the reserve areas on a larger scale (De Wet 1995:45). As the new focus was on short-term stabilisation the requirements for the planning were downgraded. Instead of detailed surveys with interdisciplinary teams as described by the 'New Era' document, planning was based on reports of so-called 'ad-hoc committees' who paid only short visits to the respective areas. The schemes were to be implemented faster than in previous times, as homelands had to be prepared for the resettlement of thousands of families from 'black spots', private farmland owned by white South Africans and urban areas (Hendricks 1989:319).

Despite this obvious change in policy, the new Minister of Native Affairs, Hendrik Frensch Verwoerd appointed the *Commission for the Socio-Economic Development of the Bantu Areas within the Union of South Africa* in 1950. Under Chairman F. R. Tomlinson, the Commission was appointed in order to

> 'conduct an exhaustive inquiry into and to report on a comprehensive scheme for the rehabilitation of the Native Areas with a view to developing within them a social structure in keeping with the culture of the Native and based on effective socio-economic planning.' (Union of South Africa 1955:xviii)

The Tomlinson Commission submitted its final report in the mid-1950s, which was then the most comprehensive survey ever conducted to address the situation of black South Africans (Yawitch 1982:27; Hendricks 1989:310; De Wet 1995:45). The report not only dealt with issues of land use but also with mining, secondary and tertiary industries, urbanisation, ecclesiastical development, health, welfare and education. The Commission's main line of argument was that a solution to the 'native question' in South Africa could be achieved by either complete segregation or integration; however, the latter option 'could only lead to problems and dangers of the gravest intensity' (Union of South Africa 1955:105).

The Tomlinson Commission – despite recognising that land was insufficient to 'produce the food requirements of the population' (Union of South Africa 1955:74) – abided by the narrative that the limiting factor in the rehabilitation of the farming areas was the bad land use practices of the black farmers (Union of South Africa 1955:77). As other state officials had done before, the team under Tomlinson stressed the negative effects of the 'cattle complex'. While white farmers relied on profitable animal types, black farmers would always prefer cattle. The Commission referred to the example of the Thaba Nchu district, where cattle outnumbered sheep enormously (17 cows per sheep). In contrast, 'European farmers' would own an almost equal ratio of cattle and sheep (Union of South Africa 1955:79). Irrespective of the fact that the figures on Thaba Nchu's cattle-to-sheep ratio provided by the Tomlinson Commission did not correspond to those mentioned in the reports of the culling officers, there was certainly a predominance of cattle in the black communities. However, the Commission did not consider that sheep had been identified as the main threat to the *veld* and thus had been systematically reduced during the culls in the 1940s. Hence, the higher amount of cattle was at least to some extent an outcome of government's aversion to small stock.

Several ideas that had been formulated in precedent reports and policy statements are recognisable in the report of the Tomlinson Commission. For instance, it was suggested that the division of land into residential, arable and grazing areas was to be continued. Most central in the Tomlinson report was the notion that rural populations were to be divided into (i) full-time farmers, who would be entitled to use an 'economic farming unit', and (ii) non-farming people, who were to be relocated to newly established rural townships to gain an income from off-farm occupations. In contrast to previous 'betterment' approaches, the Commission insisted that 'the number of families to be settled as farmers shall not exceed the number of [economic farming] units' because 'sub-maintenance farming cannot be allowed' (Union of South Africa 1955:113-114). However, it was not defined in the report how these two groups were to be chosen (Yawitch 1982:28).

The Commission stipulated that an 'economic unit' should generate an annual gross income of £60 to enable a family to live from land use alone. Whether survival on this amount was realistic or not is addressed below. The authors of the reports suggested that in cases where the allocation of 'economic farming units' was not possible, settlement should be based on half units; but in mixed-farming areas, such as the Thaba Nchu district, not less than 1.7 ha arable land were to be allocated. Moreover, it was recommended that non-farming families might settle temporarily in a residential area or, if they wanted, immediately in a 'Bantu town' (Union of South Africa 1955:118). With these suggestions, the Commission undermined its own principal recommendations. As I will show later, in practice both the curtailment of 'economic units' and the settlement of non-farmers in residential areas were the rule rather than the exception.

Whereas the idea of 'economic farming units' and the division of land according to its purpose were seconded by the National Party government, other suggestions were not accepted (Hendricks 1989:310; De Wet 1995:48). The Tomlinson Commission had recommended the sale of 'economic units' to individual land users. In cases where the owner would over-utilise his land, the SANT should have the right to cancel the title and sell it to another person (Union of South Africa 1955:118). The proposition of freehold title deeds was rejected by government officials because it would not be in accordance with the 'culture' of black people. Hendricks (1989) argues that the idea of freehold 'economic units' would have counteracted the 'retribalisation policy' based on Bantu authorities as it impeded the chief's control over land (ibid. 312).

The required budget estimated by the Commission was cut from £104,486,000 to £36,600,000 with the consequence that several development ideas were not considered at all from the start. These were for instance the creation of credit facilities by establishing a land bank-like organisation, the improvement of education facilities, and the development of mining as well as secondary and tertiary industries in the homelands. Hence, the basic prerequisite for the creation of income alternatives for non-farming families was not implemented (De Wet 1995:48–49). From the late 1950s onwards, the government invested more money in the development of border industries that adjoined the homelands and created a five-year plan for the Bantustan development in 1961. *Inter alia* this plan proposed the investment of £38 million for the creation of non-agricultural villages for the 'surplus population'.

One Job is not Enough: Income Needs and Realities

In order to determine what a viable farming unit should be, the Commission referred to the evidence of several witnesses. The respondents had different opinions about how much annual income had to be generated from land use in order to supply a household. Figures varied between £100 and £200 per annum, but a majority defined an annual gross income of £120 as necessary (Union of South Africa 1955:113). This is more or less in line with the estimations of South African academics. For instance, Krige and Krige (1956) estimated that a family of five persons needed an income of £150 per year from subsistence farming to fulfil their needs in the early 1950s (ibid. 25), while Simkins (1981) calculated that about £100 per annum was necessary for a family of the same size in the mid-1950s (ibid. 282). In any case, all estimates and calculations are much higher than those suggested by the Tomlinson Commission. Wolpe (1972) had pointed out that the apartheid state aimed to prevent the movement of black South Africans to the cities, and to maintain a 'reservoir' for cheap-labour power in the reserves. This could only be achieved by enabling rural households to engage in agricultural production 'at a level which, while not too low to contribute to the reproduction of migrant workers as a class, [was] yet not high enough to remove the economic imperatives of migration' (Wolpe 1972:437). Therefore, the creation of a full-time farmer class was not a serious aim of the government.

Assuming that Simkins' calculation represents the minimum, a family needed for survival, then the average annual incomes gained from various occupation types (Table 4) show how difficult it was for black South Africans to gain enough money to feed a family. The first four rows depict the average incomes in typical occupation spheres, based on estimates for 1949. It should be noted that these incomes varied greatly. The last five rows of Table 4 indicate data that were collected in an agro-economic survey conducted by the Tomlinson Commission in the mid-1950s. The study included 880 families in mixed-farming areas with high and low rainfall; 200 in supervised and non-supervised irrigation schemes; and 100 families in purely pastoral areas. The data refer to the families' incomes in cash and in kind.

Table 4 Average annual income for different occupational groups as indicated by the Tomlinson Commission[211]

Occupation type	Average Annual Income
Farm worker	£83
Mine employees	cash: £47 and 9 shillings cash & in kind: £90
Secondary industry employees	£110
Commerce employees	men: £99 women: £55
Agro-economic survey data	
Mixed farming in high rainfall area	£29
Mixed farming in low rainfall areas	£9
Pastoral farming	£40
Irrigation farming (under white supervision)	£110
Irrigation farming (not under white supervision)	£28

The Commission estimated that in 1951 about 503,000 men had temporarily left the South African reserves in order to work for cash. This was about 12% of the total population living in these areas. Women also temporarily migrated for work, which they often found on white-owned farms during the harvest seasons from March until November. Moreover, the domestic service sector was an important field for female labour migrants, particularly in the Orange Free State (Wells 1983:59). There the average monthly income in 1952 was £8 and 13 shillings (Union of South Africa 1955:38). According to the Commission's data there were considerably fewer female than male labour migrants. In May 1951, during the peak season of female migration, about 66,000 women were working outside the reserve areas (Union of South Africa 1955:94).

[211] Data source: Union of South Africa (1955:35-37; 85-86).

5.4 'They Wanted Us to Be in Lines': Resettlement in Thaba Nchu

In Thaba Nchu, the mid-twentieth century marked a shift in the history of environmental governance as interventions now targeted a more comprehensive reorganisation on the regional level. First, the use of agricultural land was again restructured by dividing land and herds into 'economic units'. Second, the former residential patterns were dissolved and households resettled in newly erected grid-patterned 'Trust villages' as local residents call these villages constructed by the SANT. Hence, for the first time 'betterment' directly affected not only farming households but all resident families, and also all the villages across the whole district. Moreover, the forced resettlement transformed communities more significantly as people were evicted from their places of belonging, where they had lived for decades and where their ancestors were buried.

When considering resettlement in the Thaba Nchu district it is necessary to differentiate between various forms that evolved. First, centralised villages were erected on land that was bought by the SANT. As the land had previously been privately owned, these villages were often constructed on barren ground, though not necessarily unused land. In the district, such villages had already been erected on SANT land in the 1940s. Second, centralised villages were constructed on old reserve land which had been previously inhabited. Here, all existing households were removed to newly grid-patterned settlements that were often constructed in part on the ruins of former homesteads. The aim was not only a concentration of the homesteads but also the ordering and numbering of houses and kraals, as this facilitated control. The creation of planned settlements on reserve land concentrated on the area of the Sediba reserve, whereas the old locations in the Thaba Nchu reserve, namely Mokoena, Ratlou and Moroka, remained more or less unchanged (Murray 1992:157). Third, rural non-agricultural villages were constructed at the edge of Thaba Nchu town. Residents in these new townships were allowed neither to keep animals nor to cultivate fields. In the following I will present in more detail how the replanning of land use and residential patterns affected people's lives on the newly acquired SANT land, in the old reserves and in the urban areas.

5.4.1 Early Centralised Villages on SANT Land

From the late 1930s onwards, the SANT purchased the 'released areas' that had been identified in the Native Trust and Land Act of 1936 and that were to be added to the 'scheduled areas' determined by the first land act of 1913. In the Orange Free State about 68,523 ha of privately owned land had been identified as 'released areas'. With the exception of one farm close to the Witzieshoek reserve, later Qwaqwa homeland, all of this land was situated in the Thaba Nchu district (Murray 1992:128–29). After the passage of the 1936 act the SANT began to purchase white-owned and black-owned farms located in the 'released areas': two larger pieces of land south and north of the Thaba Nchu reserve and a smaller portion adjoining the Sediba reserve (ibid. 130, figure 4.1). Many of the farms owned by Barolong were mortgaged (Chapter 3.1) and the owners were often unable to pay the required interest on the mortgagees. The SANT, which was entitled to redeem such mortgage bonds, refused to do so in most cases. Sooner or later the owners of those farms had no choice but to sell their land to the SANT. Besides these black-owned farms situated in the 'released areas', some Barolong also owned land in white farming areas. These so-called 'black spots' were soon to be eliminated (Murray 1992:136–44).

In the early 1940s six centralised villages were established on the SANT land, i.e. Moroto and Tiger River situated north of the Thaba Nchu reserve, and Gladstone, Tweefontein, Rietfontein and Klipfontein situated south of Thaba Nchu (Figure 1). Moroto and Tiger River were inhabited by a considerable number of Xhosa families and other non-Barolong who had lived there when the farm still belonged to white landowners (Murray 1992:167). That the newly acquired land was already occupied by several households was not an untypical case. The four centralised villages south of Thaba Nchu should make provision for about 500 families (ibid. 167). As the land purchased by the SANT automatically became a 'betterment' area, people living in the 'Trust villages' experienced the rigours of livestock culling and the control of land use from the very beginning. Another wave of land purchase was conducted in the late 1940s. Further grid-patterned villages were erected soon, namely the southern Woodbridge and Commissiedrift as well as the northern Merino and Paradys. People who were settled in these new villages were mainly (i) those who previously had been living on these former private farms; (ii) the 'surplus families' of the already densely populated Trust villages built in the early 1940s; and (iii) people from the overcrowded Thaba Nchu reserve (ibid. 171).

The construction of planned villages continued during the next two decades. Two of my research sites – Rooifontein and Middeldeel – are typical examples of 'Trust villages' constructed on SANT land. The farm Middeldeel No. 709 was purchased by the SANT in the late 1940s. By then, only three families lived on the 428 ha. While on the southern edge of the farm the villages of Paradys and Merino were erected soon after the purchase of the land, Middeldeel remained unsettled until the 1960s. The first families who moved to Middeldeel in 1963 came from Rooibult, a location in the old Sediba reserve. Additional households from Rooibult followed in 1964 together with other families that had been evicted from white-owned farms.

According to an older woman interviewed in 2011, at least some of the early arrivals moved from Rooibult to Middeldeel on their own initiative.[212] There was a rumour that people would have to leave their former homes and as a result some men discussed whether it would be better to leave immediately. A few of them might have hoped that they would find better conditions at the new farm. Rangelands in Rooibult were the most severely degraded in the Sediba reserve.[213] It is not clear how many families belonged to that group of initial migrants in 1963, but oral histories suggest that it were not much more than a handful.[214] The woman from Middeldeel remembered their arrival as follows:

> Paulina Mokgothu was born in 1944 in Rooibult as one of seven children. Her father was one of the first men who decided to leave their homes and move to the farm at Middeldeel. Before the family left Rooibult he went to the new home destination, where he found nothing but barren ground. There were neither streets nor houses nor fences. Using corrugated iron from their former home, he built a little shack in order to provide his family with a roof over their heads. Then he went back to Rooibult, and together with his family packed their household goods on oxen wagons and moved to Middeldeel. After their arrival, the family constructed a bigger mud house on a numbered site that was pointed out to them by white officials (Interview with Paulina Mokgothu, Middeldeel, 10 September 2011).

---

[212] Interview with Paulina Mokgothu, Middeldeel, 10 September 2011.
[213] SAB BAO 5761 H128/1666/197 part 1, Herwinningsverslag: Sediba Gebied, Thaba Nchu, 22 May 1958.
[214] Interview with Paulina Mokgothu, Middeldeel, 10 September 2011.

As indicated above, the family of Paulina Mokgothu belonged to a minority that left their homes at least to a certain extent on their own initiative. In contrast, the majority of the households in the old Sediba reserve remained in their homes until they were told to resettle or removed forcibly.

5.4.2 Replanning and Resettlement in the Sediba Area

The first planning of a spatial reorganisation and rehabilitation was done initially only for the Sediba irrigation scheme in 1956.[215] The report was compiled by an ad-hoc planning committee that would meet 'from time to time'[216] and consisted of the Native Commissioner and his assistant, several agricultural officers, and the agricultural foreman. Their aim was *inter alia* to define an appropriate 'economic unit' for the farming households. There were several possibilities, as 'economic units' could consist either of (i) a piece of dryland, an irrigation plot and livestock; (ii) an irrigated plot only along with livestock; (iii) a mixed system, with some farmers having a piece of dryland, irrigated land and livestock and others having only one type of land and livestock. At the beginning the report summarises basic information on the Sediba location and its resources, encompassing an area of about 1,665 ha. The grazing area of 686 ha was divided in six camps ranging from 57 ha to 193 ha. The dryland fields (534 ha) consisted of 5.14 ha plots which had been surveyed a couple of years previously, to the disapproval of the villagers (Murray 1992:174). Moreover, some residents had access to the Grobler irrigation scheme, which by then had 111 ha. Thus, the total area used for cultivation was only little smaller than the rangeland area. According to the report, the actual settlements extended over an area of 333 ha; but this was not a realistic estimation as the ad-hoc committee counted all land that was not explicitly demarcated as grazing camps or fields as 'residential sites'.

Generally, the condition of the rangelands was defined as good. While the quality of the dryland fields was not mentioned in the report, the irrigation scheme was said to be mismanaged, which had led to a significant decrease in soil fertility. Officials considered inadequate fertilisation to be the reason for its bad condition.[217] This was a constant problem in the irrigation scheme, as people were not allowed to drive their animals onto the land

---

[215] SAB BAO 5761 H128/166/197 part 1, Ad-hoc Verslag Sediba Besproeiingsskema, Thaba Nchu, 28 February 1956.
[216] Ibid. 1.
[217] Ibid. 3.

after harvest. Hence, the natural fertilisation as practiced in the dryland fields was not possible in the scheme. Moreover, the dung produced in the kraals was mainly used as fuel, since firewood was scarcely available in the grassland biome.

Table 5 Population and land use statistics of Sediba location: comparison of 1948 and 1956[218]

	1948	1956	Difference 1948/1956
Number of families	205	234	+ 14%
Livestock owners and access to land (Dryland and/or irrigated land)	177 (86%)	145 (62%)	- 24%
Livestock only	10 (5%)	43 (18%)	+ 13%
Land only	14 (7%)	11 (5%)	- 2%
Neither land nor livestock	4 (2%)	35 (15%)	+ 13%

When comparing the population and land use statistics of 1956 with data from 1948 (Table 5) it is striking that the number of families increased to a significant extent, which was certainly not only the result of a natural increase, but rather of the immigration of 'surplus people' to Sediba. Moreover, as the table indicates, access to land had become more exclusive. The proportion of families having access to both livestock and fields decreased by almost one quarter within a period of less than ten years. On the other hand, the number of families who owned livestock only and were not able to fulfil part of their basic needs from cultivation increased; the same goes for the number of families without access to either land or livestock. The figures must be interpreted with caution, as the official definition of 'a family' is not made clear and the accuracy of the counting remains sometimes dubious. Nevertheless, the figures indicate a clear trend away from subsistence farming. I suggest two main reasons for this trend. First, many of the

---

[218] Data for 1948 are drawn from: SAB NTS 7526 717/327 part 2, Senior Agricultural Officer, Thaba Nchu, to the Chief Native Commissioner of the Western Areas, Potchefstroom, 25 October 1948, attached list. Data of 1956 are drawn from SAB BAO 5761 H128/166/197 part 1, Ad-hoc Verslag Sediba Besproeiingsskema, Thaba Nchu 28 February 1956, p. 3.

newly arrived families had no access to natural resources as rights to land had already been allocated to the established residents long before then. Second, livestock culling had increased the burden on farming families and made the cultivation of land more difficult. Several families affected by culling had presumably given up their land due to lack of draught power. This trend was not halted but accelerated with the allocation of 'economic farming units', as I will show below.

The ad-hoc committee recommended that the Sediba location was to be reorganised as an 'irrigation unit'. In consequence, all dryland fields were to be transformed into grazing land and cultivation was only to be allowed on the irrigable plots. According to the planning report, about 80 families out of the existing 234 would remain in Sediba, each of them using an 'economic unit' of 1.28 ha of land and 5 large stock units. The 'surplus population' was to be resettled somewhere else. Former dryland farmers were to move to the Rakhoi area (Figure 1) where they could continue with dryland cultivation and animal husbandry. However, it was never considered whether Rakhoi would have enough space for 65 families. Non-farming families were to be removed to a non-agricultural village, which was not specified further in the recommendations. The 80 families who were allowed to remain in the Sediba location were to be resettled in two centralised villages, each of the latter having three newly subdivided grazing camps.[219]

The plans of the ad-hoc committee were never put into practice. The proposal that only 80 families should remain in the location was unacceptable to the NAD.[220] New calculations were made, and the determined carrying capacity was raised in order to make a settlement of about 160 families in Sediba and Rakhoi possible.[221] This shows how flexible such ostensibly fixed figures as the carrying capacity actually were. As soon as more space

---

[219] SAB BAO 5761 H128/166/197 part 1, Ad-hoc Verslag Sediba Besproeiingsskema, Thaba Nchu, 28 February 1956.

[220] SAB BAO 5761 H128/1666/197 part 1, District engineer, Bloemfontein, to the Chief Native Commissioner of the Western Areas, Potchefstroom, 24 May 1956; SAB BAO 5761 H128/1666/197 part 1, Chief Native Commissioner of the Western Areas, Potchefstroom, to the Secretary of Native Affairs, Pretoria, 28 May 1956.

[221] SAB BAO 5761 H128/1666/197 part 1, Secretary of Native Affairs, Pretoria, to the Chief Native Commissioner, Potchefstroom, 19 October 1956, p.1.

for resettlement was needed they could be changed. However, all these discussions and recalculations were a vain endeavour. In 1957, the Chief Native Commissioner demanded that instead of a replanning of the Sediba location only, the whole reserve and its adjacent SANT farms (about 20,568 ha) were to be considered. This included six locations, namely Sediba, Talla, Rooibult, Bofulo, Kgalala and Morago, each of them consisting of several scattered settlements.

In 1958 a rehabilitation report for the Sediba reserve was compiled by a similar team of officials.[222] As previously, the report contained a small section which dealt with the ecological conditions. The soils in the irrigation scheme were described as bad, showing indicators of mismanagement. In contrast, the soils of the dryland fields were regarded as generally highly fertile, without any signs of erosion. Grazing lands were also in a comparatively good condition as soil erosion along rivers had been stopped. Only in the Rooibult area were sheet and gully erosion visible to a great extent.[223] Clearly, the reasoning for the rehabilitation was not primarily based on ecological concerns (Naumann 2014:51).

In total 1,232 families – about 6,400 people – were to be resettled in the new Trust villages. The Grobler irrigation scheme should provide 82 'economic farming units', each consisting of 1.28 ha of an irrigated plot – subdivided in 14 beds – and only two dairy cows. As no oxen were allowed, the irrigation farmers were expected to hire tractors provided by the tribal authority or the government.[224] Besides the 'irrigation units', the report planned for 509 'economic units' in the dryland area, each consisting of 7.7 ha of arable land – of which one third had to lie fallow – and ten cattle. It was estimated that with this 'economic unit' each household would generate a gross annual income of about £61 and 15 shillings.[225] Ten head of cattle was not much, but could be enough to plough and to provide some milk for the family. However, in order to make enough rangeland available for more than 5,000 cattle a reduction of the cultivated area – by then about 4,700 ha – by almost a fifth of its size was regarded as necessary by officials (Naumann 2014:51).

---

[222] SAB BAO 5761 H128/1666/197 part 1, Herwinningsverlag: Sediba Gebied, Thaba Nchu, 22 May 1958.
[223] Ibid. 3.
[224] Ibid. 8-13.
[225] Ibid. 6, 13.

The planning of the Sediba reserve area envisaged only 591 'economic units', thus more than half of the residents were regarded as 'surplus people', and for who a solution had to be found. The ad-hoc committee suggested that only farming households with fields and livestock should be considered for the 'economic units'. People without access to land and those who only kept cattle were to be resettled. The latter group consisted, according to the official opinion, mainly of labour migrants who had been the main activists in the revolts against the rehabilitation efforts (Chapter 5.5.2). The committee expected the assistance of the Barolong authority with regard to the selection of proper farming families. 'Surplus households' were to be resettled to a site west of Thaba Nchu, where Selosesha is located today. The committee stressed that the establishment of these non-farming settlements should be accelerated since a further delay would hinder the implementation of 'betterment' in the rural areas. Altogether, the costs for the rehabilitation of the Sediba reserve were estimated at £16,365; this included no compensation for the resettlement of homesteads.[226]

The Secretary of Bantu Administration and Development was not satisfied with the envisaged number of farming families, not least because the space in the planned townships was restricted. The Pretoria official therefore recommended in a comment on the report that the 'economic units' of the dryland area should be subdivided into 300 full- and 412 half-'economic units', and 172 plots of 0.85 ha. Thus, a total of 884 dryland farmers and additional 82 irrigation farmers would have been accommodated in the area.[227] With this proposition the idea of providing full 'economic units' to a limited number of farming households was called into question, and in the end, was never put into practice (Naumann 2014:52). That the 'betterment' was 'loosely planned', as Hendricks (1989) puts it, was therefore not the main reason for its failure. Rather, the whole idea of a 'full-time' farmer class was incompatible with the needs of apartheid planners, whose plans required space for 'surplus people' and the maintenance of a cheap labour 'reservoir'.

The actual resettlement of households started in Sediba in the mid-1960s. Some residents moved immediately; others refused to move for some time.

---

226 Ibid. 13, 17-18.

227 SAB BAO 5761 H128/1666/197 part 1, Secretary of Bantu Administration and Development, Pretoria, to the CNC, Bloemfontein, 16 February 1959.

New plots were laid out in a standardised grid pattern. In some cases – when new settlements were erected on the foundation of old ones – this led to the absurd situation in which a household had to demolish its homestead just to erect it a few metres away in line with the neighbouring plots. Houses that were built in these plots had to be constructed in the centre of the plot about 4.5 m away from the front fence.[228] Although from time to time it was adumbrated by officials that households would receive compensation,[229] informants interviewed in 2011 stated that they never did receive any money to refund their removal costs.

Disrupting Neighbourhoods Based on Kinship

Several studies have shown that resettlement had long-term effects on social ties that existed between community members (McAllister 1989; De Wet 1995; Fay 2012). But the impacts on social structures varied from case to case. In the villages on newly purchased Trust land, all arriving households were newcomers in the initial stage. In contrast to that, the situation in the old reserve locations was more complex. Families removed from white-owned farms, 'black spots' and other Trust villages were resettled between families that had lived in the area for at least three generations. In the case of Phatlha (Figure 12), a former settlement of the Sediba location, established families had to move to new plots designated for dryland farmers – later called 'Sediba Trust' – only to discover later that the new village for the irrigation farmers – later called 'Sediba Scheme' – had been erected on their former homesteads. Many newcomers were resettled in Sediba Scheme, which did not ease the already strained atmosphere. Newcomer households were observed critically and the situation was fraught with conflict.

> 'You see, they [government officials] removed us from our own place. Now I can't understand why they took us from our residential place and brought us here and took people from the farms and took them to our own place? They should have left us there, because we grew up there and we used to love that place. Moreover, there are the graveyards and the ancestors, where we were

---

[228] SAB BAO 4831 F53/1666/1 part 1, Minutes of Meeting, Petra Clinic, 18 April 1964, p. 2.

[229] For instance, SAB BAO 4831 F 53/1666/1 part 1, Minutes of Meeting, Petra Clinic, 18 April 1964, p. 2.

born.' (Interview with Lydia Pobe, Sediba, 4 September 2011; in Naumann 2014:52)

Today intra-community tensions are still visible in Sediba. For instance, during a transect walk with an informant descended from an old established family I was told that people in Sediba Scheme had only been entitled to a maximum of three cows and a small plot because they were 'too lazy'.[230]

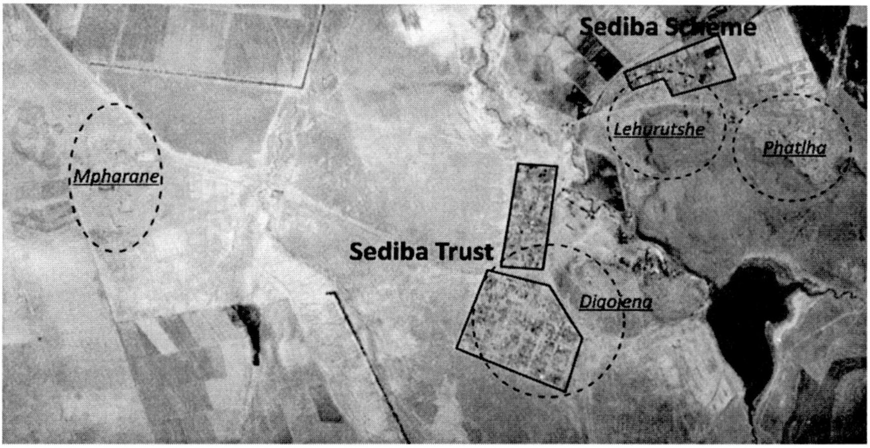

Figure 12 Sediba Trust and Sediba Scheme in 1992. Dashed rings indicate pre-betterment settlements; continuous lines the new Trust villages[231]

The people living in the Thaba Nchu villages were the ones most directly affected by the resettlement. Sooner or later labour migrants also noticed the transformations taking place in their rural home areas; in particular, those who worked far away were often confronted with a *fait accompli* when they came home. Lydia Pobe, who worked in Johannesburg between 1953 and 1981, still maintained close ties to her birthplace, Phatlha. She was in Johannesburg when she received a letter from her uncle (MB) and village headman Monchusi, 'Listen my daughter, we have moved from Phatlha. We are on the other side at Digojeng [in Sediba Trust].'[232] During Christmas time she went home for a long weekend and took a look at the

[230] Informal interview with Zacharia Mokgobo, Sediba, 19 July 2011.
[231] Image source: aerial photograph 007/00036, job 969, 18 August 1992, provided by National Geo-Spatial Information; additions made by author.
[232] Interview with Lydia Pobe, Sediba, 4 September 2011.

new streets. Monchusi had reserved a plot for her, though she and her husband were not living in the district and came only for Christmas and Easter holidays. While they were staying in Johannesburg the headman's son lived on the plot. When Lydia's husband passed away in 1977, she stayed on in the city for a couple of years. In 1981, she moved back home to her relatives and the graves of her ancestors.[233] The example of Lydia Pobe shows that migrants maintained close ties to their place of belonging and at least some wanted to move back to their homes, with or without access to livestock and land.

Figure 13 Homesteads of the Monchusi family group in a section of Sediba Trust (source: Google Earth; additions made by author)

Lydia's account is interesting for another reason: it sheds some light on the question of how the forced removal affected the spatial distribution of family groups. In Chapter 3.3, I have illustrated that pre-'betterment' settlement patterns were mainly based on kinship. Often one or more agnatic family groups lived together. At first sight, it seems reasonable to suggest that families were not able to sustain these patterns after their removal into

[233] Interview with Lydia Pobe, Sediba, 4 September 2011.

the concentrated settlements. Oral accounts of village elders and archival sources indicate that many people had no choice about where to move. Apparently, this was not always the case, as influential people could pursue their own interests when new residential plots were allocated. In the case of the Monchusi family this is still visible in the settlement patterns today. Figure 12 shows the Sediba location. The dashed rings indicate the location of the former pre-'betterment' villages, the continuous lines the new concentrated Trust villages. Headman Monchusi gathered his relatives in the rectangular part of Sediba Trust (Figure 13).

Today, of the 13 households named Monchusi nine are situated in the same part of the village. A closer look reveals that the former headman had managed to concentrate his relatives predominantly in one corner of the village section, situated next to one of the village's grazing camps. In the case of some other family groups, too, one can detect such patterns. Thus, the allocation of new residential plots was not only done by chance or according to the 'first come, first served' principle. Data rather suggest that influential men who were loyal to government officials – such as headman Monchusi, who gave information about planned protests against 'betterment' to the Native Commissioner (Chapter 5.5.1) – pursued their own interests when new residential plots were allocated.

5.4.3 Replanning of the Urban Locations

On 14 February 1964, a planning committee submitted an agricultural planning report for the Thaba Nchu locations Moroka, Ratlou, Mokoena and Motlatla, as well as for farm land that was bought by the SANT in 1938 and 1939 and adjoined the settlements.[234] Between March 1962 and the day of handover, the committee had worked for 130 days on the planning of the rehabilitation of these areas, including surveying, searching for possible sites, planning the villages and compiling the report. In total, a population of 12,150 people, or 2,427 families, lived in the planned area. Of those a majority of 61% (1,492 families) had no rights to land or stock at all, only 24 % (580 families) were both livestock owners and cultivators, and about 15% (355 families) had rights to cattle or land.[235] Hence, the situation in terms of land use activities differed significantly from that in

---

[234] SAB BAO 5761 H128/1666/214, Rehabilitasie: Beplanning van Thaba 'Nchu (Middel Gedeelte): Groep Trustplase, 14 February 1964.
[235] Ibid. 10-11.

the Sediba reserve. Households in the old Thaba Nchu locations were much more dependent on incomes from wage labour and were thus not as strongly affected by the implementation of the 'betterment schemes' as people in Sediba (Murray 1992:157). Furthermore, the ecological condition in the Thaba Nchu area was different. The soil fertility in the agriculturally used areas was low and large parts showed signs of erosion. The condition of the grazing land was also depicted as poor. Moreover, termites were destroying up to 50% of the grazing land. They were regarded as a major problem in the rangelands, but no solution was offered in the report.[236]

West of Thaba Nchu town a 1,634 ha area was designated for the erection of the above mentioned 'rural town' Selosesha. This place was reserved for non-farming families. When the report was completed an area of about 347 ha had already been surveyed and laid out for about 900 new residential sites. The remaining 1,287 ha that were to become part of Selosesha included parts of the Ratlou, Mokoena and Moroka locations that were already densely settled.[237] Hence, the planners strove to combine the township with the old locations to form a large non-agricultural urban area. The existing settlement structures of the old Thaba Nchu locations were not to be changed significantly as they were already concentrated (Murray 1992:157).

Additionally, the report envisaged the formation of eight villages on the adjoining farmland bought by the SANT, each with its own residential area, four grazing camps and arable fields. One of them was my research site, Rooifontein. Households in these new villages were to receive an 'economic unit' consisting of a 0.2 ha residential plot, 12 large stock units and 7.7 ha of land, of which one third had to lie fallow.[238] In these villages the landowning households from the urban area were to be resettled where they could continue with agricultural production.[239] Needless to say, there was not enough space in the rural areas for all those who wanted to continue their farming activities. People who did not leave the urban locations

---

236 Ibid. 1-6.

237 Ibid. 3, 15.

238 Ibid. 12.

239 SAB BAO 5761 H128/1666/214, Chief Bantu Affairs Commissioner, Potchefstroom, to the Secretary of Bantu Administration and Development, Pretoria, 11 June 1965.

were forced to give up both land and livestock. To the 'dismay and bewilderment of the Tribal Authority they were cajoled into signing documents declaring that they were selling their stock voluntarily, hence they forfeited all their rights'.[240]

Selosesha quickly developed into a large township inhabited by wage labourers. Non-farming households from other rural villages did not find space in that area. Besides Selosesha other concentrated settlements, called Bultfontein No. 1, 2 and 3 – known today as Zones 1, 2 and 3 – were erected on a farm of the same name in the late 1960s and 1970s. All these households were strongly dependent on incomes from wage labour (Murray 1984:46–47); yet many people were living a life of 'abject poverty' due to unemployment.[241] Moreover, crime and theft were serious problems in the townships, as an informant in Sediba remembered:

> 'There was too much crime. You would go out of your house, and say you go to work [and] leave the children behind. If they go to school, some people would come and find that there is no one left at home. They would steal everything from you.' (interview with Dichutso Mosotho, Sediba, 8 August 2011, translated by Tsametse Pobe)

Many rural households that refused to resettle were removed to a Bultfontein location in the late 1960s and 1970s. Often this was only a temporary place of residence until the people eventually were resettled to Botshabelo (Figure 1). Botshabelo (Sesotho for 'place of refuge') was created in 1979 when about 40,000 illegal squatters, mainly Basotho-speaking people, were resettled from a farm in Thaba Nchu which was then already part of Bophuthatswana. In the following years, hundreds of thousands of people were brought to the relocation site; within six years, Botshabelo had developed into the second-largest township in South Africa, with a population of about 500,000 by the mid-1980s (Murray 1987:318).

People were brought by lorries to Bultfontein and Botshabelo and had to live in tents at first. 'There was hunger and not even firewood', recollected an elder woman in 2011. The Franciscan priest Cosmas Desmond, who

---

[240] SAB BAO 5761 H128/1666/214, Barolong Tribal Authority, Thaba Nchu, to the Bantu Affairs Commissioner, Thaba Nchu, 16 August 1965, p. 2.
[241] Ibid. 3.

visited several resettlement areas throughout the country in the 1960s, described the situation in these townships as follows:

> 'The lorries arrived at Morajo [Morago] on 20 April [1969], loaded the people and their belongings and dumped them either in Trust villages or at Bultfontein. In many of the villages I saw groups of demoralized-looking people living under a few pieces of corrugated iron. These were the people from Morajo. This was about five weeks after the removal and none of them seemed to be making any effort to build a house. One or two of those who went to Bultfontein were loaned tents, but the others were given nothing at all. They did not receive any compensation and some of them complained that their furniture and fittings were ruined in transit. [...] No sanitation was provided and in May 1969 there were very few latrines in evidence. There was one borehole with a windmill, which was the only amenity provided on an otherwise completely bare stretch of open veld. No emergency rations were provided, there was no school, no shops, no clinic – nor any sign of these being provided – and no transport to Thaba Nchu [which was about 9 km far away]. The people have to do their shopping and even buy wood for their fires in Thaba 'Nchu [...] At Bultfontein the people have no land or cattle; just the residential plots [...] Many of them managed to leave their cattle with friends in the Trust villages or on farms.' (Desmond 1972:210–11)

Desmond's account of the situation in Bultfontein corresponds with the memories of older inhabitants that I interviewed in 2011. A forced removal to one of these townships was the worst thing that could happen to a household. Nevertheless, the last sentence in the quote also reveals an interesting fact, namely that the evicted people managed to keep a few animals back in their former homes.

5.4.4 Changes of Land Use from a Bird's-Eye Perspective

The long-term changes in land use patterns resulting from 'betterment' can be visualised by taking a look at aerial photographs provided by the South African mapping organisation 'National Geo-Spatial Information' (NGI). From the 1950s onwards flights were made over the Thaba Nchu area or parts of it. Eleven photograph series at varying scales are available for the region. As Nancy Jacobs (2003) did for Kuruman, I will analyse some of these to provide a visualisation of the effects of the state intervention in the mid-twentieth century. Together with informants I visited several sites on the ground in order to avoid a misinterpretation of regional aerial photographs. There I marked GPS waypoints that enabled me to trace the same areas again in satellite images provided by Google Earth. Additionally, I found in the archival documents several maps of the research area that facilitated a better understanding of the aerial photography. Finally, the general surveyor maps of the Thaba Nchu area were a valuable aid in interpretation, as the maps are very detailed (1:50,000) and even contain information about anti-soil erosion work.

Figure 14, 15 and 16 show the area of Morago in the Sediba reserve between 1950 and 1992 (for the precise location of Moroga see also Figure 1). A mountain top, indicated by an 'M' in all three images, serves as a point of orientation. This mountain is part of a ridge, which appears as a dark grey region in the lower portion of the images, and which formed the southern boundary of the old Sediba reserve (Murray 1992:4). The land south of the ridge belongs today to my research site, Middeldeel, and is used for grazing. In 1950 the farm had already been purchased by the SANT, but no location had been erected there yet. I would like to call the reader's attention to the land north of the ridge; i.e. the land used by residents of Morago location. In 1950 people still lived in scattered settlements, hence no concentrated residential sites are evident. Rather, here and there one can detect homesteads, with their typical rectangular enclosures, on the mountainside. One such enclosure is visible just to the right of the letter 'K' in the image below (Figure 14).

Morago was early shaped by *donga* erosion, which was already being addressed by soil conservationists in the 1930s. The arrows indicate a zone where contour lines had been constructed to prevent the runoff of water and soil. In contrast to the mountain ridge, the area at the foot of the mountain appears as a lighter grey, which suggests that the vegetation cover was less dense on the lower parts of the land. However, it is important to stress that the photograph was taken in September, and thus at a time of the year when grasses are naturally on the decline. Moreover, the preceding rainy season had been a comparatively dry one in Thaba Nchu, and only in March and April was heavy precipitation recorded.[242] North of the degraded area, the numerous small fields under cultivation are most striking.

[242] SAB BAO 5761 H128/1666/214, Rehabilitasie: Beplanning van Thaba 'nchu (Middel Gedeelte): Groep Trustplase, 14 February 1964.

Obviously, a great proportion of the land used by Morago residents was still used for ploughing purposes in the early 1950s. This was not only the case in this village but can also be seen in aerial photographs of other parts of the Sediba reserve (Naumann 2014:50).

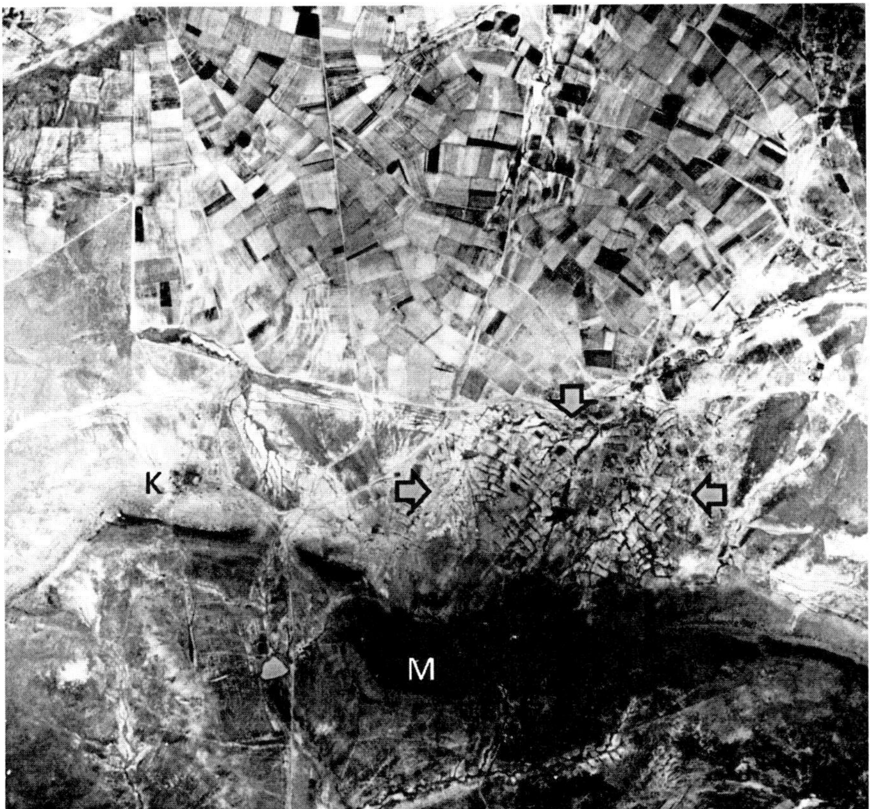

Figure 14 Morago area in 1950: shaped by soil erosion measures, scattered settlements and intensive cultivation[243]

---

243 Aerial photograph 002/00281, job 246, September 1950, provided by National Geo-Spatial Information; additions made by author.

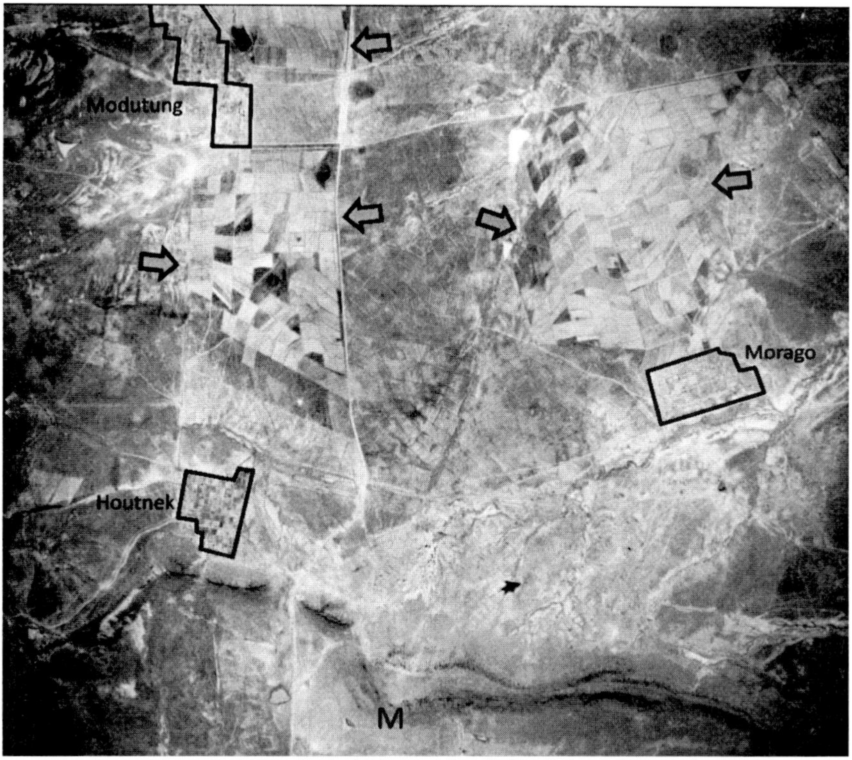

Figure 15 Morago area in 1968: centralised villages and reduced cultivation land[244]

18 years later, in 1968, the situation appears very different. Figure 15 indicates that land use patterns had changed tremendously within less than two decades. Residential patterns had been reorganised into the typical grid-patterned villages. Most radical are the transformations evident with regard to cultivation. First, the fields – indicated with arrows – appear to be laid out in a more ordered way than in the past. This was the result of new ploughing techniques, which had to run along the contours of the land. Second, the total area of land under cultivation had diminished considerably. In fact, in the late 1960s large parts of the land were used for grazing. In the image, these former fields still bear traces of ploughing activities,

[244] Aerial photograph 001/00033, job 612, 27 September 1968, provided by National Geo-Spatial Information; additions made by author.

but they were subsequently turned into rangeland. Since the rehabilitation measures did not include the plantation of valuable fodder grasses, the quality of those rangelands was presumably rather poor and dominated by pioneer species.

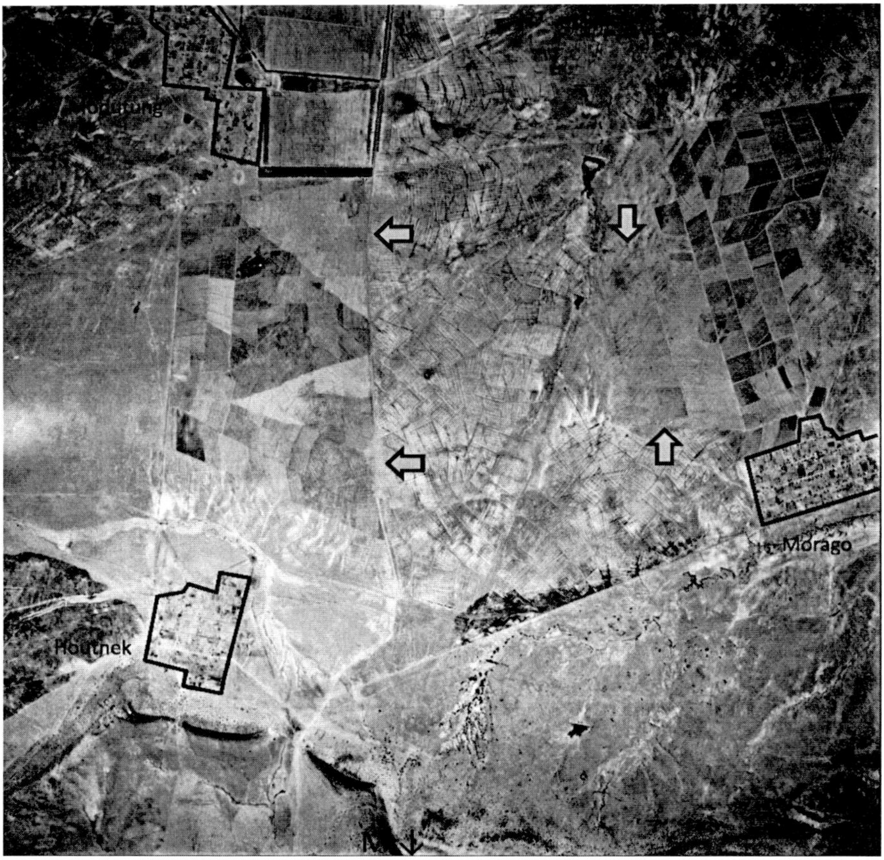

Figure 16 Morago area in 1992: increasing number of fields lie idle[245]

The last image of the Morago area (Figure 16) was taken in 1992 from a lower altitude. The aerial photograph suggests that some fields in that area

---

245 Aerial photograph 007/00036, job 969, 18 August 1992, provided by National Geo-Spatial Information; additions made by author.

were still cultivated. However, the total area of land under cultivation had lessened yet again. In this image, larger parts of the former fields are lying idle (indicated by arrows) and have also turned into rangeland. Today, there are no longer any fields left in the Morago area.

5.5 Resistance to Politics and Politics of Resistance

Resistance to 'betterment' evolved in many rural areas in the South African Union, though its form, intensity and duration varied (Hirson 1977; Beinart and Bundy 1980; Yawitch 1982; Lodge 1984; McAllister 1989; Delius 1996). People expressed their dissatisfaction by not attending official assemblies and arranging their own meetings, by writing protest letters to high ranked Native Affairs officials, and by disobeying rules and laws as well as deliberately destroying conservation work. Apart from the All African Convention, which printed pamphlets against rehabilitation schemes (Beinart and Bundy 1980:302), resistance movements against 'betterment' in the 1930s and 1940s were not strongly connected to national political organisations. Moreover, resistance was rather characterised by spontaneous actions, e.g. the defiance of the livestock culls (Yawitch 1982:18).

In the 1950s and 1960s resistance to 'betterment' blurred into resistance against the Bantu Authorities Act (Yawitch 1982:22). Chiefs and tribal authorities became the targets of protests, as they were now central actors in the enforcement of rehabilitation schemes and regarded as 'instruments of the South African state' (Beinart and Bundy 1980:305–06). At the same time, protest movements were increasingly supported by national political organisations, and became better organised and more violent, which was for instance reflected in the rural revolts in Mpondoland (1960-1961) and Tembuland (1962-1963) where attempts were made on the lives of chiefs and headmen (Beinart and Bundy 1980:304–10; Lodge 1984:261–94).

Despite widespread opposition, actions that ran counter to 'betterment' were mostly unsuccessful (Beinart and Bundy 1980:313; Yawitch 1982). In some cases, the implementation of the schemes decelerated because government officials faced strong opposition. This was also true for the Sediba area where the enforcement of the resettlement started about six years after the rehabilitation planning had been done. But in the long term, resistance attempts failed in Thaba Nchu as they did in other rural areas of South Africa. Reasons for this were manifold: among others, agriculture was losing its importance as a mode of production, and the movements often remained localised and did not have access to the same means of communication as did the dominant classes (Beinart and Bundy 1980:313). In the following I will present further reasons that were also relevant in the Thaba Nchu case.

Major instances of resistance to 'betterment' in Thaba Nchu have already been elaborated by Colin Murray (1992), and there is no need to extensively repeat all the examples he depicted. Nevertheless, there are some events documented in archival sources that he either left out or only touched on briefly – perhaps for reasons of space. In addition to these aspects, I will try to provide insight into the local dynamics of the protest and attempt a characterisation of the various groups of actors involved, based on the interviews that I conducted in 2011.

5.5.1 Unrest and the Government Response in the Sediba Area

In contrast to the Thaba Nchu reserve, Sediba was a trouble spot in terms of resistance to 'betterment' throughout the 1940s, 1950s and 1960s (Murray 1992:172). Yet to understand the unrest only as a sign of opposition to rehabilitation would be too simple. As elaborated above, resistance to 'betterment' blurred with resistance to the Bantu Authorities Act of 1951. To understand the political dynamics in the Sediba area, it is necessary to recapitulate its distinct history. As pointed out in Chapter 3.2.3, the area was a former cattle-post inhabited by supporters of Samuel Lehulere Moroka, who claimed to be the legitimate successor of the deceased Chief Moroka in 1880. However, he was not recognised as chief by the colonial administration. Under the administration of the Orange River Colony the Sediba area, which was then already a 'native reserve', received its own reserve board. When in the late 1920s the idea of a combined local council for Thaba Nchu and Sediba under the provision of Act No. 23 of 1920 emerged, Sediba residents and board members consistently stressed that they were not in favour of a consolidation of the two reserves.[246] In 1939 the chief of the Barolong received civil jurisdiction over Sediba, but still in the mid-1950s some people did not accept him as their leader (Murray 1992:173). After the Sediba reserve board was abolished in 1953, inhabitants of the old reserve complained that they were ignored by the leaders in

[246] See for instance SAB NTS 8142 2/341 part 1, Minutes of the Meeting of the Sediba Native Reserve Board, 19 July 1927; SAB NTS 8142 2/341 part 2, Minutes of the Meeting of the Sediba Native Reserve Board, 17 March 1939.

Thaba Nchu. Bearing in mind that the chief and the tribal authority supported most 'betterment' measures enforced by the state, it is not surprising that Sediba residents' opposition to 'betterment' blurred with their reservations about the Thaba Nchu authorities.

Open rejection of the rehabilitation measures was expressed shortly after the introduction of livestock culls in the early 1940s. In 1944, 131 inhabitants of the Sediba reserve wrote a letter to the Secretary of Native Affairs pointing out the severe effects the culls had on the population of the reserve, which was largely dependent 'upon the farming products as means of gaining a livelihood'.[247] The major point of objection was that even young animals were counted on the stock lists that each livestock owning household possessed. Hence, people had to cull their calves in order to save oxen 'to manage to pull the ploughs'. Moreover, the writers complained that 'the poorest people, that is those who own nothing as far as stock is concerned, were each allowed to own one head of cattle and in some cases two. This brings us to a question of: What will a man, with a family, do with one or two head of cattle?'[248] Obviously, livestock owners were the initiators of this letter; hence, it represented the views of a group of comparatively wealthy farmers.

The opposition to culling culminated in 1950, when about 800 livestock owners of the Sediba reserve – in other documents a figure of 650 stock owners is mentioned – did not present their animals to the culling officers.[249] Stock owners had been annoyed by the officers' arbitrariness for years. At the previous culls the households' herds had been reduced to a fixed number of animals. When farmers afterwards paid *bogadi*-cattle or lost animals due to diseases they were not allowed to purchase new ones in order to replace the stock. Hence, they had the impression that it was the intention of the culling officers to reduce their herds to nothing.[250] When another cull was announced in November 1950, several headmen, members of the reserve board and inhabitants held an assembly which was *de*

---

[247] SAB NTS 7526 717/326 part 2, Letter to the Secretary of Native Affairs, Pretoria, signed by 131 inhabitants of the Sediba reserve, 1 October 1944.
[248] Ibid.
[249] SAB SAP 481 15/20/50, Acting Commissioner of the South African Police, Pretoria, to the Secretary of Native Affairs, Pretoria, 23 December 1950.
[250] SAB NTS 7526 717/327 part 3, Attorney H.M. Basner, Johannesburg, to the Chief Native Commissioner, Potchefstroom, 9 March 1951.

facto illegitimate, as Chief Moroka did not attend. Participants decided that no villagers should present their animals to the culling officers. This plan was reported to the Native Commissioner by the above-mentioned headman Monchusi of Sediba location. In turn, another assembly was held under Chief Moroka which was attended by 300 people, including all headmen and board members. The Native Commissioner, who was also participating, warned the people to obey the law and bring their animals for culling. But the majority of the audience expressed their unwillingness to cooperate and only 20% of the livestock owners presented their animals to the culling officers in the following days. Immediately afterwards evidence was taken from Chief Moroka and one of his councillors, headman Monchusi and several 'native' rangers in order to discover the main culprits of the protest.[251]

Four members of the then still-existing reserve board as well as three headmen were identified as the agitators. Arrest warrants were issued for all of them, except for one headman who agreed later to let his animals be culled. On 21 February 1951, they were imprisoned.[252] In jail, they were expected to sign a statement in which they had to declare that they would bring their animals for culling, and be obedient and loyal to the Native Commissioner and the chief of the Barolong. Moreover, they had to agree that in case of non-compliance they would be arrested and subsequently removed from the reserve.[253] Despite the fact that all the prisoners signed this statement, government authorities approached them about a month later again in order to convince them to leave the reserve voluntarily. All six men refused to do so and one vividly expresses the fear of removal by comparing it with a death penalty:

---

[251] SAB NTS 7526 717/327 part 3, Evidence of Headman Andries Monchusi, Sediba, 9 December 1950; Evidences of Chief Councillor Walter Zacharias Fenyang and Chief Albert Setlogelo Moroka, Thaba Nchu, 9 December 1950, Evidences of Native Rangers David Liseko, Zacharia Setouto and Tom Kato, 15 December 1950.

[252] SAB SAP 481 15/20/50, Acting Commissioner of the South African Police, Pretoria, to the Secretary of Native Affairs, Pretoria, 23 December 1950; Deputy-Commissioner South African Police, Bloemfontein, to the Commissioner South African Police, Pretoria, 21 February 1951.

[253] SAB NTS 7526 717/327 part 3, Statement signed by Hosia Mochiwane, Jeremiah Maibi, Shadrach Malebo, Joseph Malebo, Levi Mholo and Michia Malebo, Bloemfontein Gaol, 4 March 1951.

> 'The Government is throwing a knife in front of me and it is asking me to kill myself. If I do not kill myself, then the Government is going to kill me in a most cruel way. [...] When the Government says it is going to take me by force and throw me out of my land of birth, imposing a deportation upon me which is just as bad as a death sentence, then I will be exposed to robbers and murderers. [...] Now that I have no power to do anything, it rests with the Government who brought me into this jail to exercise its powers. I have no power in this world for protection, because the Government who was supposed to protect me has thrown me away. [...] I have been told [...] that I am accused for not bringing my stock for culling and also other information that I have done other things which have not been told to me. I am surprised that I can be deported without a hearing to prove whether I am guilty or not.'[254]

Although the Native Commissioner and the chief of the Barolong were in favour of deportation, Pretoria officials decided that the six men were to be allowed to move back to their homes. The example shows that the chief wanted to suppress resistance in Sediba by all means necessary. Moreover, it illustrates that the government authorities used perfidious means to put pressure on opponents by uttering threats that had no legal basis. This was mainly because they feared another 'Witzieshoek', a reserve that was situated in the eastern Orange Free State. There 'betterment' and, in particular, culling were applied with the same vigour as in the Thaba Nchu district. Resistance to livestock limitation culminated in Witzieshoek in 1950, when 14 reserve inhabitants and two policemen were killed and almost 100 people injured during a shooting incident (Hirson 1977:124–25).

People in Sediba not only objected to the culling but also to governmental interference with regard to arable land. Before the determination of 'economic units' in the late 1950s, the government had the policy of providing each holder of a residential site with a 5.1 ha plot, of which 3.4 ha were intended for the cultivation of summer crops and 1.7 ha for winter plantations.[255] In 1941, when government authorities started to subdivide dryland into regular plots, the headman of Sediba location encouraged the villagers

---

[254] SAB NTS 7526 717/327 part 3, Statement of Michal Malebo, Bloemfontein Gaol, 17 April 1952.

[255] SAB NTS 308 10/54, Assistant Native Commissioner, Thaba Nchu, to the Secretary of Native Affairs, Pretoria, 8 October 1941.

not to accept the new fields. This form of protest eventually cost him his position (Murray 1992:173). In consideration of the fact that in the following years no demarcation of arable lands is mentioned in the archival files of Thaba Nchu, it appears that the department's intervention momentarily halted. State regulation of cultivation reached another peak in 1949, when demarcation of 5.1 ha fields continued. Again, several people refused to accept their new plots by leaving them uncultivated;[256] but the means of protest were not restricted to defiance. In September and October 1952, a small group of cultivators ploughed over contour walls erected between the fields in the Sediba reserve. In March 1956, two weeks after numerous residents had again ignored the culling order, about 200 people from the Morago, Kgalala and Talla locations ploughed their old fields again (Murray 1992:176–77). Protestors demanded a hearing with the Chief Native Commissioner in order to express their concerns. On 27 March 1956, a public meeting was held at Petra clinic, which was situated between Talla and Morago location. The major complaints brought up were that (i) fields had been demarcated without the agreement of the residents, (ii) culling had formerly been claimed to have to be enforced for only three years, and (iii) the Sediba reserve board had been replaced by the Barolong tribal authority in 1953 (Murray 1992:177).

The protocol of the meeting reveals the paradoxes 'betterment' caused. A farmer from Morago location argued:

> 'Our complaints with regard to the fields are the demarcated plots. We want our old lands as formerly. I had a field of three morgen [2.6 ha] myself, and I was happy with it. [*Chief Native Commissioner:* But now you get six morgen!] I was happy with my three morgen. Yes, I was absolutely content. […] I want only my old field back. My old field is not yet allocated to someone else. It still lies as in the past. I did not *want* to have a demarcated field.'[257] [translated from Afrikaans]

Before the arable lands had been surveyed, each landowning household ploughed only fields sized according to their specific needs. Some households required larger fields – because they had to feed many dependents

---

[256] SAB NTS 7692 387/332, Additional Native Commissioner, Thaba Nchu, to the Chief Native Commissioner, Potchefstroom, 7 May 1953.

[257] SAB NTS 7692 387/332, Notule van 'n vergadering gehou te Petra-Kliniek, Sediba, 27 March 1956, p. 2.

and had the labour force to do so – whereas other families had neither the need nor the means to cultivate a large area. Other protestors perceived that the government's aim was to weaken farmers:

> 'What makes me sad [...] is that we are people that have been working well. The government sent people to teach us how to farm. We learnt well and had a good life. When we became strong men, the government came again to do away with our wealth. I am very poor. We are indigent. We had full spans of oxen. Today I have only four cows. [...] Leave us alone as we were with our cows and our fields. [...] Also in Transvaal, there are people who refuse to leave their fields to let them cut. The government will leave them alone. The government should leave us alone too, so that we can die here. If you go to one of the grazing camps now you will see that only 50 cattle graze there and still a lot of grass is there. [...] If you release us now, I would be able to buy 6 cattle from my own harvest next year.'[258] [translated from Afrikaans]

Besides, the participants of the meeting extensively discussed the abolishment of the Sediba reserve board and criticised the Barolong tribal authority that was established in its place. Although the decisions of the reserve board were not always appreciated by the locals, it was nevertheless an institution where Sediba inhabitants could bring up complaints, worries and questions directly, as each village had a board member. People received a response comparatively promptly after a meeting and therefore felt that their opinion had been heard. Moreover, the board made minutes of the meetings in which all resolutions were recorded and, when necessary, could be retraced later on. In contrast, the Barolong tribal authority was less accessible to Sediba inhabitants. First and foremost, the meetings took place in Thaba Nchu, which was far away from the northern locations. Moreover, ordinary people were not allowed to speak at those meetings. The farmers of Morago complained that the questions of the residents were not answered and people doubted that their wishes and objectives were even presented at the authority's assemblies.[259]

---

[258] SAB NTS 7692 387/332, Notule van 'n vergadering gehou te Petra-Kliniek, Sediba, 27 March 1956, pp. 8-9.
[259] Ibid. 4.

Complaints submitted at the meeting remained unheard. Instead 'the wheels of the legal machinery continued to grind' (Murray 1992:178). Between May and December 1956, 600 convictions – resulting in either fines or heavy labour – were made, involving 194 residents of the Sediba area.[260] People had been charged for not presenting their animals at the culls, damaging of contour banks, and maltreatment of animals, and were punished with fines ranging from £5 to £10, or else with several weeks of hard labour.[261] Many people were legally advised by Johannesburg-based lawyers Basner, Kagan and Millin (ibid. 178) who were engaged by a Talla resident who was an ANC member (for the connection between the ANC and the protestors see below). The government's clampdown was effective and by the end of July 1956 without exception all residents had brought their animals to the livestock cull.[262]

For the next years unrest seemed to ease (Murray 1992:178). However, developments in other African countries influenced the people in Thaba Nchu, at least according to some government officials. Thaba Nchu's Native Commissioner Welman worried particularly about the events in Ghana, which became independent in 1957, and about the 'fiasco' in the Belgian Congo.[263] In the latter case, heavy riots in 1959 led to a hasty initiation of reforms for the country's independence, which followed only a year after the unrest (Marx 2004:260). The developments in the Belgian Congo especially encouraged black people and initiated a '*veld* fire', as Welman put it:

> 'Natives who until recently were very sympathetic to apartheid policy take over the slogan "Africa for Africans" today, and announce this in broad daylight. It seems to me that we have lost the race against time. The *veld* fire of the pan-African Bantu nationalism spreads like cancer in our whole national existence and there wait pitch-dark days for the Whites in South Africa. It appears to me that the only way to turn something around will be

---

[260] SAB NTS 7692 387/332, Native Commissioner, Thaba Nchu, to the Chief Native Commissioner, Potchefstroom, 15 December 1956.

[261] SAB NTS 7692 387/332, Native Commissioner, Thaba Nchu, to the Chief Native Commissioner, Potchefstroom, 18 May 1956, pp. 1-2.

[262] SAB NTS 7692 387/332, Native Commissioner, Thaba Nchu, to the Chief Native Commissioner, Potchefstroom, 28 July 1956.

[263] SAB NTS 7692 387/332, Native Commissioner, Thaba Nchu, to the Chief Native Commissioner, Bloemfontein, 19 January 1959.

> to think big and to hand over whole provinces and cities to the Bantu and to tell him: "Here is your territory. Here is the White's territory. If you want to stay here you are nothing. If the White wants to stay in your territory he is nothing.'[264] [translated from Afrikaans]

Refusal to Resettle

Unrest reached another peak in the mid-1960s, when the resettlement of households in centralised villages was under way in Sediba. On 13 March 1964, the residential certificates of all people still living in the old settlements were withdrawn, because officials wanted to speed up the relocation process. Accordingly, the people also *de facto* lost all rights to grazing lands and fields, as those rights were linked to the holding of a residential permit.[265] This matter was intensively discussed at tribal authority meetings and during a special assembly that was held again at the Petra clinic on 18 April 1964. Native Commissioner Welman, six members of the tribal authority, and the headmen of Kgalala, Sediba, Bofulo, Moroto and Morago locations were present along with about 200 residents.[266] Welman had initiated the meeting because *inter alia* an anonymous letter written by the 'Sediba Reserve People' was sent to the Minister and the Secretary of the Department of Bantu Administration and Development (BAD). A major complaint was that the Native Commissioner had threatened the Sediba people at a public meeting on 14 March 1964, i.e. one day after the cancellation of the residential permits. He was accused of having said that 'the villages of the people who didn't move to the new places will be destroyed... the 14th April will be a turmolious [sic] day'.[267] All people would have to apply for a plot in the new residential areas by then.[268] Even though the old settlements were still intact on 18 April, people feared for their homes. At the Petra Clinic meeting one Morago resident stated:

---

264 Ibid.

265 SAB BAO 4831 F 53/1666/1 part 1, Minutes of Meeting, Petra Clinic, 18 April 1964, p. 1.

266 Ibid.

267 SAB BAO 5761 H128/1666/197 part 1, Anonymous letter to the Secretary of Bantu Administration and Development, Pretoria, 25 March 1964.

268 SAB BAO 5761 H128/1666/197 part 1, Minutes of Meeting, Molatswaneng School Morago, 14 March 1964, p. 1.

> 'We are people that still live at the old-old places, since 1882. When we think back to 1882 – and we hear that we have to leave these old places – it will be difficult for us and we will bury ourselves in these new places. We will never come out again. To bury oneself is very difficult [...] You [the Native Commissioner] have the power to take us away; but we don't want to go there.' [269]

Whereupon the Native Commissioner answered:

> 'Did you wear a tie in 1882, jackets, shoes etc.? Why do you do it today? Has there not been progress in between? We just want to make the residential area compact in order that you are close to the grazing land and fields etc. When you live concentrated then we will save 300 morgen [256.96 ha] of grazing. We try to help you but you do not want to play along. This is why I ask why do you wear shoes and so forth, why don't you wear what you have had worn in these times?'[270]

Apart from using an inappropriate analogy – a change in clothing is hardly comparable with leaving one's home – in this statement the Native Commissioner provides a figure that is questionable. Old settlements were often located in mountainous country, and hence not in the best grazing areas anyway. To argue that a relocation of households would result in more than 250 ha of grazing land was therefore an overly positive calculation. Another problem was that those who were prepared to resettle did not find the best conditions when they moved, as one resident of Kgalala described:

> 'The residential areas are very bad. I left [the old settlement] and I took plot no 173 and 174 but others had been moved in there. What should I do? I am not a striker, (...). At first, I assisted widows and other people who suffer to move. And now I get only bad sites in the gravel pit.'[271]

Like other people who did not remove their homesteads on time he was reprimanded by the Native Commissioner. According to him it was the

---

[269] SAB BAO 4831 F 53/1666/1 part 1, Minutes of Meeting, Petra-Clinic, 18 April 1964, p. 4.
[270] Ibid. 4-5.
[271] Ibid. 7.

man's own fault that he did not find an appropriate plot, as one should mind one's own business first before helping others.

The meeting in 1964 was not successful from the officials' point of view, as some residents still ignored the order for removal. At tribal authority meetings, the matter was discussed constantly. In June 1965 one councillor lobbied for the deportation of the protest's 'ring-leaders'.[272] Oral histories indicate that this was done, at least in the case of some of the protesting households. The government 'hurt the protestors until they could feel the pain', as the 63-year old Andries Kgware expressed it.[273] According to his recollections even a pregnant woman had been arrested. She gave birth to her daughter in jail. Her husband, who had allegedly been one of the agitators, worked outside the Thaba Nchu district and had apparently not been found. Hence, the authorities arrested his wife in order to impose pressure upon him.

As I have shown, the complaints brought up by Sediba farmers went up in smoke – both those against 'betterment' and those against the tribal authority. This was not only because the Sediba protestors did not get any support. Additionally, Native Affairs officials always argued at these official meetings that 'betterment' opponents were only a small fraction of the society, and hence deprived their complaints of their legitimate basis.[274] This leads to the questions of how many people actually agreed with the protestors, how many were loyal to the ruling authorities, and how these different groups behaved towards each other. These questions will be addressed in the following subchapter.

5.5.2 Protestors vs Loyalists: An Attempt at Classifying Actor Groups

> 'There were two parts … the protestors and the people who have agreed with the government.' (Interview with Andries Kgware, Sediba, 17 August 2011, translated by Tsametse Pobe)

---

[272] SAB BAO 4831 F 53/1666/1 part 1, Minutes of Meeting Barolong Tribal Authority, Thaba Nchu, 9 June 1965, p. 6.

[273] Interview with Andries Kgware, Sediba, 17 August, 2011, translated by Tsametse Pohe.

[274] For instance SAB NTS 7527 717/327 part 4, Notule van Vergadering gehou in kantoor van die Bantusakekommissaris, Thaba Nchu, 26 January 1960, p. 10.

The archival sources that deal with the various protests between the 1940s and 1960s in the Sediba reserve do not provide a clear picture of the number of people involved. First and foremost, this is due to misrepresentations by officials who adjusted the figures according to their own requirements to provide a basis for their argument. When in 1950 the culling order was ignored for the first time, sources mention that 650 or even 800 persons had been involved and had to be prosecuted. However, when officials debated with protestors these figures were never mentioned, as they would have run counter to the official argument that the protest was undertaken by only a few 'troublemakers'. Thus, the information on protest participation in archival sources must be interpreted with caution.

Archival documents suggest that the opponents to 'betterment' during the 1940s and 1950s were for the most part labour migrants who were still involved in farming activities. The 1958 rehabilitation report for the Sediba reserve describes the inhabitants as generally loyal to the government, except for a number of migrants who kept livestock. They would be 'without exception' those who had been rebellious, who did not accept the cutting of the fields, and hence were not engaged in cultivation. Indeed, informants interviewed in 2011 stressed that not all reserve inhabitants acted in concert. However, oral histories indicate that the official account was an over-simplification of the situation. In fact, non-migrating households and active cultivators were also involved in the protests against the 'betterment schemes'. An example is the family of the above cited Andries Kgware, who were briefly introduced in the vignettes in Chapter 3.3.5. His parents were comparatively wealthy as they had a little shop and were successful farmers.[275]

It is perhaps no surprise that mainly the wealthier farming households opposed 'betterment'. They were the ones that had the most to lose, i.e. livestock and land. In the later years of protest, the motivation might have changed, as objection was increasingly raised to the Barolong tribal authority, which did not only affect land users. That labour migrants played a critical role in the protest campaigns was not only the case in Thaba Nchu. In the Transkei too, migrant workers with access to grazing and arable land made up a large proportion of the opponents. 'Betterment' meant a threat to their livelihood, which was based on incomes from wage labour and farming. An abolition of the latter would result in proletarianisation

---

275 Interview with Andries Kgware, Sediba, 17 August, 2011.

(Beinart and Bundy 1980:303; Lodge 1984:28). From the perspective of farming migrants in Sediba, this would have meant a decline in their households' resilience, because they wanted to maintain their agricultural basis and they knew about the terrible conditions that prevailed in the semi-urban townships, such as Bultfontein.

A second reason why labour migrants in particular belonged to the group of activists was their engagement with the ANC, which is illustrated in detail by Murray (Murray 1992). A local branch of the organisation had been founded in Sediba in 1945. Soon they explored the topic of livestock culling as one major issue that had to be addressed. Leading actors in the Sediba unrest were often active ANC members. Many of them were also labour migrants, as they were the most important contact persons between the Sediba protestors and the provincial ANC-branch in Bloemfontein (ibid. 174–81). The protestors' close ties to ANC politics were also reflected in the means of resistance that they chose. Instead of protesting against 'betterment' regulations, they disobeyed the rules, taking a similar line to that of the ANC's Defiance Campaign of 1952 and 1953.

> 'Everybody's protest was [that] they wouldn't agree with [it]. They wouldn't show that they were on protest or marching, no. They would disagree with every rule that they got, that they had to change this into that. They didn't want any changes because ... they were used to living with horses, sheep, cows. That was their life. It was like that. They wouldn't let anybody change it.' (interview with Andries Kgware, Sediba, 17 August 2011, translated by Tsametse Pobe)

However, it seems that the ANC-branch at the provincial level was not always in accordance with the events, as a protocol of an ANC meeting in Bloemfontein suggests. In 1951, when the alleged agitators of the culling defiance were released from jail, the ANC held a reception in their honour. The then ANC president Dr J.S. Moroka, who was also attending the meeting, stated that he personally knew the six men that had been imprisoned, and that he had decided not to take action against their imprisonment because they had held meetings from which he had been excluded. He admonished the audience that the ANC had been founded because black South Africans had wanted to unite each other in 1912, but that they still would not work hand in hand in 1951.[276] The reason why Dr Moroka was not inducted in the plans of the Sediba protestors might be rooted in his personal position. Besides being the ANC president between 1949 and 1952, Moroka was also a representative of the landowning Barolong elite, who were regarded with suspicion by Sediba residents. Moreover, it has to be considered that Moroka, who was, together with Walter Sisulu, speaking for the repeal of several unfair laws such as the stock limitation (Lodge 1984:42), had in fact proposed himself that the 'betterment' regulation should be implemented in the Thaba Nchu area.[277]

In opposition to the protestors were people who were generally loyal to the government authorities. Often these people were employed by the government, such as councillors of the tribal authority, headmen and local rangers. However, some headmen also resisted 'betterment', though this normally meant that they were dismissed sooner or later. Several tribal authority councillors were farm owners. Hence, their loyalty to the government was motivated by the wish to maintain certain privileges. On the other hand, not all loyalists worked for the government. I assume that at least parts of the land- and stockless households also belonged to this group, as to them 'betterment' promised an improvement. Besides resource pressure and livestock inferiority, the official reason for culling was the promise that stockless people would be able to obtain grazing rights as soon as the stock numbers of the large herd owners were reduced. As I argued in Chapter 4.4.5, there are justified doubts that stockless people actually benefitted from culling, but this might not have prevented them from opposing the protestors.

To increase the pressure upon the protestors, Native Affairs officials tried to pit loyalists against 'troublemakers'. At a public meeting in Morago in 1964, Commissioner Welman said that residents could find out very quick who was destroying conservation work, and that they should work together to bring such people down. Moreover, he stressed that the whole location could be fined for such actions (e.g. for cutting the fences).[278] Such threats

---

276 SAB SAP 481 15/20/50, Deputy Commissioner of the South African Police, Bloemfontein, Commissioner of the South African Police, Pretoria, 11 July 1951.
277 SAB NTS 8141 1/341 part 2, Minutes of Meetings of the Thaba Nchu Reserve Board, 15 March 1939.
278 SAB BAO 5761 H128/1666/197 part 1, Minutes of Meeting held at Molatswaneng School, Morago, 14 March 1964.

certainly worked, as the following two vignettes from Sediba location suggest.

> Andries Kgware attended the school of the Lutheran Church in Sediba location. In 1961, when he was still in Standard Five, he and his siblings had to leave the school according to his father's decision. Previously, the Kgware children had experienced hard times since they were bullied by teachers and other children. The official reason was that their father's sheep would have destroyed the school's vegetable garden. However, Andries was convinced that it was other people's sheep that had eaten the plants in the garden. His father was allegedly inculpated because he was one of the protesters against the 'betterment'. In the aftermath of this event, Andries and his sisters were beaten up by classmates. At that point, the father decided to take his children out of the school. Andries started to work as a gardener for the Department of Agriculture and his sisters went to another school in Thaba Nchu. (interview with Andries Kgware, Sediba, 17 August 2011, 16 September 2011)

> In 1956, Merriam Molehe was born in Mpharane, where she lived in one household together with her parents, three sisters and two brothers. She remembers the 1960s as a difficult time. Her family had no cattle, as the animals had been confiscated since the family belonged to the group of protestors. Moreover, they had no access to fields anymore. The family survived by helping other households during the harvest and by sending their sons to work in town. In the mid-1960s the government told the families in Mpharane that their homesteads were in an area now defined as a grazing camp. They were expected to move to the new village, Sediba Trust. By 1967, most of the approximately 20 households of Mpharane had moved, but four families still refused to resettle. One of them was Merriam Molehe's family. The pressure on Mpharane residents was high. Frequently, officers of the South African police would come and threaten them. Sometimes they arrested fathers and mothers and left the children behind; but pressure was not only exerted by officials. Merriam, who attended the same school as Andries Kgware, remembered that she and the other children of Mpharane had to pass a gate on her way home from school. Boys from the other settle-

> ments were waiting there each day, and shouted 'bana ba strikers' [children of the striker] when they were approaching. Often, they were beaten by those boys. Years after the erection of the new settlement, the last families of Mpharane eventually left their homes. Merriam's parents were allowed to move to a residential site in Sediba Trust but still had no access to arable and grazing land. (interview with Merriam Molehe, Sediba, 17 August 2011)

The two cases illustrate the wide gap and social tensions existing between the loyalists and protesters in Sediba location. Conflicts were not only enacted between adults but, too often, also between the children. In the end, the government and the loyalists had the greater power, and the means to suppress the protestors. All of the protestors eventually either moved to a new Trust village or settled in a semi-urban township.

5.6 Preliminary Summary

Thaba Nchu was a rather typical case of 'betterment' implementation, which began in the initial phase with reclamation measures aimed at the restoration and improvement of natural resources, particularly the soil and grazing. However, during the second phase of 'betterment', illustrated in this chapter, a shift towards stabilisation and control is evident. The shift in 'betterment' policies was described for other regions too (Beinart and Bundy 1980; Yawitch 1982; Lodge 1984; Hendricks 1989; Delius and Schirmer 2000). The stabilisation phase that reached its peak in the 1950s and 1960s was characterised by methods of surveying, inventorying, zoning, mapping and standardisation. The centralisation and geometric arrangement of settlement structures made rural populations more 'legible', to use Scott's (1998) term, and facilitated state functions, such as taxation, but also – and this was perhaps more important – established control over people. The reorganisation of rural communities also included a division of society and nature (i.e. grazing and cultivated areas) into 'discrete categories' (Neumann 2005:121). Each category was linked to different sets of institutions, e.g. rotational grazing management in the rangeland area and the division of standardised residential plots in the 'human area'. As I have shown in the previous chapters, the categorisation of land types originated in the first half of the twentieth century. Thus, resettlement only completed the division of land that had been previously prepared. In order to enforce the institutions on the local level, the state employed local residents. Rangers had the right to sanction norm deviance (Chapter 6.4.3); thus, they were an extended arm of the state authority. Moreover, with the imprisonment of protesters, the government used increasingly coercive means and 'legitimate violence' (Peluso 1993:199) in order to gain social control.

All government-imposed interventions and institutions seemingly had one thing in common: they were based on 'western' mathematics, natural sciences and technology, i.e. allegedly objective knowledge systems. By taking recourse to 'objective' measures, the state tried to legitimise intervention. The concept of 'economic units' can be interpreted as the state's attempt to define social-ecological resilience: officials determined how much could be extracted from natural resources in order to provide households with sufficient income and prevent an overutilisation of the ecological system. However, as the examples of the 'economic units' and the carrying capacity show, such measures were flexible and could be changed when necessary in order to achieve a political aim.

What conclusions can be drawn with regard to the effects of 'betterment' on the social-ecological system? This question can only be answered by considering the different types of intervention separately. The early conservation measures against soil erosion – i.e. the filling of *dongas*, the scarifying of bare patches and the construction of contour lines – had only limited effects on the system. Certainly, these measures did not change social relations between farmers or lead to institutional transformations. Whether they did indeed halt a further increase in soil erosion is difficult to estimate with hindsight. Yet archival data indicate that they had less destructive effects than, for instance, the erection of grazing camps or the establishment of the Grobler irrigation scheme. Both the fencing-off of rangelands and the irrigation scheme were examples of bad planning. The former did not result in an improvement of the *veld* but rather in its deterioration. The latter quickly led to sheet erosion processes, and in the long term to severe salinisation of soils, as I will show in the next chapter. Pre-'betterment' local land use practices can be described as forms of 'integrative management' that aimed at a high utilisation of natural resources. For instance, after the harvest, cattle were driven on the fields where they would eat the unused remainders of the harvest and at the same time naturally fertilise the soils. With the clear definition of areas for residence, grazing and cultivation, the use of land was separated into distinct spheres. This development not only transformed local institutions of land management, but also had negative effects on the ecological system, as the example of the irrigation scheme shows: the land there was constantly in need of additional fertiliser.

I propose that the above-mentioned interventions did not massively reduce the resilience of households and communities. This development was rather a result of livestock culls and resettlement. Ironically, culling was the only measure that presumably had a positive effect on the natural resource base, as it decreased the grazing pressure. However, high grazing pressure was the result of ethnic segregation rather than poor management practices. Nonetheless, the effects of culling on the social system proved destructive. As herd sizes were reduced drastically, culls resulted in the medium-term in a decline of cattle lending. The standardisation and reduction of cattle herds to a minimum, therefore, undermined patron-client relationships. However, these relationships were important for agricultural production as they enabled poorer people without livestock to gain access to cattle for ploughing.

Finally, resettlement physically disrupted social relations throughout whole communities and disconnected many households from their natural resource base by relocating them in non-agricultural townships. For these the resettlement was connected with an immediate decrease of their households' resilience as the townships were places of extreme poverty, unemployment, theft and violence. The farming households that were relocated within the rural areas faced often a restriction of their field sizes. This also decreased their resilience, though less suddenly than in the case of the households who had to leave to the urban townships.

6. RURAL DEVELOPMENT IN BOPHUTHATSWANA, 1970S-1994

I illustrated in the previous chapter that the 1950s and 1960s marked a critical shift for rural households, since 'betterment' not only changed resource access and land use practices but also disrupted long-established settlements and detached people from their former homes. The following chapter will show that transformations did not stop after the resettlement of households had been accomplished but continued into the 1970s and 1980s. With the establishment of the Republic of Bophuthatswana, new administrative bodies emerged, which reframed the conceptualisation of the Thaba Nchu social-ecological system. Despite a long agro-pastoral history, in the second half of the 1980s the area was defined as a district suitable for pastoral production only. By re-evaluating the social-ecological system the Bophuthatswana authorities created a new narrative that is still relevant today. Hence, for a better understanding of the conception of the recent system, an elaboration of the agricultural policies and debates in Bophuthatswana is essential. In addition, the following subchapters will concentrate on the question of which actors tried to maintain agricultural production in a system in which cultivation had become uneconomic, and how they did so. Moreover, I will show that certain aspects of the social-ecological system remained unchanged in Bophuthatswana, such as the governmental control of land use.

6.1 'Independence' of a Dependent Homeland

The Homelands Constitution Act of 1971[279] paved the way for the self-governance of the Batswana homeland, i.e. Bophuthatswana. The Tswana territorial authority was replaced by a legislative assembly, which had more legislative powers than its predecessor. The assembly consisted of 72 members, four of whom were representatives from Thaba Nchu, i.e. the chief and three appointed members of the tribal authority (Erasmus and Krige 1998:11). In 1972, Bophuthatswana was granted self-governing status. The newly established Bophuthatswana National Party (BNP) won the elections against the Seoposengwe Party. Lucas Manyane Mangope, chief of the Bahurutse ba Manyane and leader of the BNP, became the chief

[279] Republic of South Africa, Act No. 21 of 1971: Bantu Homelands Constitution Act.

minister of Bophuthatswana. In this phase of self-governance, the departmental administration of the homeland – which resembled that of the Republic of South Africa – was far from being dominated by black people. In fact, many leading positions in the departments and other governmental bodies were occupied by white officials. In the Bophuthatswana Department of Agriculture, for instance, technical posts, and those of extension officers and veterinary officers, were mainly held by whites (Butler et al. 1977:43–44).

In 1977, one year after the Transkei, Bophuthatswana was the second homeland that was declared 'independent'. Ciskei and Venda followed in 1979 and 1981 respectively, whereas the other homelands remained merely self-governing. With the formation of these 'pseudo' nation-states, the apartheid strategy of racial segregation was completed by inventing national differences between white and black South Africans (Ferguson 2006:56). National symbols such as anthems, flags and mottos were produced by the BAD in Pretoria (ibid. 57). However, in a similar manner to the other 'independent' homelands, Bophuthatswana was never acknowledged as a 'real' state by the international community (Halbach 1988:4–5; Drummond 1991:371; Jones 1999:593, 597; Ferguson 2006:57).[280] Initially, Bophuthatswana was fragmented into 26 pieces of land. Hence, consolidation of the land blocks was one of the major aims of the legislative assembly (Jeppe 1980:54). Although several land pieces were united by eliminating 'black spots' and incorporating white-owned farmland, Bophuthatswana never became a consolidated block of land but consisted of seven parts, of which Thaba Nchu was the most geographically isolated. The territorial fragmentation of Bophuthatswana had negative effects on the state's efficiency, as it complicated communication, administration, economy and development (Jeppe 1980:167).

'The ethnic state idea' (Cowley and Lemon 1986:252) was never put into practice in Bophuthatswana. In 1980, only 42% of the Setswana-speaking

[280] Ferguson (2006) argues that the absence of international recognition of the homelands facilitated a more rigorous critique of their poverty and political economy. He contrasts the homelands with the case of Lesotho: the small kingdom is recognised as a 'real' nation-state by the international community, but is in fact similarly dependent on the South African economy and marked by poverty. However, as Lesotho is a legitimate state, the economic situation is not understood through its relation to South Africa – as in the case of the homelands – but as a locally made problem.

people (2.4 million in total) were living in 'their' homeland, and at least one third of the republic's 1.3 million inhabitants were non-Batswana. During the 1970s population figures for Bophuthatswana had been increasing at an annual rate of about 5.7%, which indicates a massive population influx due to forced removals (ibid.). Thaba Nchu belonged to the districts with a very high proportion of non-Batswana residents. In 1972, about 42% of the population of the district were not Setswana-speaking.[281] On the other hand, the ideal of an economically 'independent' state was never achieved either. Historically, the development of Bophuthatswana's economy was strongly connected to the Republic of South Africa, as labour migration remained the main source of income for the inhabitants of the Bantustan, and thus affected all 'spheres of life' (Beuster 1983:1).

After 1977, the Bophuthatswana territory that had been previously owned by the South African government and South African Development Trust (SADT), which was the new name of the SANT, was transferred to the new republic. Authority over all Bophuthatswana state land was vested in the president, who was responsible for protecting the land rights of Batswana groups and communities and supporting their material and moral welfare (Jeppe 1980:62). The land question was not solved, however, as more additional land had to be purchased, for instance in order to replace the 'black spots' in white farming areas. Both the South African and Bophuthatswana government agreed that the SADT would buy the additional land, make plans and settle it, and then transfer it to the government of Bophuthatswana (Jeppe 1980:63); in other words, in these cases the previous procedure of 'betterment' was continued. However, Thaba Nchu was an exception in this regard as it was the only district where up to 99% of the land had already been subjected to agricultural replanning between the 1940s and 1970s (Jeppe 1980:150, table 11). Hence, it was the only district where 'betterment' was *de facto* completed.

6.2 Re-aligning Agriculture: Actors, Policies and New Concepts

Agricultural development was the prime focus of the Bophuthatswana policies. An increase in agricultural production, particularly in the cultivation of crops, was not only important in order to feed the Bantustan population and make the republic 'independent' of expensive food imports, it also

[281] SAB SON 1606 D12/2/4/2/4/3 vol1, Government Anthropologist Breutz to the Secretary of Bantu Administration and Development, 15 April 1972, p. 1.

constituted the basis for the development of the rural areas. The country's agricultural potential was – at least according to Bophuthatswana officials – excellent, and far from being fully exploited (Beuster 1983:1–2, 6 of appendix). Of all land suitable for cultivation only about 18% was used for the planting of crops (Jeppe 1980:116). Nevertheless, compared to other Bantustans, such as KwaZulu, in large areas Bophuthatswana had arid conditions that hindered agricultural production (Butler et al. 1977:200). The districts of Ganyesa and Tlhaping-Tlharo, which were situated in the very dry, western part of the Bantustan, were basically unsuitable for dryland cultivation. Hence, the Department of Agriculture discouraged the local farmers from ploughing even small patches, and promoted animal husbandry alone as a farming practice (Beuster 1983:22–23 of appendix). On the other hand, the districts of Ditsobotla and Molopo were situated at the boundary of the maize triangle, which was one of the main cropping areas in Southern Africa, whereas Thaba Nchu was located in an excellent wheat production zone (ibid.7–8 of appendix). Moreover, Thaba Nchu was described as being better suited to agriculture and animal husbandry than other districts in Bophuthatswana (Butler et al.1977; Beuster 1983). First and foremost, this was said to be the result of the higher rainfall, with an annual mean of 600 mm. Yet a few years later Thaba Nchu's suitability for agriculture was re-evaluated with a completely contrary outcome (Chapter 6.3).

Besides the Department of Agriculture – which was responsible for extension services, agricultural research and planning – other state corporations and boards, namely the Agricultural Advisory Board, the Economic Development Corporation of South Africa, the Bophuthatswana National Development Corporation, the Tswana Agricultural Corporation and the Agricultural Development Corporation of Bophuthatswana (Agricor), played central roles in the homeland's agricultural sector (Jeppe 1980:83–84). The parastatal development corporation Agricor was, together with the Department of Agriculture the main body that administrated agricultural production in the Thaba Nchu district. Agricor was established in 1978 for the purpose of guiding the Bophuthatswana Bantustan to self-sufficient food-production within the shortest possible time (Drummond 1990:337). Agricor's aim was to increase not only the agricultural production, but also the socio-economic development of rural communities in order to improve the inhabitants' quality of life (Agricor 1986:4). The corporation introduced the concept of *temisano*, meaning 'farming together', which formed the basis of the development activities. The concept included four major components:

(1) 'Agricultural production' was the main focus of Agricor's work. In practice that meant that possible project sites were identified in the rural areas. After a planning phase that included technical and economic studies in order to analyse feasibility, negotiations were to be initiated with the residents of the area to ensure that everybody agreed with the project. Once the project started, Agricor secured the financing. The overall aim of such projects was to 'generate self-sustaining incomes comparable to those earned in other sectors of the economy' (Agricor 1986:4).

(2) 'Community development and social planning' was another focus of Agricor, which took into account that a successful development of the rural areas could only be achieved by generating jobs which were not directly linked to farming. Emphasis was placed mainly on the collection of socio-economic data in the rural areas, which were then used to formulate recommendations for possible development projects (ibid.).

(3) Another component of *temisano* was the 'training' of farmers and their families in order to ensure that they successfully run the agricultural projects once the management and administration were handed over to them (ibid.).

(4) The last component was the creation of 'agro-businesses', i.e. 'jobs and incomes for those not accommodated in the farming projects' (ibid.). Work could be provided through the establishment of agro-industries, e.g. the canning and processing of plants, or through cottage industries, e.g. backyard candle factories.

The idea of 'farming together' reflects the general policy of the Bophuthatswana government with regard to agricultural development. An increase in production was to be achieved by transforming the individually practiced supplementary agriculture into a cooperatively managed commercial agriculture in which small farming units were consolidated into larger, viable units. By forming cooperatives, farmers would reduce their production costs, have better access to credit facilities and control each other's production, which in turn would increase their collective output (Beuster 1983:26-27 of appendix). The agricultural cooperative supplied its members with implements, seeds, fertiliser, and diesel, marketed the products of the cooperative, and rendered other services to its members, such as the ploughing of land. A distinction was made between primary and secondary agricultural cooperatives. The former consisted of ten or more farmers and operated mostly at the level of agricultural production. The latter subsumed two or more primary agricultural cooperatives and focussed mainly on marketing, supply and service aspects (ibid. 27 of appendix). As I will show in the following subchapters, the formation of cooperatives was also encouraged in Thaba Nchu, though their economic success was very limited.

6.3 Agricultural Production in Post-'Betterment' Thaba Nchu

Between 1986 and 1987 a feasibility study was done for the Thaba Nchu area which aimed to identify possible projects for regional development in order to reduce the region's high unemployment rate and economic dependence.[282] The study was commissioned by the Bophuthatswana Department of Economic Affairs and the Public Service Commission, and carried out by the town and region planning bureau Oakenfull, Hall and Kleynhans, which was based in Sandton, Johannesburg.[283] The planning bureau was asked to integrate both the government's and local people's perspectives on development in its report. The planners therefore collected information from various government departments regarding existing and proposed development projects and the general potential of the region. They also held two workshops – one in the rural areas north of Thaba Nchu and one south of the town – at which attendants were asked to compile a priority list of development needs. These two workshops with an unspecified number of participants were the only occasions where the 'people of Thaba Nchu'[284] were asked about their opinion on regional development.

The most important priorities identified by the rural inhabitants were improving the water supply, creating employment opportunities and the need for a better distribution of schools, pre-schools and kindergartens. In the southern villages, better access to medical services and additional arable

---

[282] NWA OPD 9/6/3 (34) part 1, Thaba Nchu Budgeting and Development Plan: Report on Phase 1 of the Plan, Vol. 2, June 1986, p. 1.
[283] NWA OPD 9/6/3 (34) part 2, Thaba Nchu Budgeting and Development Plan: Consolidated Report for Phases 1 and 2, December 1987.
[284] Ibid. 15.

land were other aspects perceived as pressing. The improvement of markets, shops, sport and cultural facilities, and housing were not identified as urgent needs in either workshop. In addition to the workshops, a survey of rural and urban households was conducted in order to assess their socio-economic situation. Finally, the planning committee held meetings with the town councils in Thaba Nchu and Selosesha whose views on regional development requirements were also recorded and incorporated in the report.[285] The development report differs from previous 'betterment' reports mainly in the fact that it was at least partly based on social as well as natural scientific surveys. As the report offers for the first time detailed data that allow a better understanding of the past conditions of the social-ecological system, the major findings will be summarised in the following.

6.3.1 Ecological Conditions

A reconnaissance survey was undertaken by the Institute of Pedological Research of the University of Potchefstroom in order to determine the potential of the arable soils. Scientists argued that the soils in the Thaba Nchu region did not have a high potential in terms of an economically viable cash crop production. Some patches in the area would be suitable for the production of wheat and vegetables, but in order to determine where these sites were located detailed soil surveys had to be conducted. Moreover, scientists suggested that areas that were not suitable for cultivation were to be replanted with *Medicago sativa* or perennial grasses, and hence transformed into pasture land.[286]

A reconnaissance survey of the rangeland vegetation done by the Department of Botany of the University of Potchefstroom identified the following six *veld* types in the Thaba Nchu area:[287]

(1) Grassland dominated by the climax species *Themeda triandra* was found on deep soils over vast areas in the northern part and in smaller areas in the southern part of the Thaba Nchu district. Only limited sites that consisted of this *veld* type were severely overgrazed and trampled.

(2) Throughout the region grassland with *Eragrostis curvula* and *Eragrostis plana* was found, typically on poorly drained soils and

---

[285] Ibid. 13-18.

[286] Ibid. 50-52.

[287] Ibid. 53-55.

along floodplains and watercourses. The *veld* condition of these areas was highly varied due to management differences in the past. A small area was heavily overgrazed and characterised by an encroachment of the Karroo bush.

(3) In the southern parts of the districts grasslands dominated by *Cymbopogon plurinodis* were found on deeper soils. Many of them had been ploughed before. Again, this *veld* type was highly varied in terms of quality. Isolated sites close to watercourses or on slopes were severely overgrazed.

(4) Hills and ridges with shallow rocky soils made up another *veld* type which was found throughout the region. Vegetation in these areas was rather vulnerable and prone to overgrazing.

(5) *Veld* characterised by open trees and a shrub layer was found on higher mountains and ridges.

(6) Old abandoned fields constituted another *veld* type which was found in substantial parts of the region. Succession stages in these areas varied depending on how long the field had been abandoned.

Based on this botanical survey it was recommended that the natural rangelands be subdivided by taking into account their botanical composition, direction of slope, degree of steepness, general *veld* condition and degree of erosion, as well as other factors such as water availability and animal types. A rangeland management based on a rotational grazing system was suggested, with a slow rotation interval and long resting periods. That this type of management had actually been enforced since the 'betterment' was not mentioned. Moreover, the report recommended that old abandoned fields and cultivated plots with a low potential were to be sown with pasture species, e.g. *Medicago sativa*, *Digitaria eriantha*, *Chloris gayana* or *Eragrostis curvula*. In areas that were heavily degraded a withdrawal of animals for a period up to five years was suggested. Bare patches were to be mechanically scarified, fertilised and fenced off to allow the growth of pioneer plants.[288]

The scientists triangulated the data of the soil and vegetation surveys with aerial photographs from the 1980s in order to gain some sense of degradation trends. Data indicated that the soil erosion had increased between 1981 and 1985. The main problem was the development of gullies or *dongas*.

---

[288] Ibid. 56-57.

About 4.8% of the Thaba Nchu area were affected to such an extent that the land was not usable. Furthermore, scientists also detected signs of sheet and rill erosion. The latter was characterised by the formation of small run-off rills, which were mainly found in unprotected or old abandoned fields. The degradation of vegetation and soils accelerated rainwater runoff and thereby lowered the infiltration rate. As a consequence, groundwater supplies were reduced or had even dried out. Furthermore, the soil erosion also contributed to the silting of dams, and hence to the reduction of surface water supplies. According to the scientists a rehabilitation of soils was therefore necessary to avoid further degradation of land and water resources.[289]

6.3.2 Moribund Agriculture?

The largest part of the total Thaba Nchu area was defined as agricultural land (Table 6). About one quarter of it was registered under freehold by the mid-1980s. In total 165 freehold farms or plots existed, of which 42 were owned as undivided shares (i.e. belonging to several owners) and eight were leased to white farmers from South Africa.[290] According to the report only the latter plots and 20 Barolong-owned farms were used efficiently. However, production on these farms could be increased if access to markets and extension services was improved. The rest of the freehold land was inhabited by scattered households and often heavily overgrazed, and hence not regarded as productive.[291] Besides privately-owned farms, there existed the so-called ‘Sepane farms’ situated east of Thaba Nchu and north of Botshabelo, which had been purchased by the SADT in 1979, as well as the ‘Excelsior farms’, which were situated south of the Sediba location and had been bought in 1983 (Murray 1992:208). Both land blocks were incorporated into Bophuthatswana in 1989 (Erasmus and Krige 1998:12). Bophuthatswana leased these farms to individuals. Some of them

---

[289] Ibid. 58.

[290] The report stated that in Thaba Nchu there were 170 pieces of land under freehold, of which five were registered to the government of Bophuthatswana. It is not clear why these were not counted as state land. As I only regard land that is owned by either natural persons, companies or organisations as private land (Barker 1998:5), I excluded these five farms from my figure.

[291] NWA OPD 9/6/3 (34) part 2, Thaba Nchu Budgeting and Development Plan: Consolidated Report for Phases 1 and 2, December 1987, pp. 56, 62-63.

successfully purchased the land after 1994.[292] In total, these farms covered an area of 13,742 ha, and consisted of rangeland along with a number of smaller irrigation plots that were lying idle at the time of the report was compiled.[293]

Table 6 Land use types in the Thaba Nchu district in 1985[294]

Land Type	Area in ha	Area in %
Rural settlements	1,726	1.3
Urban and peri-urban settlements	2,850	2.3
Eroded areas	6,213	4.8
Arable land	19,250	14.9
Rangeland	98,943	76.6
Woodlots[295]	140	0.1
Total	**129,122**	**100**

Although the agricultural production on private land was depicted as rather meagre, the authors' evaluation of the communal areas was even worse. The rural villages were described as 'essentially non-agricultural'[296] as the majority of the households would rely on off-farm incomes. Several drivers were identified for this trend. Firstly, many families had no rights to arable or grazing land as the 'economic farming units' determined in the 'betterment planning' had long since been distributed. As only households with rights to arable land were allowed to keep livestock, all others had to rely on incomes from other sources. Secondly, population growth had resulted in a splitting up of 'economic units' into half units and sometimes into even smaller plots. As a consequence, the plots available for cultivation were too small to allow production 'at subsistence level and wage earning [became] essential'[297] even for households that had rights to land. Thirdly, the incomes generated from the non-agricultural sector were higher than those

---

[292] Interview with S. Matsididi, Khumo Flat, 16 September 2011.

[293] NWA OPD 9/6/3 (34) part 2, Thaba Nchu Budgeting and Development Plan: Consolidated Report for Phases 1 and 2, December 1987, p. 62.

[294] Data source: NWA OPD 9/6/3 (34) part 2, Thaba Nchu Budgeting and Development Plan, p. 62

[295] Smaller woodlots had been established throughout the area during the 'betterment' in the 1960s (Chapter 4.4.4).

[296] Ibid. 63.

[297] Ibid.

that could be achieved with agricultural production in the communal areas. Hence, rural households preferred to engage in wage labour (Naumann 2014:54). The absence of a work force due to labour migration contributed to the trend of many plots being left uncultivated.

Despite having identified the main problem, i.e. insufficient plot sizes, the authors concluded that agricultural productivity could only be increased when 'agriculture becomes more acceptable as an occupation'.[298] They also stated that 'the rural people represent a large but ineffectual resource which desperately needs motivation, guidance and organisation'.[299] This fitted with the view of some officials, who regarded the people's attitude towards agriculture as an important factor that had affected the country's agricultural sector negatively. As the managing director of Agricor put it:

> 'Agriculture in the less developed areas is today often associated with an image of poverty, hardships, subsistence economy, dirty work and very little scope for real progress. This poor image has in many instances caused the status of agriculture and those involved in it, to be extremely low – especially in the eyes of the younger generation.' (Beuster 1983:6 of appendix)

Only the private farmers and about 300 members of the 14 'primary cooperatives' were said to have an interest in farming. The agricultural production of the latter, however, though well-equipped technically, was not economically viable, as they only produced crops in dryland fields.[300] That the planning committee had already written off the cooperatives in 1987 is surprising, since their economic activities continued to be encouraged by the Department of Agriculture and Agricor until the early 1990s, yet this also happened in other districts, as the example of Ditsobotla suggests (Francis 1999).

There was still some cause for optimism in Thaba Nchu's communal areas in the form of the irrigation schemes in Feloane and Woodbridge, as well as the Grobler irrigation scheme in Sediba. The agricultural production practices of the land users varied, but the report came to the conclusion that 'in the main the land is reasonably well used producing vegetables for the region'.[301] As I have no detailed data on the land use history of the Feloane

---

[298] Ibid. 64.
[299] Ibid. 73.
[300] Ibid. 73-74.
[301] Ibid. 62.

and Woodbridge schemes, I can only refer to Sediba here. With regard to that case, I have shown in Chapter 4.4.2 that agricultural officers often described the Grobler scheme as badly managed, and – in contrast to the dryland fields – as affected by soil erosion. As the scheme had not been efficiently used since its origin in the 1930s it is questionable whether this had changed in the twenty years between 'betterment' and the mid-1980s. Rather it seems more likely that the feasibility study overrated the success of the schemes.

In fact, a more detailed soil analysis of the irrigation schemes conducted a year later arrived at a very different evaluation. In 1988, scientists of the UFS recommended that the land of the three schemes should not be used for irrigation on a large scale as the soils would be prone to salinisation. The worst soils were located in Feloana and Woodbridge, but parts of the Sediba scheme were also affected.[302] The overly positive evaluation of the irrigation schemes stated in the 1987 development report must be understood in the context of the agricultural policies of the Bophuthatswana government. Irrigation promised an increase in production and was regarded as 'modern' and 'advanced' and therefore favoured by officials.

After considering all the various kinds of survey data, the authors of the feasibility study concluded that the potential for cultivation would be very limited due to the area's unpredictable rainfall and harsh winters. In total only about 400 ha in the whole district were suitable for proper irrigated cultivation, which was an 'extremely disappointing'[303] insight. The authors thought, however, that the establishment of irrigated gardens of 10 to 20 ha could be an option for the rural communities. Each resident household would receive a so-called 'comma plot' of 0.05-0.1 ha in these gardens, which could be used for vegetable production. Water would be provided through boreholes. This idea of irrigated gardens was never put into practice during the Bophuthatswana era, but recent agricultural projects that I will describe in Chapter 7.2.1 show obvious similarities to that approach. Crop production in drylands was to be discouraged in most villages, except at those sites where soils were in a good condition and situated 1,600 m

[302] NWA, Agri Info 1/26, Besproeiingsgeskiktheid van Gronde in die Feloane-Sediba- en Woodbridge Gebiede te Thaba Nchu, Bloemfontein, September 1988, pp. 8-9.
[303] NWA OPD 9/6/3 (34) part 2, Thaba Nchu Budgeting and Development Plan: Consolidated Report for Phases 1 and 2, December 1987, pp. 70-72.

above sea level. Here, according to the authors, only suitable crops such as wheat, sunflowers and sorghum should be planted, using improved technical standards.[304] Surprisingly maize, which had been promoted by the former NAD for decades, was not even mentioned as a suitable crop for dryland cultivation.

6.3.3 'Households at Risk'[305]

During the 1970s and in the first half of the 1980s the Thaba Nchu district experienced a significant population increase of about 3.14% (Chapter 7.3.1). The new Bultfontein townships and Seloseha in particular grew in terms of household numbers. To investigate the economic situation of the households in the district a survey was undertaken, which included 54 households in the rural villages north of Thaba Nchu town, 27 in the southern villages and 179 in the urban and peri-urban areas where no agriculture was practiced. On average, the households surveyed had about 6.2 members. The proportion of people between 16 and 49 years was higher in the urban (49.7%) than in the rural areas (42.8%). The situation with regard to people of 50 years and older was different. About 18.3% of the rural populations and only 13.3% of the urban fell in that group.[306] This indicated that the young people were moving to the urban areas; presumably not only in order to find wage labour, but also to escape the elders' authority.

Women between 16 and 49 years of age outnumbered men of working-age in the ratio of 1.29:1. This was a result of male migrants leaving to search for jobs outside the district. However, compared to other areas this was still a moderate difference. In Venda for instance, in the mid-1980s about 30.9% of the total population were women between 15 and 64 years of age, whereas only 15.9% of the total residents were men of the same age group (Lahiff 2000:74). Why was Thaba Nchu different? I can only suggest possible explanations. One reason could be changing patterns of labour migration, i.e. in the female out-migration increased. This would also explain why the women-to-men ratio for the people over 66 years of age in the district was more skewed in favour of females, with 1.83 women per man. In the older age group labour migration was dominated by men. The new migration patterns are also reflected in the oral histories I collected in 2011.

---

[304] Ibid. 77-78.
[305] Ibid. 18.
[306] Ibid. 19.

Many women worked in the domestic service in Bloemfontein or other cities, in the retail or manufacturing sector. For Kuruman too it has been established that women were increasingly drawn into wage labour as the twentieth century progressed (Jacobs 2003a:139). Still, the proportion of female-headed households in Thaba Nchu was high in the mid-1980s – about 38% in the rural and 45% in the urban areas[307] – which suggests that migration of men was still predominant. Unfortunately, there are no precise statistics on the gender of household heads before the 1980s, thus it is difficult to say whether female headship was increasing or not.

About 46% of the working-age population were employed, almost three quarters of them living in the urban areas in Thaba Nchu. 12% of the people between 16 and 49 years of age were still studying. About 41.65% were unemployed, of which 60% indicated that they were actually searching for work. The unemployment rate was more severe in the case of women. Whereas in the villages twice as many women were unemployed as men, in the urban areas there were up to three times as many. About 34.4% of those employed were working inside the region in Thaba Nchu or Selosesha. The remainder either had to commute daily (26.1%) to Bloemfontein or other towns in the Orange Free State, or were living as migrants (39.5%) in the Orange Free State and Transvaal provinces or in other parts of Bophuthatswana. The main occupational spheres were the domestic sector (28%), followed by the building sector (14%), retail and service sector (14%), manufacturing (13%), mining (11%) and public and private administration (11%). Interestingly, of all survey participants only one was employed in the agricultural sector, which had been such an important working area in the past.[308]

Based on the survey data, it was estimated that the households had a mean annual income of about R3,300 in the urban areas and only slightly less in the rural villages (about R3,000).[309] As households had on average 6.2 members, this would be a mean annual income of about R480 per person in the rural areas. Thus, the average household's income was lower than that of households in low-income countries as defined by the World Bank

---

[307] Ibid.

[308] Ibid. 21-24.

[309] Ibid. 25. The Pound was replaced by the Rand when South Africa became a republic and left the Commonwealth in 1961.

(World Bank 1986).[310] Interestingly, the difference between the incomes of the urban and rural households in Thaba Nchu seemed to be only minor. This contradicts the general observation that particularly rural households in the homelands became impoverished (Halbach 1988:238). However, as the survey only presented mean values it is impossible to make any assumptions about the distribution of incomes and the socio-economic stratification of households.

Table 7 Income sources and their proportions for rural and urban areas in Thaba Nchu[311]

Source of Income	Proportion of total income (Rural Areas)	Proportion of total income (Urban Areas)
Remittances from outside (cash and goods)	60.6%	61.7%
Pensions	7.7%	9.1%
Agriculture	7%	-
'Penny Economies'	2.8%	1.1%
Others	21.9%	28.1%

Table 7 displays data on the composition of incomes in the Thaba Nchu district. The figures suggest that income diversification was more important for rural households than for their urban counterparts. The majority of the rural and urban incomes were derived from remittances and goods sent from labour migrants and commuters working outside the district. The second largest category is 'other incomes', which consisted of wages earned within the Thaba Nchu district and social welfare payments other than old-age pensions, i.e. disability grants and child-support grants. In theory black South Africans had been entitled to a pension since the 1940s. However, as the brief history of social welfare in Chapter 8.4.3 will show, gaining access to pensions was particularly difficult for people living in the rural areas. Despite the fact that the Thaba Nchu villages had a higher pro-

[310] An annual per-capita income of 400 US$ was defined as the threshold below which a country was defined as low-income (World Bank 1986:ix). In the mid-1980s the Rand: Dollar exchange rate was 2:1 (Odendaal 2013) thus, annual per-capita income in in Thaba Nchu was about 240 Dollar.
[311] Data source: NWA OPD 9/6/3 (34) part 2, Thaba Nchu Budgeting and Development Plan, p. 25.

portion of people older than 66 years, pensions made up a smaller proportion of the mean incomes of rural households than of their urban counterparts. A small percentage of the incomes was derived from so-called ‘penny economies’, such as the selling of beer, sewing, and making bricks and repairs. Finally, agricultural production contributed to the household income even less than old-age pensions. [312]

Even when assuming that the amount of agricultural income was perhaps underestimated, the data suggest that the decoupling of rural livelihoods from land had been initiated. Table 8 shows the distribution of ‘economic farming units’ in all 37 rural communities in the Thaba Nchu area. In total, 3692 residential sites existed. Less than one third of the rural households had a full ‘economic-unit’. It is doubtful whether a household could survive on the basis of a full unit; but clear is that almost 70% of the rural population were unable to do so. More than one third had no rights to land at all.

Table 8 Distribution of 'economic farming units' in all rural communities in 1975[313]

Access to Land	Number of households	Percentage
Full ‘Economic Farming Units’	1,125	30.5%
Half ‘Economic Farming Units’	1,270	34.4%
Landless and Stockless	1,297	35.1%
Total	**3,692**	**100%**

Finally, the authors of the report did not identify the rural households as the most vulnerable, but rather those who were living at the fringes of the urban areas. In the newly developed townships such as the Bultfontein Zones (Chapter 5.4.3) people were not yet integrated into the wage labour market, but also had no access to arable and grazing land. They were not

[312] NWA OPD 9/6/3 (34) part 2, Thaba Nchu Budgeting and Development Plan: Consolidated Report for Phases 1 and 2, December 1987, p. 25

[313] SAB SON 1606 D12/2/4/2/4/3 vol1, Ondersoek na die ekonomiese en werkgellentheidsbasis van Thaba Nchu, January 1976, p. 25.

only at risk economically but also with regard to other aspects such as infrastructure and social services.[314]

In summary, sixty years of environmental governance and land use regulation in the Thaba Nchu district have had mainly negative effects on the social-ecological system. The worst implication for rural households was an increasing decoupling of their livelihoods from the natural resource base. The people of Thaba Nchu had to base their incomes on multiple pillars in order to compensate for the decline in self-produced food. Labour migration was the preferred way out of poverty but not all people were successful in finding jobs. Qualitative data indicate that growing numbers of females were engaged in migration too, leaving behind villages that were increasingly dominated by children and elders. The interviews I conducted do not clearly reveal whether the women's decision to engage in wage labour was only an economic one. However, Bozzoli's (Bozzoli 1991) account of Phokeng women indicates that the desire to defy patriarchal control was just as much a motivation as the cash income.

6.4 Land Use and Institutions on the Local Level

The mid-1980s feasibility study drew a bleak picture of the agricultural production at the local level. However, to indicate that households had no interest in farming would be a misleading generalisation. In fact, several households were still involved in land use activities. As animal husbandry was not very capital-intensive once a herd was established, livestock ownership was not restricted to the wealthier households. Although I do not have figures for the distribution of livestock, it is certain that more rural households in the communal areas were involved in cattle farming than in cultivation. The plantation of crops, in contrast, had become a highly mechanised business that required either the purchase or renting of expensive machinery, particularly tractors. Hence, engagement in ploughing activities in the Bophuthatswana era was mainly restricted to the wealthier families (Naumann 2014).

6.4.1 Cooperative Cultivation and the Myth of Upgrading

Two groups of cultivators were planting crops in Sediba during Bophuthatswana times: individuals that ploughed the small fields that were part of

---

[314] NWA OPD 9/6/3 (34) part 2, Thaba Nchu Budgeting and Development Plan: Consolidated Report for Phases 1 and 2, December 1987, p. 33.

their 'economic unit', and a primary cooperative (Chapter 6.2) called *setlamo* (Setswana for 'united group') that cultivated cash crops (Naumann 2014:53). The group was established[315] at the urging of Bophuthatswana officials who argued that it would upgrade the local community. Like other cooperatives, the Sediba *setlamo* had an elected management committee which consisted of a chairman, a secretary and a bookkeeper. The group initially had about 20 participants; most of them were men. The decision to join the group was mainly motivated by the hope for an improvement of one's livelihood by becoming engaged in commercial agriculture:

> 'Actually, the *setlamo* came with the upgrading. We thought that the upgrading would be good. And it should be good actually [...] Because we learnt something from farmers. [...] You see that, what is around us, you call it commercial farms. They have farmed with cattle, sheep and maize. When he [the farmer] harvested we saw how many tractors, how many tons came out and were taken to the cooperative.' (interview with Frans Kganare, Sediba, 7 November 2011, parts translated from Afrikaans by the author)

As a first step, suitable sites were identified where the *setlamo* would establish several larger fields (see Appendix 11.2) and which were to be cultivated mainly with maize, wheat and sunflowers. Within the prospective area some individual farmers still used land. The *setlamo* members and government officials entered into negotiation with them. Individual farmers were argued into leaving their fields and taking over lands in another area.

> 'Fortunately, all people who have been here [individual farmers], all of them got lands too. In other words, here was nobody that had no place to plough after the *setlamo* had begun.' (interview with Frans Kganare, Sediba, 7 November 2011, parts translated from Afrikaans by the author)

Joining the *setlamo* promised progress, but not joining the group also had advantages. Most obvious was that the *setlamo* had to produce cash crops that were sold in town, whereas individual farmers had more power of de-

---

[315] Informants could not remember the exact year of the establishment of the *setlamo*, but it was in the early 1980s at the latest (interview with Frans Kganare, Sediba, 7 November 2011).

cision over what to do with their harvest, i.e. selling or using it for consumption. Households who could afford it were engaged in the *setlamo* and continued to plough their former lands, as the following example shows:

> Frans Kganare (66 years) worked for more than 20 years in the gold mines in Odendaalsrus. His wife and his father lived in Sediba while he was migrating to the northern Orange Free State. As other labour migrants did, he invested parts of his wages in land use activities. His household was one of the rare exceptions that could afford to buy a tractor on their own. Kganare joined the *setlamo* a couple of years after its formation while he was still working in Odendaalsrus. Due to his work duties, he was not able to attend all the meetings of the group. Yet he was an active and conscientious member, as proved by his records of the *setlamo* finances that he still keeps in his house today. Despite his dedication to the *setlamo*, Frans said to his father: 'Pa, when I join the *setlamo*, you should not join the *setlamo*. You should always plough your own fields with my tractor.' Although Frans Kganare believed the *setlamo* would come along with upgrading, he had an interest in maintaining the independence of his family. Indeed, according to him his father's fields were the last ones that were still in use in the mid-1990s. Hence, individual farming 'survived' longer than the cooperative effort (from interview with Frans Kganare, Sediba, 7 November 2011).

The example shows that wage labour had become a determining factor of a household's ability to engage in cultivation. By investing salaries in agricultural capital migrant households could maintain the production of crops. In contrast, those who did not have sufficient cash often had no choice but to abandon their former fields as cultivation had become too expensive. Ploughing with oxen was basically impossible as cattle numbers had been reduced radically in the previous decades, and the dependence on heavy machinery made the cultivation of small fields inefficient. People were neither able nor willing to pay the costs as the risk of harvest failure in a semi-arid environment was too high: 'sometimes you pay for the tractor and you don't benefit anything due to the drought' (interview with Johannes Mihi, Sediba, 12 September 2011).

In the Sediba *setlamo* each member had to pay a monthly contribution. Though this fee was adjusted according to the member's financial situation, the person had to have a certain financial stability to be part of the cooperative. Hence, the *setlamo* was not an option for the poorest households. The monthly contributions were used to pay for a tractor that was provided to the group by the government. Seeds, fertiliser, diesel and the rental of other machinery were paid with loans taken out from Bophuthatswana Agribank. After the harvest was sold to a cooperative in town the outstanding loans were immediately deducted from the revenues, which resulted at best in a 'zero total invoice' (Naumann 2014:53). Similarly, to the informants in Sediba, interviewees in Rooifontein recollected that the farmers never made much money from the harvests (community meeting Rooifontein, 23 September 2010; informal interview with Benjamin Koloane, Rooifontein, 26 April 2011). The question is, why did people still continue with farming when they must have realised that it was economically unsuccessful? One explanation is that agriculture was still a part of people's identity as agro-pastoralists – at least for the older generation. Hence, they tried to maintain a practice that was important to fulfil their 'normative needs' (Crane 2010:n.pag.).

When the officials of the Department of Agriculture demanded that the Sediba group cooperate more closely with the farming groups of Talla and Rooibult, the original *setlamo* was enlarged. Yet this did not solve the financial problems but only made the intra-group cooperation more difficult. Together with eleven primary cooperatives in Thaba Nchu, the Sediba *setlamo* was cancelled in 1991 (Department of Agriculture and Natural Resources 1991:11, table 7.2). The tractors of the still heavily indebted group were sold and the money distributed to the members. Each one received a few hundred Rand – much less than they had invested in the years before (interview with Frans Kganare, Sediba, 7 November 2011). The economic failures of the cooperative farming projects in Thaba Nchu were not exceptional cases but were rather typical, as the high number of cooperative cancellations in the early 1990s (Department of Agriculture and Natural Resources 1991:10–12) and scientific accounts (Francis 1999:55–56) of other districts suggest.

6.4.2 Persisting Pastoral Production

In contrast to the cultivation, animal husbandry was not on the decline. The trend that was already visible in the 1960s – i.e. that an increasing number of households abandoned their fields but maintained their livestock herds – continued in Bophuthatswana. This had several causes. First, livestock husbandry was less work-intensive than cultivation as animals only had to be driven to the grazing camps in the morning and collected there in the evenings. Second, it was less expensive to maintain an existing herd than to continue with the cultivation of fields. Concurrently the price of cattle increased which made cattle farming more lucrative.[316] Third, animal husbandry was less risky than cultivation. A drought could be buffered with the purchase of additional food or the sale of animals; though the latter alternative was not preferred by people with small herds, as my observations in 2011 suggest. And fourth, cattle ownership was important not only from an economic perspective but also from a social one: livestock was used for rituals and as bridewealth, and determined a man's status in the community.

Government authorities continued to control animal numbers by conducting regular livestock counts and pledging farmers to keep stock cards on which all livestock sales, purchases, births and deaths were recorded. The herd size that farmers were entitled to did not increase as the system of 'economic units' was continued. People that had grazing rights were allowed to keep about five or ten cattle depending on whether they were entitled to a half economic unit or a full unit. Agricultural officers in Bophuthatswana were still allowed to cull a household's surplus animals.[317] Yet oral histories indicate that culling was not strictly enforced by agricultural officers in Thaba Nchu. None of the persons interviewed in 2011 could remember forced sales of livestock during that time. This was not the case in all Bophuthatswana districts, as indicated by the example of Kuruman, where in 1983 about 20,000 donkeys were eliminated by the Bophuthatswana government (Jacobs 2001, 2003a).

The costs connected with the usage of the communal grazing were low. As with the case of the arable lands, households with access to rangeland had to pay an annual fee, the amount of which is unfortunately not specified in the Bophuthatswana documents that I traced in the archive. However, the development report of 1987 indicates that this was not a great financial burden for livestock keepers – which, according to the officials, resulted in

---

[316] NWA OPD 9/6/3 (34) part 2, Thaba Nchu Budgeting and Development Plan: Consolidated Report for Phases 1 and 2, December 1987, p. 64.

[317] NWA OPD 6/7/2/5 part 3, Barolong Tribal Authority Council Meeting, Thaba Nchu, 13 May 1977, p.1.

overgrazing and resource degradation.[318] Indeed, the surveys conducted by the natural scientists suggest that soil erosion increased and that the vegetation was of low quality in parts of the district. However, with regard to the latter aspect, it should be considered that the former fields that had been transformed into rangelands during 'betterment' had never been replanted with perennial grasses; neither had they been allowed to rest for an extended period. Hence, the low grazing quality was not only the result of high grazing pressure.

6.4.3 Rangers: Contested Guardians of the Veld

The control of livestock husbandry on the local level, particularly the rotational grazing management, was enforced by so-called rangers (Naumann 2014:53–54). In Thaba Nchu, rangers already controlled the grazing areas in the 1940s.[319] During 2011 I interviewed two former rangers in different villages in the Thaba Nchu area.[320] According to them, the rangers were initially accountable to the Department of Agriculture, and later – in Bophuthatswana – to Agricor. Each ranger was responsible for several villages whose rangelands he patrolled regularly. On these patrols rangers controlled the infrastructure, i.e. fencing, wind pumps, and watering points, and impounded livestock that were wrongfully grazing in camps that had been closed for a resting period. Confiscated livestock were brought to the governmental farm Boston Skiet or to a fenced-off area close to Thaba Nchu town. When cattle owners wanted their animals back, they had to pay a fee of about R2.50 per head (rangers, Thaba Nchu area, 7 June 2011 and 10 June 2011). The rangers recorded the observations made during their daily patrols and regularly reported them to Agricor:

> 'Complaint about Mr M., headman of village X: Mr M does not have a good communication with the rangers. […] The camps are not in a good condition. They don't seem to have good people

[318] NWA OPD 9/6/3 (34) part 2, Thaba Nchu Budgeting and Development Plan: Consolidated Report for Phases 1 and 2, December 1987, p. 64.

[319] In 1944 rangers are mentioned for the first time in the minutes of the Sediba reserve board, SAB NTS 8142 2/341 part 2, Minutes of Meetings of the Sediba Native Reserve Board, 22 May 1944.

[320] Although they were willing to have conversations with me about their former jobs, the rangers requested strict anonymity. Hence, in the following section I will not mention any village or persons' names in order to conceal any information that could reveal the identity of the informants.

> to fix and protect. I am complaining because of broken fences all over the camps of his village. He [the headman] does not encourage his community to fix [the fence] where it is broken. I told him to come and collect the fence and poles from me at X. Above all things, in his community cattle are grazing in the wrong camps of the village: Last but not least, we did not agree that they can use all camps. All of them are open, which is wrong. He made a conflict against his community and the rangers.' (ranger report to Agricor, March 1990, translation from Setswana)

It was particularly rangers' power to control livestock movements and to confiscate animals that led to Thaba Nchu farmers' dislike of them. In Transvaal rangers were regarded as spies of the white officials, and beaten up or even killed by local residents (Delius 1996:120–27). Although there are no known murders of rangers in Thaba Nchu, the relation between livestock owners and the *veld* guardians was also fraud with conflict:

> 'I am so worried about the people of headman T. from village X. […] We [the rangers] have camped their cattle [because they had grazed in the wrong camp]; so they [the livestock owners] went there and destroyed the fence and took their cattle. I recognise that they do not obey the laws and do things which are not right. […] On today's date, they came to us with their headman T. by car. They came with their fighters with weapons. We recognised that they were ready to fight with us.' (ranger report to Agricor, July 1990)

Apparently, the situation eased; at least no serious fight is mentioned in the reports of this ranger. However, it shows that livestock owners were willing to resort to violence. A ranger interviewed in 2011 stated that he was once beaten up in the 1980s. Three people that he had seen before at the camps where the confiscated cattle were kept caught him in town. Two held his hands; the third stole his money. When he tried to escape they hit him on his head and stabbed him with a knife. He was seriously injured but survived the attack (ranger, Thaba Nchu area, 7 June 2011).

Strikingly, local residents today would give a completely different account when asked about the system of rangers. As long as the guardians of the *veld* existed, infrastructure would have remained intact, livestock theft would have been a minor issue, and the rotation of livestock would have saved the vegetation. In hindsight, for many farmers, the positive aspects of the ranger system seemed to outweigh the negative. However, the reports of the ranger to Agricor cited above also suggest that not everything functioned well, despite regular controls. Rangers could not prevent fences being broken and remaining unrepaired. They had no absolute control over camps that were closed in order to rest the vegetation. Yet livestock theft was certainly made more difficult due to the regular patrols.

6.5. Preliminary Summary

At the beginning of this chapter, I pointed out that the agricultural potential of the different areas of Bophuthatswana was highly varied. The Thaba Nchu district was initially described as one that had better conditions for cultivation than other Bophuthatswana districts, mainly because the mean annual rainfall was considerably higher. However, ten years later the development report regarded the agricultural potential of the area as rather low and stated that most parts of the region would be unsuitable for cultivation. What reasons can be found for this change of mind? Had the soil quality deteriorated due to bad farming practices? Was the previous evaluation of the feasibility of dryland cultivation in that region too positive? These questions cannot be answered with absolute certainty. It is possible that the soil quality of the arable lands had changed during the second half of the twentieth century. The detailed soil survey of the irrigation schemes done in 1988 suggests this hypothesis. Yet it must be considered that the salinisation that the survey identified as a major problem in those schemes is a phenomenon particularly found on irrigated soils. In other words, one cannot draw conclusions with regard to the quality of the soils in the dryland areas. The statement that the semi-arid climate would hinder efficient dryland cultivation is not convincing. For instance, on the so-called 'Excelsior farms' that adjoin the Sediba location in the south, farm owners still cultivate maize, sunflowers and *Medicago sativa* on dryland fields today. For these reasons, I suggest that the main causes of the changed evaluation of Thaba Nchu's agricultural potential were not environmental factors, i.e. soil quality and climate, but rather changing political motives.

When the Thaba Nchu area was subjected to 'betterment', this was done in order to prepare the rural areas for the resettlement of thousands of households, to allocate pieces of land to as many families as possible, and to promote subsistence agriculture on a minimal scale. The main aim was not to secure rural livelihoods, but to prevent further migration to the urban areas. Moreover, ensuring migrants had access to land allowed their wages to be kept low, as their households would still have an income from the land. In the young republic Bophuthatswana, the focus shifted, and subsistence agriculture lost its importance. Commercial crop production became the focus of attention, as a homeland had to be fed and its independence from the South African economy strengthened. An increase in agricultural production required large-scale irrigation schemes and the use of tractors and other advanced machinery. Thaba Nchu, however, consisted mainly of small fields, and – compared to other district such as Taung where a 3,500 ha irrigation scheme was situated (Stevens 2006:283) – was not suitable for irrigated crop production. From the government's perspective, the region had no agricultural potential, and thus was declared unsuitable for cultivation. Until today the Department of Agriculture defines the area as usable only for pastoral production, which ignores the long history of dry-land cultivation that was efficient enough to increase rural households' food security.

However, the new perspective on Thaba Nchu cannot solely be ascribed to a policy change, but also to the effects of 'betterment'. Although Thaba Nchu was the only district that was almost 100% replanned, the communal areas were said to be heavily degraded and 'essentially non-agricultural'. After the mid-1960s, many arable fields had been abandoned for various reasons: (i) fields were too small to produce efficiently; (ii) off-farm incomes were significantly higher than those gained from agriculture; (iii) due to the lack of draught animals ploughing had to be done with tractors, which increased the costs of cultivation. An indulgent evaluation of 'betterment' would draw the conclusion that the schemes were not successful, whereas a less lenient examination would argue that the replanning had systematically undermined small-scale agriculture; or, in social-ecological terms: 'betterment' decoupled rural livelihoods from the land. Though it is impossible to say how agriculture would have developed under a system involving less control and coercion, the formation of a comparatively wealthy land user class was visible in pre-'betterment' times. Some of them remained engaged in agriculture in the new Bophuthatswana state by joining cooperatives. By the eve of the first democratic elections, however, most of them had discontinued.

Finally, it is important to stress that generally the means of intervention were no less coercive in Bophuthatswana than they had been before. This is illustrated for instance by the Kuruman 'donkey massacre' (Jacobs

2001). In Thaba Nchu too, Bophuthatswana officials continued to violently enforce their politics, though not with regard to animal numbers. In 1982 two communities at the foot of the mountain of Thaba Nchu were destroyed in order to establish a game reserve and luxury hotel with an attached gambling casino. Resident households were resettled to other villages without proper compensation (Murray 1992:201–03). On the other hand, the methods of official surveys appear to have been more participatory, i.e. it involved holding workshops to generate wish lists of local populations with regard to development projects. However, not much had changed in the broader narrative and the power of authorities to 'communicate truth' (Jacobs 2001:492); even when such 'truth' was a misconception of reality. Instead of arguing that land was insufficient for economically viable production in Thaba Nchu, the authors of the feasibility study considered the main reason for agricultural decline in the area to be the attitude of the rural people. As in 'betterment' times, it was the farmers, and not the conditions under which they had to produce, that were presented as the root of the problem.

7. ENVIRONMENTAL GOVERNANCE IN POST-APARTHEID SOUTH AFRICA

Whereas the previous chapters have illustrated the historical continuities and changes in human-environment relations, I will focus now on recent configurations of the social-ecological system. With the end of apartheid in 1994, Thaba Nchu – like other South African homelands –underwent significant political, economic and social transformations. Today, the area is part of the Mangaung metropolitan municipality, which encompassed 45 wards in 2010. The Thaba Nchu area consists of six wards, each of which is represented by an elected ward councillor. According to the constitution of the Republic of South Africa, local governments are in charge of the development of the municipality. Hence, as an executive authority, the municipality plans, budgets and manages the social and economic development of the region and the provision of services (Mangaung Local Municipality 2010:1). In national censuses, the Mangaung municipality achieves comparatively good results in terms of regional development. For instance, on the municipality level, more than 85% of the population has piped water inside a dwelling or yard and almost 80% of households have weekly rubbish removal services. Yet these data do not reflect the great differences within the municipality. In rural Thaba Nchu there is no rubbish removal at all, and many households have inadequate water supply (Chapter 2.2). Moreover, development projects for formal housing focus on the urban areas of Thaba Nchu. Therefore, many rural communities feel neglected by politicians.

Whereas the local government system provides a new institutional setting for public authority, other political institutions survived in the Thaba Nchu area after the end of apartheid: the region falls under the authority of the Barolong boo Seleka Traditional Council, which is essentially the successor of the former tribal authority but with less decision-making power. The traditional council administrates the communal land and still issues PTOs for the residential sites (Chapter 2.2). Thus, in terms of land allocation, not much has changed on the community level; however, the situation is different with regard to land management. The incorporation of Bophuthatswana into South Africa brought about the collapse of state-implemented institutions for rangeland management and the discontinuation of subsidisation for agricultural cooperatives. However, the end of apartheid also opened up new opportunities for both communities and households. An ambitious land reform programme was initiated by the new ANC government with the aim of breaking down racial inequalities in terms of access to land.

In the following sub-chapters, I will explore how the land reform took effect in the region. Moreover, I will show that the state has largely withdrawn from the strict regulation of land use, but still runs agricultural development projects in the area. As before, these projects are based on science and technology and, I argue, ignore socio-economic conditions at the local level. Finally, a pressing question is asked: what happened to the narratives that underpinned apartheid policies? Recently, scientists have challenged the view that overgrazing is the main cause of resource degradation in communal rangelands and have called for a more critical judgement of vegetation and soil conditions in communal areas (for instance Rowntree et al. 2004; Vetter 2013). However, narratives of degradation still persist in debates, as an example of a report concerned with resource use in Thaba Nchu will show. This report was compiled by scholars of the UFS in the mid-1990s in order to inform government officials about the then-current land situation in Thaba Nchu. It argues, *inter alia*, that demographic development is a major threat to the ecological system. I will critically explore whether these assumptions are valid by analysing past and recent population dynamics in rural Thaba Nchu.

7.1 Compensating Past Injustices: Land Reform in Thaba Nchu

To overcome the injustices of the racialised land distribution created during apartheid – i.e. the provision of only 13.7% of the country's land for black South Africans – an ambitious land reform programme was initiated in 1994. Much scholarly attention has focussed on the success of the reform and its impacts on rural livelihoods (Lahiff 2002; Hall 2009a; Hebinck and Shackleton 2011b). Up to this date, conclusions are often quite sobering. For instance, a large quantity of agricultural projects failed due to various factors such as inappropriate planning, group conflicts and lack of capital (Hall 2009b:43). On the other hand, the process of redistribution of land to black South Africans has been criticised for being too slow (Du Toit 2011). In the following, I will briefly outline the main components of post-apartheid land reform in South Africa and then shed some light on the importance of land reform in the Thaba Nchu area and the aspirations of communal farmers with regard to tenure security.

The three components of the South African land reform can be subsumed under the terms 'restitution', 'redistribution' and 'tenure reform'. The Restitution of Land Rights Act of 1994[321] addresses people who had been displaced and dispossessed after 1913. Compensation for their forced removal is achieved by either restoring rights to the land under claim, or by providing alternative land or a financial relief. Until the end of 1998, which was the deadline for submissions of land claims, more than 68,000 such claims were submitted to the Commission on Restitution of Land Rights by individuals and communities in urban and rural areas (Lahiff 2002:10). The Restitution of Land Rights Amendment Act of 2014[322] extended the cut-off date for the submission of land claims until 30 June 2019. In the former Thaba Nchu area too, a restitution land claim was settled recently, which will be illustrated below.

The aim of 'redistribution' is to make land available for formerly disadvantaged groups of people, i.e. the landless, labour tenants, farm workers and emerging farmers, for agricultural and non-agricultural production as well as residential use. In the first five years the policy aimed particularly at the poorest by providing a Settlement Land Acquisition Grant (SLAG) of R16,000 to households with a monthly income under R1,500. In most cases several households pooled their grants together and purchased white-owned farms (Lahiff 2002:10). In 2001, the policy shifted towards a more neo-liberal approach (Hebinck and Shackleton 2011a; Lahiff 2011). To apply for a grant within the Land Redistribution for Agricultural Development (LRAD) programme, people had to contribute capital in cash or in kind, e.g. agricultural machinery. LRAD grants were much higher than SLAG, and ranged from R20,000 up to R100,000 depending on the individual contribution, which varied from R5,000 up to R400,000 respectively. It is important to note that the LRAD programme addresses individuals, and not households, as done by SLAG. When a group wants to apply for an LRAD grant they have to establish a legal entity, i.e. a communal property association (CPA), a trust or a closed cooperation (Lahiff 2002:10; Jacobs, Lahiff, and Hall 2003:4–5; Hebinck and Shackleton 2011a:19).

---

[321] Republic of South Africa, Act No. 22 of 1994: Restitutions of Land Rights Act.
[322] Republic of South Africa, Act No. 15 of 2014: Restitution of Land Rights Amendment Act.

The last pillar of land reform, i.e. the tenure reform, is aimed at the protection of residential rights of people living on private or state-owned land, and a 'reform of the system of communal tenure prevailing in the former homelands' (Lahiff 2002:10). The slow process of tenure reform created several acts that addressed different aspects of tenure insecurities, i.e. the Land Reform Labour Tenants Act,[323] which protects the rights of labour tenants, or the Communal Property Association Act,[324] which enables groups of people to form a legal entity and obtain private rights to land (ibid. 10–11). A recent attempt to establish secure tenure rights of people in former homeland areas was undertaken with the Communal Land Rights Act of 2004,[325] which facilitates the transfer of title deeds to communal land from the state to communities. Parts of the land may also be subdivided and handed over to individuals. Communities that receive a title deed have to establish committees for the administration of the land. However, administration may also be vested in the hands of traditional councils in cases where they exist (Cousins 2009b:12–13). The act was not a success, as shortly after its passing four communities successfully undertook legal proceedings against the act and challenged its constitutionality (Cousins 2009a). A major critique was that the act undermined small, independent communities that fall under the jurisdiction of new traditional councils. To this date, the debate over tenure security in communal areas has not come to an end, and critical voices call for a complete rethinking of tenure reform and a shift away from privatisation-oriented approaches (Cousins 2009a; Lahiff 2009).

7.1.1 Land Reform in the Free State Province

The land reform process in South Africa is slow; however, it seems to be even slower in the Free State province (Bureau for Food and Agricultural Policy 2013:12). In 2013 a land audit found that about 86% of the province's commercial agricultural land (12,473,175 ha) was still owned by white farmers. Only 2.96% of the land was owned by previously disadvantaged people or still in the process of redistribution, i.e. not yet transferred to the new farmers. The remainder was owned either by foreign investors

---

[323] Republic of South Africa, Act No. 3 of 1996: Land Reform Labour Tenants Act.

[324] Republic of South Africa, Act No. 28 of 1996: Communal Property Associations Act.

[325] Republic of South Africa, Act No. 11 of 2044: Communal Land Rights Act.

(0.24%), the state (0.89%) or various other actors (4.27%), such as mining companies, other private companies and churches. For about 5.31% of the land, the audit could not gain information regarding ownership (ibid. 8).

That the land reform process in the Free State province seems exceptionally slow has several explanations. First, the province has a comparatively low number of restitution claims, with only 41 claims for rural areas until 2013. In other provinces such as KwaZulu-Natal or Limpopo more than 2,000 claims have been made. As restitution claims depend on the demand, i.e. 'the greater the extent of forced removals, the greater the extent of restitution' (ibid. 13), the small number of claims in the Free State indicates that less removals had taken place. However, only removals after 1913 are considered for restitution, therefore no claims are possible for the dispossessions that took place when Thaba Nchu was annexed by the Orange Free State (Chapter 3.1). Second, in contrast to other provinces a vast majority of the claimants in the Free State (about 76%) had been compensated financially and not with land (Bureau for Food and Agricultural Policy 2013:13).

An evaluation of the entire redistribution process is difficult as data for the provincial and local levels are difficult to obtain (ibid.16). In the Free State province, only 427 people got access to about 70,000 ha of formerly white-owned land through redistribution programmes between April 2009 and March 2013 (ibid. 17). Compared with the national level, the province's share of redistribution cases (8%) is still lower than its share of commercial agricultural land (12%); in other words, other provinces perform better in terms of redistribution. According to the Bureau for Food and Agriculture Policy the reason for this is not clear (ibid.).

7.1.2 Striving for Tenure Security in Thaba Nchu

The issue of tenure security is discussed in the research sites, though some people have more interest in the topic than others. In particular those who want to expand their cultivation efforts from their gardens to larger areas were arguing for the allocation of title deeds. Two gardeners, Kesiilwe Theko and Patrick Malebo, who were counselled by the farmer's association Grain SA (Chapter 7.2.1) emphasised during our conversations that a farmer could only cultivate successfully when he or she obtained a title deed for the land. One said that 'title deeds make people strong', as they would enable them to receive loans from the Land Bank (informal interview Kesiilwe Theko, Sediba, 3 April 2011). Kesiilwe, and his friend Patrick, wanted to obtain the rights to an individual field of several hectares in the area where people had ploughed in the past. In 2011 a title deed was merely a dream. For Patrick, the situation was even worse as he and his parents had never had rights to a field in the village in the past. He tried to find ways to circumvent his current situation of being landless. He approached the headman and explained that he wanted to erect a fence around his grandparents' old field. The headman agreed to raise the question at the next meeting of the traditional council and asked him to apply with an official letter. Patrick succeeded, and the council officially granted him permission to use the old field and to fence it in. He was very happy, as he now had the right to exclude people and their livestock from the site. Up to November 2011, when I left, he had not yet built a fence, and thus also had not started with cultivation.

The example of the two men shows that demands for land and tenure security do not necessarily refer to a large farm – an observation that is also made at the national level by others (for instance Hall 2009b:69). Some rural villagers stated that they wished to have their own farm, but this was not always met with sympathy by other people. For instance, at a community meeting with representatives of the Department of Agriculture and Rural Development a brainstorming session was conducted and people were asked to collect ideas for community and livelihood improvements. Most points raised concerned infrastructure and services, fences and watering points, access to electricity and water for household consumption, and medical care, as well as employment facilities. Only one young man stated that he would like to acquire a farm; a statement that was met with a shaking of heads by the elders of the community.

Still, a few people who live in the research sites are members of a trust or CPA that constitutes a legal entity and owns land. Most of these landowners were found in Rooifontein. In several cases, farms that were incorporated in the 1980s into Bophuthatswana (Sepane and Excelsior farms; Chapter 6.3.2) had been bought by individuals who were better-off and lived in Thaba Nchu town. The exact number of farms that are redistributed through SLAG or LRAD grants could not be established. The Thaba Nchu-based extension office of the Department of Agriculture has only an incomplete list. In 2011, the list contained 39 farms owned by trusts or individuals. Not all of them were recently distributed, but some were *de facto* old Barolong farms that had been owned throughout the apartheid period by black South Africans.

I conducted interviews with land reform beneficiaries who were farming close to Botshabelo. One farm was owned by a CPA which consisted of 16 members, some of them former farm workers. The land, which had previously been owned by a white farmer, was transferred to the CPA in 1998. Of the 13 men and 3 women who were part of the group, several were related, and all knew each other before they claimed the land. As required for a CPA, the group had a steering committee with a chair, vice chair, secretary, treasurer and a contact person. The latter was the only person that lived permanently on the land; the other beneficiaries lived in Botshabelo. The redistribution was one of the earlier projects of SLAG. People pooled their grants and bought a 639 ha farm for more than R230,000. Moreover, they were supported with infrastructure and received a wind pump, two reservoirs, and fences, and also a water point in each of the nine grazing camps. Like the communities, the farmers faced problems with theft of infrastructure. Parts of the fences that ran along the highway were stolen, as were essential parts of the wind pump and the water points. The water situation was thus comparable with that in the communal area. Animals had to drink in open water in the *veld* or at a river.

When I talked to two of the CPA members in 2010, the farm was only being used for cattle production. In total about 52 cattle were grazing there; 28 belonged to individuals and the remainder to the entire CPA. An irrigated vegetable field of 2.5ha had been used by them in the past, but was not cultivated anymore due to the lack of adequate machinery. The CPA earned almost no money with their agricultural production. From time to time they sold cattle and used the money to pay back a land bank loan which they had taken out to purchase a herd. The only income they earned from their farm in 2010 was the rent that a telecommunication provider paid for a radio tower erected on their land. The example of the CPA illustrates the typical problems indicated at the beginning of the chapter. Lack of money and implements made a successful agricultural production difficult, and the interviews suggest that the beneficiaries could not significantly improve their livelihoods with the land.

7.1.3 Restitution of Resettled People: The Sediba Land Claim

For the majority of the inhabitants of the former Sediba reserve the redistribution programme had not brought changes to their lives. More important for them is the Restitution of Land Rights Act of 1994. A land claim was lodged for the former Sediba reserve in 1997 by a resident of Morago village, and finally accepted and gazetted in 2007 by the regional Land Claims Commission. Only in 2012 was the claim settled and more than R42,000,000 offered as financial compensation to about 2,000 beneficiaries from several villages for resettlement and dispossessions during 'betterment' (Department of Rural Development and Land Reform - Communication Services 2012; Moiloa 2012; Commission on Restitution of Land Rights 2013:18). Amongst others, residents of Sediba, Talla and Middeldeel were among the beneficiaries. When I visited Sediba in July 2012, the money had not yet been distributed. However, families were making plans about how to use the restitution payment. Of those I asked, many stated that they planned to renew their houses, spend it on their children's education, and erect tombstones for their deceased ancestors, as it was they who had suffered the most under the apartheid regime. Whether the financial compensation will have long-term outcomes for rural people's livelihoods and land-based activities remains to be seen; but certainly, the money is a positive turn in the history of the former 'native reserve'.

7.2. State Intervention with Regard to Land Use

As noted above, the municipality is accountable for the delivery of services and socio-economic development in Bloemfontein, Botshabelo and Thaba Nchu, as well as in the adjoining rural areas. Strikingly, in an integrated development plan for Mangaung, the authors state that the municipality 'has little experience in dealing with the problems of rural areas' (Mangaung Local Municipality 2010:46). Therefore, it is not surprising that the strategies for rural development remain vague in the report. The aim of the municipality is to 'facilitate an increase in economic opportunities and activity in rural areas' (ibid. 46), support rural small-scale land users and assist emerging farmers in becoming commercial farmers (ibid. 47), but how this should be achieved is not clarified in the development plan. In other words, the local government apparently has no clear road map for the communal areas.

In addition to the Mangaung municipality, there are government departments, research organisations and other non-governmental actors engaged in agricultural development in rural Thaba Nchu. The cooperative production of cash crops is no longer subsidised by the state, as was done during the Bophuthatswana era. Instead, policies and interventions for agricultural development focus on the improvement of 'food security', the new buzzword used by officials. I will present one of these projects in more detail below. With regard to animal husbandry, a sheep and wool project has been started in selected communities in the Thaba Nchu area. Thus, after decades of disapproval of sheep production by agricultural officials, the issue is again being promoted as it was in the 1930s (Chapter 4.4.5). Apart from this project, there has been no state intervention with regard to the production of grazing livestock in most communities since 1994, although some chicken and pig projects have been initiated. After the incorporation of Bophuthatswana into South Africa, most bureaucratic institutions of rangeland management were abandoned. However, it seems that the collapse of the institutions has had no dramatic effect on the ecological system so far.

7.2.1 Government Efforts to Increase Food Security

Being a former homeland area, Thaba Nchu is regarded as a prime site for food security programmes and other development efforts in the Free State. Two scholars of the UFS told me at the beginning of my research that it seems that Thaba Nchu is a kind of 'experimental site' for agricultural development. Thaba Nchu residents, according to the UFS scholars, were 'over-researched' and a bit tired of the ever-new ideas that were promoted. Whenever government and research organisations had explored new farming methods or better management ideas they tested them in Thaba Nchu. Over the past 15 years the region had experienced the introduction of infield rainwater harvesting (IRWH), greenhouses, and community gardens, to mention only the main projects related to cultivation. A community garden was established in Sediba on the site of an old vegetable garden situated next to the irrigation scheme and supplied by the Grobler Dam. In the early 2000s the garden had been relaunched as a cooperative project where people were expected to cultivate together and share the profits gained from the harvest. When I visited Thaba Nchu for the first time, the garden was no longer operating. Conflicts between the participants, the exclusion of certain members by other members, and the unredeemed promise of the

Department of Agriculture to provide a tractor were three reasons for failure mentioned by the village residents.

Greenhouses were introduced by the Department of Social Welfare in 2006. In each community, a handful of families were selected to receive a greenhouse in order to plant vegetables, but the project was not successful in the long term. In many cases the greenhouse films had been broken after a couple of years and the owners did not have the money to renew them. Moreover, the vegetables in the greenhouses had to be watered manually, which caused severe problems in those villages with an uncertain water supply. In 2011, the majority of the greenhouses were broken, used as kraals or storerooms.

The Infield Rainwater Harvesting Project

Infield rainwater harvesting was introduced in selected Thaba Nchu communities in 2003. In short, the technique requires the creation of runoff strips leading to catchment basins where rainwater is stored and plants are cultivated (Kundhlande et al. 2004:58–59). After introducing the technique at four test sites, it was expanded to about 1,000 households in the whole Thaba Nchu region (Water Research Commission 2009:30). The main aim was to improve food security, a concept that was introduced by the World Food Conference in 1974 (Esterhuyse 2012:95). The project started promisingly and agricultural economists projected significant increases in farm incomes: 'by adopting the simplest form of in-field rainwater harvesting [...] farmers can increase their income by about R800 per hectare in the case of maize production' (Kundhlande et al. 2004:xix). The project, which started in the backyards of the villages, was also to be applied on cropland in the long run and 'experts [were] convinced that with IRWH farmers can e.g. produce enough maize for own consumption and also earn cash with the sale of surpluses' (Water Research Commission 2009).

In order to strengthen the project institutionally, several levels of cooperation and support were established. In each village, rainwater associations were established in order to combine the labour and knowledge of the participants of one community. Moreover, the Thaba Nchu area was divided into three zones, each consisting of several villages. Each zone had its own committee, which would meet the other committees and share knowledge and experiences. Finally, the 'Tswelelopele Small Scale Farmers' Cooperative', an IRWH interest group at the municipal level, was formed. One of its tasks was the establishment of a close corporation that would enable the

IRWH-farmers to purchase seeds and fertiliser at better prices (Esterhuyse 2012:100). In 2011, this closed corporation was not yet registered. Technical advice for the participants was provided by the Agricultural Research Council (ARC), the principal agricultural research organisation in South Africa, which gave demonstrations of IRWH techniques and supplied seeds, fertiliser, pesticides and agricultural tools.

An evaluation of the project that took place between 2008 and 2010 made sobering observations. The village associations, zone committees and the cooperative as a whole were ineffective or misperceived by members, and showed 'evidence of overstructuring to accommodate the philosophy of participation in decision-making' (Esterhuyse 2012:100). Moreover, having been supported with agricultural capital by the ARC for three years, the participants were still expecting further assistance. All in all, none of the main aims was fulfilled at the end. The project brought no significant increase of incomes – which was actually predictable since the gardens were often very small and potential markets were too far away. Households also experienced no differences with regard to their food security. Though participants had learned new techniques for gardening in a semi-arid environment, they 'experienced the initiative as hard labour and had little hope that it would grow into a dynamic movement against poverty' (Esterhuyse 2012:103).

Esterhuyse's evaluation of the project's success corresponds with my observations. In Sediba, about 32 people were involved in the IRWH project when it was introduced to the community. Two of them also received six months' training at the GLEN agricultural college near Bloemfontein. When I worked in Sediba a couple of years later, three households still used the new cultivation method. Of the two former GLEN trainees, one was not gardening at the time of data collection. The other one, an older woman, still cultivated, though not with the new technique. It was too strenuous to prepare the garden in the prescribed way and she perceived the conventional method with beds as easier. She was still supported by GLEN and stated that she would receive regularly seeds and fertiliser. Additionally, she got two government-financed water tanks next to her house that secured the water supply in case precipitation was not enough. When I asked other former participants why they had discontinued the rainwater harvesting they often stated that they would not have the means to provide the agricultural input: 'ARC [gave] free fertiliser and free seeds before they advised us to form an organisation. After they finished their project we had nothing to sustain our aim and objectives' (interview with Johannes Mihi, Sediba, 12 September 2011).

The failure of IRWH, like the greenhouse project, was caused by its underlying assumption, one that is typical of the southern African development 'machine' (Ferguson 2003). 'Agricultural development' is synonymous with capital-intensive input, scientific cultivation techniques and – finally – commercialisation. This ideology not only resurrects the old dichotomy of 'traditional and ineffective' versus 'modern and productive' forms of land use, but also fails to consider local realities, as people may lack the money for funding. As a consequence, the poorest members of communities – who, one would assume, are the main target group of a food security programme – are basically excluded.

Hart (2011) has shown that similar problems arose in Limpopo, where a vegetable project was introduced in a small village by the extension service of the Department of Agriculture. Spinach-like plants growing wild in home gardens were an important source of food for local households. Instead of promoting the cultivation of these plants, which did not need any irrigation, the Department of Agriculture encouraged the use of exotic cash crops. The project was no success, as the water supply was inadequate for irrigation, inputs were too expensive and the new crops contributed less to households' food security than the wild plants (ibid. 330). As in the Thaba Nchu case, the 'ecological and socioeconomic circumstances (semi-arid conditions, poverty, and food insecurity) beyond the project fences [were] inconsequential, although they [were] cited as the very reasons for conceptualising and supporting the projects in the beginning' (ibid. 332).

In fairness, it has to be stressed that the state is not the only actor that promotes a costly agricultural intensification in the Thaba Nchu area

> Kesiilwe Theko, a man in his sixties, and the younger Patrick Malebo were involved in gardening (Chapter 7.1.2) and were frequently in contact with representatives of the farmers' association Grain SA. Both attended the association's information days and wanted to farm commercially. Particularly Kesiilwe, who was the wealthier one of the two men, counted on the best input: he used tested seeds of the South African company Starke Ayres, regularly applied the herbicide Roundup made by the multinational concern Monsanto, and let his soil be tested in order to identify the perfect fertiliser. His garden was certainly a success in terms of yield, yet bearing in mind the size of the cultivated

> area – less than 0.1 ha – the costly inputs were exorbitant, and not affordable for other village residents, such as Patrick. Although he also applied Roundup, he had insufficient money for industrial fertilisers, but produced his own based on a mixture of cattle or sheep manure, plant material and paper. Farming with tested seeds and chemicals was regarded by both men as 'real business' and a sign of progress. Conventional gardening was just 'to keep you busy', as Kesiilwe put it. Both complained about some of the wealthier older men that had all the technological requirements, e.g. they possessed a tractor, but did not know how to farm appropriately (based on several informal interviews and one semi-structured interview with Kesiilwe Theko and Patrick Malebo, Sediba, 6 June 2011).

A rejection of 'conventional' cultivation, or the 'old, traditional method' as it was called, was also expressed by other younger men. On one occasion, along with two former IRWH participants I visited the irrigation scheme in Sediba, where the group had cultivated some plots until 2008 but abandoned them when the supply of free material discontinued. When we arrived there, we saw a small maize field of about half a hectare. The owner, who had died a couple of weeks previously, had been part of the *setlamo* (Chapter 6.4.1) and was one of the few people in the research area that had owned a tractor. The two men pointed at the field and referred to it as an example *par excellence* of the poor yields that would be obtained with the 'traditional method'. Indeed, the maize did not look as big as the crops found in the commercial areas, yet the field was the only one as far as the eye could see. Particularly the middle-aged people who had an interest in farming activities seemed to be attracted by these 'modern' techniques and inputs. Instead of using less productive, but also less capital-intensive means, they planted nothing. For younger people subsistence agriculture is not an appreciated way of life (see also Esterhuyse 2012:99). If they are to be engaged in cultivation at all, they want to earn money from it.

7.2.2 Ecological Resilience despite Collapsing Management Institutions?

Similarly, to pastoralists in the Richtersveld (Berzborn 2006:277), livestock owners in rural Thaba Nchu complain about a deterioration in the management of communal rangelands. In contrast to the past, when animal numbers and movements were controlled by agricultural officers, rangers and fences, today almost no institutions for the management of rangelands are found. The only rule which exists at least in theory is that of the exclusion of the animals of non-residents, which Lawry (1990) refers to as the 'minimum definition of common property' (ibid. 406). Still, I met several people who stated that illegal grazing by the animals of non-residents of neighbouring villages and private farmers would be a problem. Particularly in Rooifontein, which is surrounded by other villages, townships and private land that is not always fenced off, inhabitants reported that the use of the community rangelands by outsiders is common. Several informants, including officials from the Department of Agriculture, white commercial farmers and communal farmers, stated that illegal grazing is also practised by Lesotho herders, who drive their livestock into the eastern Free State when the rangelands in their home country do not provide enough fodder. Yet this issue seems to be more severe in the vicinity of Thaba Nchu town and the neighbouring township Botshabelo. In the more distant rural communities people do not face severe problems with illegal herds from Lesotho.

Except for Middeldeel, which received new fences from the Department of Agriculture in the course of the sheep and wool project, none of the research sites had grazing camps or working water points in 2011. In these communities, rangelands are used throughout the year and animals drink at open water in the *veld* or in dams. In the case of the latter, several informants reported that particularly during the winter months animals sometimes got stuck in the mud of the dams and died. Some people have small dams on their plots or use hand pumps for ground water in order to provide livestock with water inside the kraal. Pastoral farmers who do not pay herders prefer to graze their animals close to their homesteads – on the one hand, because the animals were closer to the kraal, which meant that less time was needed to bring them back in the evenings, and on the other hand, because they could keep an eye on them. As indicated, Middeldeel has – at least theoretically – better preconditions for practicing rotational grazing, yet during my research stay herds were not rotating for several reasons. First, only one of the four grazing camps had a working wind pump in 2011. Particularly in the dry winter months, the majority of the animals were grazing there, either for the whole day or at least in the mornings (interview with agricultural officer, 17 September 2011). Second, the gates between the camps were often left open and some sections of the fences were broken. Thus, animals could actually move between the different camps. Third, one of the camps was out of sight of the village and farmers were reluctant to bring their livestock there because of the possibility of theft. A sizeable fraction of the grazing land was therefore only used marginally. Fourth, the area used for cultivation in the past was also fenced but actually used solely by the headman of the village, who owned a large cattle herd. To sum up, in Middeldeel the use of rangelands was comparable with that of the other villages.

That rangeland infrastructure is not persistent in the long term is mainly a result of theft. With the abolition of rangers, the monitoring of management and infrastructure was discontinued, which subsequently led to a breakdown of top-down-implemented institutions. In all research sites, residents complained about the situation; several people mentioned that the only solution would be the reintroduction of rangers to patrol the infrastructure regularly. I pointed out in the previous chapter that the formerly hated guardians of the *veld* are in retrospect positively valued. The agricultural officers in Thaba Nchu also mentioned that the Department had already thought about the reintroduction of a ranger. However, if the position is re-established, then rangers would not be employed in each village. Rather there would be one person responsible for the whole communal area and based in Thaba Nchu (interview agricultural officer, 17 September 2011). I assume that in that case, there will be no or only minimal improvement in the maintenance of infrastructure in the communal areas.

What conclusions can be drawn from the management situation, particularly with regard to the aspect of resource degradation? Are the rangelands in rural Thaba Nchu prone to deterioration due to mismanagement? And what are the emic and etic perspectives on this question? The previous chapters have shown that officials had emphasised the problem of large cattle owners who sought to maximise the sizes of their herds at the expense of the community as a whole. Vegetation change and soil degradation as a consequence of overstocking and bad farming practices were the official reasons for the interventions since the 1930s, and still in the 1990s resource deterioration was regarded as a major problem in the area (see for instance De Vielliers and Booysen 1998).

When asking 34 informants of Middeldeel and Sediba about their perceptions of the rangeland condition, 41% of the informants stated that the condition was generally bad. The remainder perceived the quality of the *veld* as good (32%), as good with some degraded parts (18%) or as average (9%). Several aspects were mentioned as indicators for good rangelands:

the condition of the animals ('cattle are fat and produce calves'), green grasses, a large amount of standing biomass, access to water, and trees for shadow. Species composition – one of the main scientific criteria for rangeland condition – was considered less important as an indicator. Only one informant specifically mentioned *Themeda triandra* (*seboku*) as a sign of a good rangeland. Another one stated that spiky grasses, such as the annual *Aristida congesta* (*steekgras*), would occur on bad *veld*. Generally, knowledge of grasses was not as broad as for instance that of medicinal plants (Chapter 8.5.3). On average people interviewed for the ecological knowledge survey mentioned only 3.7 names of grasses (min: 0; max: 11), with *Themeda triandra*, *Cynodon dactylon* (*motlwa* or *mohlwa*) and *Eragrostis plana* (*modula*) being the most common. This could explain why species composition was not a central emic indicator for rangeland condition in rural Thaba Nchu.

70% of informants stated that the condition of the rangelands was better in the past. Many of them explained that the grasses did not grow as high anymore. Thus, changes in rangeland condition were also ascribed not so much to the presence or absence of certain species as to the amount of standing biomass. However, there were also people with a contrary opinion, who perceived no changes (10%) or even improvements (16.6%) in rangeland quality. One informant stated that the condition would vary annually depending on rainfall. Of those people who observed a negative change in rangeland condition, many stressed that in years of high precipitation the *veld* would be in good condition again.

With two Sesotho-speaking people who proved to be knowledgeable with regard to grasses and plants in general, I tried to assess the changes in species composition over time in more detail. Both stated that the perennial climax species *Themeda triandra* is still often found in the area but had decreased in amount in the past, yet both regarded the changes as non-permanent and were convinced that after good rainfalls the grass would again be abundant. This general statement is supported by the botanical research undertaken within the RCR project. In years with good rainfalls, only small differences in biomass could be measured among commercial farms and communal rangelands in the Thaba Nchu area. Despite several decades of high stocking rates – at least, higher than suggested by the Department of Agriculture (Kotzé 2013:227) – and two decades of continuous grazing, vegetation in the communal rangelands was not found to be seriously degraded in 2011. Thus, botanists have concluded that the ecological system is rather resilient with regard to management practices (Linstädter et al. 2016). Nevertheless, the narrative of resource degradation as a major threat still persists, as the following chapter will show.

7.3 Population Dynamics and Pressure on Natural Resources

The question of how population growth affects natural resources has been debated for a long time. Two main theoretical positions have emerged. Followers of the Malthusian approach suggest that unlimited population growth will lead to resource exploitation and degradation and consequently result in poverty and hunger. For instance, marine biologist Mora (2014) projects that

> 'in a planet with limited resources and a sensitive climate, with most of its natural resources being overexploited and its economic systems overstressed, meeting the additional demands of a growing human population without destroying the Earth and our social systems will be one of the greatest tests to humanity.' (ibid. n. pag.)

In contrast to this point of view, Boserup (1965) argues that population growth initiates innovations, transformations and labour input, and thus leads to an intensification of land use that could maintain populations and sustain natural resources. From a social-ecological systems perspective, the path of development projected by Malthusians would indicate a low level of resilience. However, Hummel (2012) stresses that 'it is not population dynamics *per se* that generates these [ecological] problems, but rather the inadequate capacity of supply systems to adapt to and cope with demographic changes' (ibid. 123). In other words, a system that is adaptive, in Boserup's terms, would be highly resilient.

The narrative of overpopulation has a long history in Thaba Nchu. Government officials have repeatedly stressed that there are too many people and insufficient resources.[326] I have shown in the previous chapters that

---

[326] Already a few years after the Natives Land Act of 1913 the problem of overcrowding was regularly discussed in the meeting protocols of the Native Reserve Boards; see for instance: SAB NTS 8142 2/341 part 1, Minutes of Meeting of the Sediba Native Reserve Board, 12 July 1918 and 17 January 1919; SAB NTS 8141 1/341 part 1, Minutes of Meeting of the Thaba Nchu Native Reserve Board, 18 January 1919.

this official concern was not unjustified, as enforced segregation accelerated the influx of people into the area. Until recently, some scholars regarded population growth as a major problem in the Thaba Nchu area, as a report by the Thaba Nchu Land Reform Research Project (TLRRP) compiled by scientists of the UFS suggests (De Vielliers and Booysen 1998). In a geographical section of the report, it was argued that the natural resources in the area, particularly soils and vegetation, were overused. The extent of natural resource degradation was not assessed through detailed research, but estimated based on previous studies: at one point, a dissertation written in the mid-1960s (Roberts 1966) was cited in order to summarise the 'present state of the vegetation and the soil erosion' (De Vielliers and Booysen 1998:41).

The authors identified rapid population growth caused by the high fertility rate of Thaba Nchu women as one main reason for resource degradation (ibid. 42). This argument was in line with a World Bank document titled *Reversing the Spiral: The Population, Agriculture and Environment Nexus in Sub-Saharan Africa* (Cleaver and Schreiber 1994) that was published a few years before the TLRRP report. Strikingly, the suggestions made by the geographers resemble those of the World Bank authors. According to them, the high fertility rate could only be changed by 'improving the legal, economic and social status of women to overcome the constraints imposed on them by their traditional roles' (De Vielliers and Booysen 1998:42) – yet the report offers no fertility rate at all. The only population figure that is included – in another part of the TLRRP report written by a social anthropologist and a geographer – is an estimate that projects a figure of 76,000 people for 1995 (Erasmus and Krige 1998:22). The data on which this estimate was based were not indicated, but presumably it used previous census data. Apart from that, apparently no data were collected by the TLRRP scientists that would allow a calculation of the fertility rate or population growth rate in general.

Besides a high fertility rate, the authors were worried about future immigration into the area:

> 'Thaba Nchu has a more particular population problem in the sense that immigration to the region may increase with the improvement of facilities like running water and electricity. Improving the quality of life of the present residents are [sic] commendable, but if this leads to immigration from other parts of the Free State, or even Africa it will be a hollow gesture which could

just aggravate the stress on the environment and the natural resources. Strict measures to prevent this is [sic] therefore essential.' (De Vielliers and Booysen 1998:43)

In their conclusion, the geographers suggest two possible futures for Thaba Nchu: either the area could turn into a successful and prosperous region or it could end as a 'disaster, a miniature Sahel' (ibid. 53). To prevent the latter development the authors recommended that residents had to drastically change their reproduction behaviour and that settlement of further immigrants should be prevented (ibid. executive summary).

With regard to the fertility rate, historical data on South African demography indicate that in fact the birth rate of black South Africans started to decrease since the late 1960s (Caldwell and Caldwell 1993; Simkins 2001).[327] Whereas in the 1940s black South Africans had a total fertility rate[328] between 6 and 7 children (Caldwell and Caldwell 1993:231), the rate had decreased to just above three by the early 2000s (Simkins 2001:130). This was mainly the result of government-initiated fertility control programmes that were expected to prevent an 'explosion' of the black population (Caldwell and Caldwell 1993). The national data put the geographer's conclusions in a different light. At least on the larger scale statistics indicate that the change of fertility patterns had already begun decades ago. This suggests that similar dynamics can also be found in Thaba Nchu. With regard to immigration of people seeking better infrastructure and services, as suggested by the above citation, it seems questionable that such movements would necessarily take place only in one direction, towards

[327] Because of its heterogeneous population, the South African 'fertility transition', as the fertility decline is commonly called, had no exact starting point. While the fertility of white South Africans began to decrease at the beginning of the twentieth century, the transition of childbearing figures of 'Asians' and 'Coloureds' started in the 1950s and 1960s respectively. Black South Africans were the last group whose fertility rate began to decrease (Simkins 2001:130).

[328] The total fertility rate (TFR) is 'the average number of children women would expect in a lifetime if reproductive levels remained constant at the levels recorded in a specific year' (Caldwell and Caldwell 1993:226). For the calculation of TRF, the age-specific fertility rate is needed, which reflects the average number of births of women of a certain age group (Pauli 2000:221). As the age-groups are normally determined at five-year intervals, e.g. women between 25 and 29 years old, the TFR is calculated by summing up all age-specific fertility rates per year of life and multiplying the result by 5.

Thaba Nchu: why would people from Thaba Nchu not also search for better living conditions somewhere else and emigrate?

Hence, the question arises as to whether the geographers' neo-Malthusian causal model – i.e. that a massive population growth due to high fertility and immigration leads to environmental degradation in Thaba Nchu – is not an oversimplification of the rather complex dynamics in the social-ecological system. In order to scrutinise their hypothesis and to allow a better understanding of recent and past population dynamics, in the following I will analyse (i) historical population data, mainly drawn from government censuses or other public sources; (ii) data from the village mapping that allow some conclusions on the migration patterns in the four villages; and (iii) data from the ethnographic census that help to more precisely calculate the fertility rate of women in the rural areas over time. Moreover, I will also pay attention to the changes in the availability of resources at the local level and the question of whether there is evidence of Boserup's ideal of adaptive behaviour in Thaba Nchu.

7.3.1 Population Growth in the Thaba Nchu District, 1904–2011

For the nineteenth century, no reliable population data are available. A census conducted in 1884 when Thaba Nchu was annexed to the Orange Free State counted 7,280 people; however, this figure included only adults.[329] Murray (1992) assumes that the total population including children was at least 12,000 people, perhaps even 15,000 (ibid. 12; 36). As these data are only rough estimates, I consider in the following only data from the twentieth century.

I subdivided the analysis of the historical population data into two time periods: from 1904 until 1946; and from 1970 until 2011.[330] The main reason is that the 'Thaba Nchu district', which was the spatial basis for the census, changed considerably during the twentieth century. The old Thaba Nchu district as it was defined in the late nineteenth century consisted of about 316,450 ha (Murray 1992:39). In 1970, the Thaba Nchu district had only 114,226 ha and hence was about one third of its former size. The area that had been incorporated into Bophuthatswana in 1977 was later extended with the addition of pieces of land (about 15,000 ha) bought by the SADT.

---

[329] VAB GS 1574.

[330] For the 1950s and 1960s I have no population data for the district.

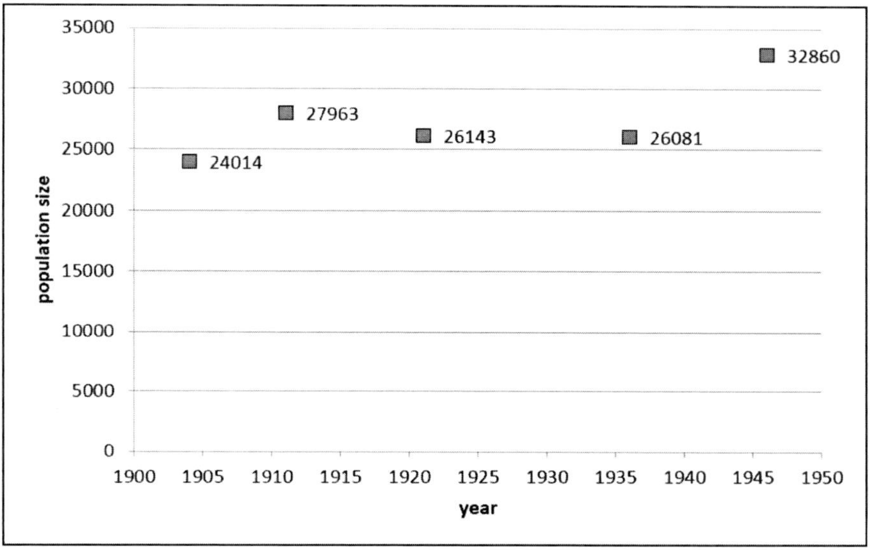

Figure 17 Population of Thaba Nchu district, 1904-1946[331]

Figure 17 shows the population data for the Thaba Nchu district between 1904 and 1946, as recorded in official sources. It is important to note that these numbers only refer to the *de facto* population; i.e. temporarily absent labour migrants were not considered for the enumeration. Over this period of four decades the mean annual growth rate was rather modest at about 0.7%. However, there were several peaks when the population numbers increased rapidly. Between 1904 and 1911 and between 1936 and 1946 the annual growth rate rose to 2.17% and 2.31% respectively. Both values could be the result of a natural population increase, i.e. due to births only (Lang 1998:387; Lang and Pauli 2002:27). However, especially for the latter time period this seems questionable, bearing in mind the broader political events. Certainly, the population increase in the post-1936 period was partly a result of an immigration of displaced people. Still, the data also

[331] Population numbers were drawn from the following sources: 1904: VAB ORC 40/28, Orange River Colony, Census, 17 April 1904; 1916: Union of South Africa (1916a: appendix 2, p. 28); 1921: SAB STK 3/21/230; 1936: SAB STK 4/21/428; and 1946: SAB STK 4/B22/428/46.

indicate that the Natives Land Act of 1913 did not cause an immediate massive immigration to the Thaba Nchu area.

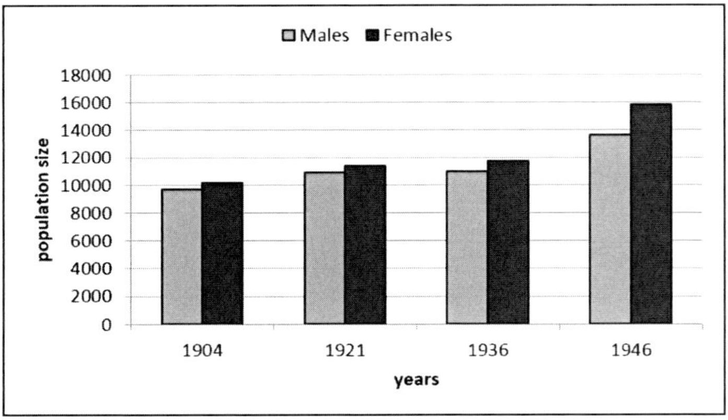

Figure 18 Gender distribution of black residents in the Thaba Nchu District, 1904-1946[332]

Despite a clear increase, the population growth in the first decades of the twentieth century was less extreme than in the second half of the century (see below). One reason for this was labour migration, which was a general trend in reserves throughout the country (Simkins 1981). Previous chapters showed that wage labour became an increasing necessity for the rural households in Thaba Nchu, with the effect that temporary emigration equalised the population influx to a certain extent. Figure 18 qualifies this statement and suggests that labour migration gained more importance in the 1930s and 1940s. I deduce this from the fact that only during that period did the female population strongly outnumber the male. This was a typical picture in the rural areas in South Africa throughout the century. For instance, in Venda there were half as many working-age males as females in 1970 (Lahiff 2000:74).

When comparing the census data with the archival sources it becomes obvious that labour migration was not the only aspect that prevented a strong population increase. Another factor was disease, which claimed many people's lives. For instance, in 1918 about 2.74% of the total Sediba reserve

[332] I have no gender-differentiated population data for 1911.

population (192 people) died because of an influenza epidemic.[333] Moreover, severe droughts resulted in the emigration of several families in search of better agricultural conditions elsewhere.[334] On one occasion in 1935, colonial officials also promoted the emigration of households to the Kuruman district as there was still space for 100 'surplus families' and about 2,200 head of cattle.[335]

Figure 19 illustrates the *de facto* population numbers since the 1970s. First, I want to look at the growth rate over the long term. Between 1970 and 2011, the mean annual growth rate was 1.68% and thus higher than the annual growth rate between 1904 and 1911. Between the 1970s and 2010s population growth also peaked at certain points. The highest increase occurred between 1970 and 1985, when Bophuthatswana was established and Thaba Nchu became part it. The homeland formation led to a population influx which resulted in an annual growth rate of 3%. But the significant increase does not mean that labour migration came to a halt. Figures on gender distribution in the Thaba Nchu area in 1970 indicate that male labour migration was still commonplace: 22,939 women and 18,285 men are reported to have been living in the whole region. After the homeland formation the annual growth rate decreased from 1.63% (1985-1996) to 0.56% (1996-2001) and finally to 0.32% (2001-2011). I argue that this is mainly the result of emigration in combination with the above mentioned 'fertility transition', which I will discuss below.

---

[333] SAB NTS 8142 2/341 part 1, Minutes of Meetings of the Sediba Native Reserve Board, 22 November 1918; Another instance that caused many deaths was a typhus epidemic in 1921, see SAB NTS 8142 2/341 part 1, Minutes of Meetings of the Sediba Native Reserve Board, 20 May 1921.

[334] For instance, in 1916 during a period of drought several households of the Sediba 'native reserve' left for Bechuanaland, SAB NTS 8142 2/341 part 1, Minutes of Meetings of the Sediba Native Reserve Board, 17 November 1916.

[335] SAB NTS 8142 2/341 part 2, Minutes of Meetings of the Sediba Native Reserve Board, 25 October 1935.

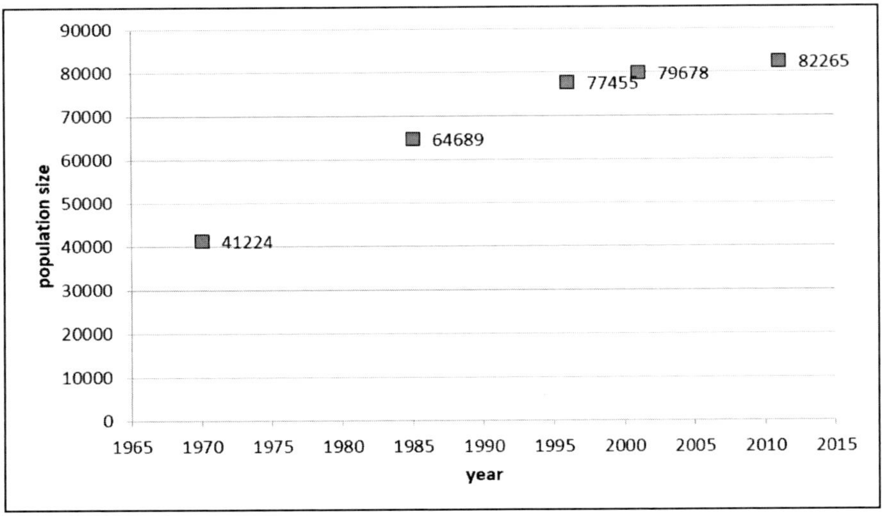

Figure 19 Population of Thaba Nchu, 1970-2011[336]

A look at the absolute population numbers indicates that the pressure on the land increased during the twentieth century. Even more striking are the results when considering the population density. As mentioned above, the old Thaba Nchu district was three times larger than the area that was part of Bophuthatswana, and this territorial decrease coincided with a population increase. Thus, population density rose from 7.6 persons per km² in 1904, to 38.4 persons per km² in 1970, and by 2011 was about 63.7 persons per km². Compared to the homeland Qwaqwa, which was also situated in the Orange Free State, population numbers in Thaba Nchu were still moderate. In 1970 the former Witzieshoek 'native reserve' had a population density of 52 people per km² (Halbach 1988:68).

[336] Population numbers were drawn from the following sources: 1970: SAB SON 1606 D12/2/4/2/4/3 vol1, Ondersoek na die ekonomiese en werkgellentheidsbasis van Thaba Nchu, January 1976; 1985: NWA OPD 9/6/3 (34) part 2, Thaba Nchu Budgeting and Development Plan: Consolidated Report for Phases 1 and 2, December 1987, 1996: Centre for Development Support (CDS) (2003); 2001: Mangaung Local Municipality (2007); 2011: Frith (2013b).

7.3.2 Fertility Dynamics Revised: Births in the Research Sites

In order to analyse the fertility dynamics, the childbirth history of 74 women between the ages of 16[337] and 86 was recorded in the ethnographic census. In total, these women gave birth to 164 children born alive. The recording of the mothers' and children's birth dates in my research area was not as problematic as it has been in other fields (see for instance Howell 1979:24; Fischer 1997:64; Bollig 2006:68). Exact ages were in most cases known. In the few instances where older women were not sure about the dates of birth of some of their children, I tried to estimate the age by gaining more information on the child's relative age, i.e. whether the child was older or younger than a sibling whose birth date was known. In about one-fifth of births (21%) the father was indicated as 'unknown', a typical answer when the mother and father of the child had never been married and the father had abandoned her and the child. Not all of the remaining 79% of births for which a father was recorded happened within a marriage. However, the father was normally the boyfriend of the mother. In some of these cases, the child grew up in the father's family.

Of the 74 women, 25 (33.8%) were younger than 29 years of age, 20 women (27%) were between 30 and 44 years old, and 29 women (39.2%) were 45 years of age or older, and thus had already completed their reproductive phase. First, I want to analyse the fertility of women of the latter group only. In total, they had given birth to 115 children that were born alive. The completed fertility rate[338] for these women aged 45 years or older was 3.97 children; with a minimum of one child and a maximum of eight children. Thus, no case of 'complete sterility' (Howell 1979:125) was found in the sample. On average, the first child was born when the women were 19.8 years old; about 44.8% had their first birth between the ages of 15 and 19, and 51.7% between 20 and 24 years of age. One woman had a comparatively late first birth when she was 29 years old. Women gave birth

[337] In ethno-demographic studies, the age of 15 is typically regarded as the beginning of the reproductive phase of a woman (Pauli 2000:220). However, in my survey no woman was 15 years old at the time of data collection (reference date is 5 June 2011, i.e. last day of census collection).
[338] The completed fertility rate is the average number of children of women whose reproductive phase is already over (Lang and Pauli 2002:20).

to children up to their early forties. The average interval between births was 3.8 years.[339]

The analysis of the fertility of women over 44 years of age provides only retrospective insights on reproductive behaviour, but does not say anything about recent developments. To do that, it is necessary to include younger women in the demographic analysis too. When looking at all women aged 15 years and older, the results are slightly different. The average age at the first birth is now 20.6 years. For women between 15 and 44 years old, the mean age at first birth rises to 21.3 years. Thus, women today tend to have their first child later than the previous generations. The total fertility rate of the 74 women is 3.3. This rate indicates how many children a woman would have at the end of her reproductive phase – based on the assumption she would 'behave' according to the calculated age-specific fertility rate. As the total fertility rate is lower than the completed fertility rate, data indicate that women tend to have fewer children today. However, compared to the data from the South African community census, the sampled women in Thaba Nchu have a higher total fertility rate than the national average. This was estimated at 2.5 births per women for 2007 (Statistics South Africa 2007:12).

As a last step, I will shed light on the question of how fertility changed over time. This could be done by comparing total fertility rates for certain decades, e.g. the total fertility rate of women in the 1950s compared with the total fertility rate in the 1960s and so forth. In order to do that one would have to base the calculations on the age-specific fertility for birth cohorts, as done by Pauli (Pauli 2000:239). Since I have only a relatively small sample of women – Pauli's included more than 300 – this method seems inappropriate for my case study. Instead, I decided to use the general fertility rate, which considers all women in their fertile phase (Newell 1994:38–39). For each year between 1961 and 2010, the number of children born to women between 15 and 44 years is divided by the number of women. The result is then multiplied by 1,000 (ibid). I summarised the annual fertility rates in ten-year intervals and calculated the average value. Births before 1961 were not included, as the sample size was too small and in several years no children were born at all.

---

[339] In the sample three pairs of twin births were found. I considered each of them as one birth for the calculation of the interval.

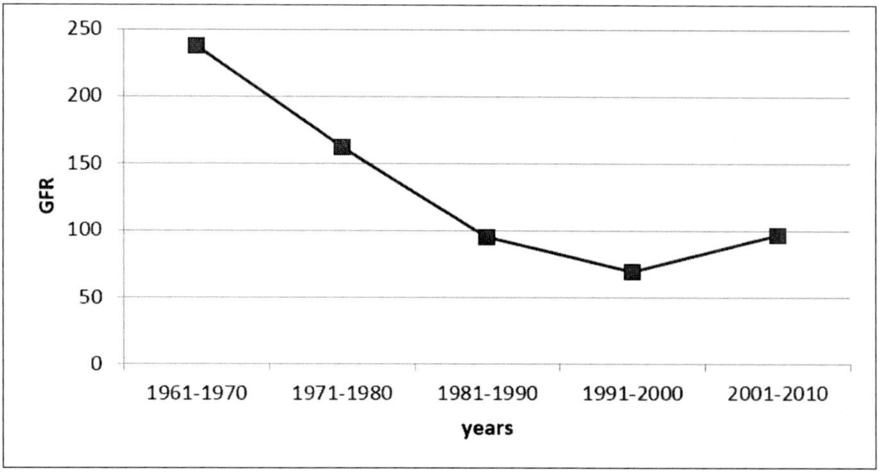

Figure 20 General fertility rate in the four research sites, 1961-2010

The results for the general fertility rate are presented in Figure 20. In correspondence with the general national trend that was observed by statisticians, the fertility rate declines in the 1960s and over the following decades. Reasons for this were the above-mentioned fertility control programmes, as well as the changes in marriage patterns depicted in the following chapter. In the 1980s and 1990s, the rate of decrease slowed down. Interestingly, the fertility rate slightly increased in the 2000s again. This rise could be connected to the introduction of the child support grant in 1998 (Chapter 8.4.3) which might have motivated women to have more children. However, a larger sample size is needed to prove whether the fertility rate is indeed rising.

7.3.3 Village Extension vs Declining Populations

Aerial photographs indicate that after the incorporation of Thaba Nchu into Bophuthatswana, the rural villages did not grow significantly in size. Comparing images from 1975 with those from 1988 or 1992 hardly any transformations in the village structures are evident. Recent aerial photographs of my research sites, however, show clear changes. Between 1992 and 2004 all four villages expanded spatially. Several new streets, residential sites and houses emerged on the fringes of the villages. At first glance, one could assume that the population has increased due to immigration. However, the village assessment as well as population figures indicate that this assumption is wrong. Firstly, it turns out that in several cases these newly established households were in fact those of residents who had been born in the village. After marriage or – in the case of single mothers – after the birth of a child, people founded their own, new household in another part of the village. Secondly, the villages experienced a period of emigration during post-apartheid times, and these emigrating families clearly outnumbered the few immigrants that settled in the villages over the past 20 years. Although the Thaba Nchu area as a whole experienced a constant population increase, the opposite is observable in the rural communities, yet this was not always the case, as the historical population numbers of Sediba village will illustrate.

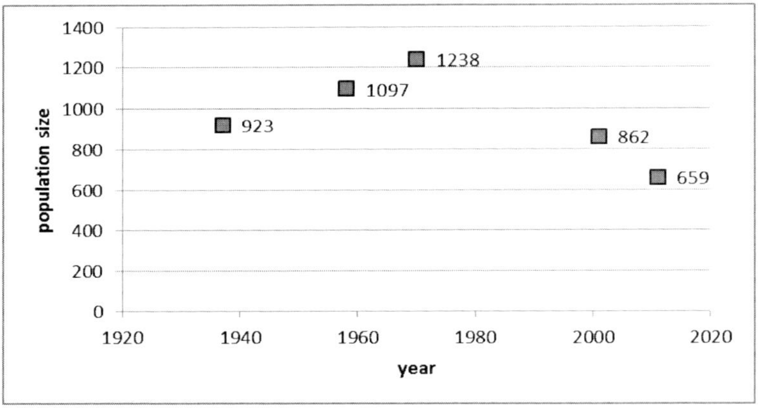

Figure 21 De facto population numbers Sediba village, 1937-2011[340]

Figure 21 illustrates the growth and decline of the *de facto* population of Sediba village between 1937 and 2011. The mean annual growth rate between 1937 and 1970 was about 0.9%. Between 1970 and 2011 the value decreased to -1.5% per year. This reverse trend was mainly an effect of

[340] Population numbers were drawn from the following sources: 1937: SAB NTS 9502 138/400/5 part 1, Memo of the Director of Native Agriculture, 3 April 1937; 1958: SAB BAO 5761 H128/1666/197, part 1, Herwinningsverslag Sediba Gebied, Thaba Nchu, 22 May 1958; 1970: Bevolkingsensus Thaba Nchu, (day & month unknown) 1970; 2001: Frith (2013a); 2011: Frith (2013b).

emigration, though I assume that the major outflow of people occurred after 1994. When the residents of former Bophuthatswana had the opportunity to move freely throughout the whole country many households left the villages temporarily or even for good. In 2011, I counted 254 residential plots on which a dwelling had been built. Only 194 of them were permanently occupied. Some of the owners of the empty houses were either dead (5 cases) or living in another household in Sediba (13 cases). In 11 cases, it was unclear to whom the dwelling belonged. However, the majority of the empty houses (31 cases) were owned by people that lived and worked in another town or city. The main destinations of these emigrants were Bloemfontein or Thaba Nchu, yet they were not visiting on a regular basis. Other families had left the Free State and were living in other provinces, e.g. Gauteng, North West (Rustenburg, Orkney), Northern Cape (Kuruman) and Western Cape (Cape Town).

A comparison of census data for 2001 and 2011 for the other research sites shows that a decrease in the population was a common trend in the villages' recent history (Table 9). Interestingly, the old 'native reserve' villages of Talla and Sediba experienced a more marked population decrease than Middeldeel and Rooifontein. In the case of the latter, the less severe population decline might be the result of the village's proximity to Thaba Nchu, which makes emigration less necessary. The question of why the population of Middeldeel decreased the least out of all these sites is difficult to answer. During the village mapping, I found that of 126 dwellings only 93 were occupied. In the majority of cases the owners lived and worked in Thaba Nchu, Bloemfontein or another town. Thus, it seems incorrect to suggest that outmigration played a minor role in Middeldeel. Rather I suggest that the main wave of emigration had already taken place by the turn of the century. Whatever the reasons for the faster or slower rates of population decrease might be, data indicate that the population pressure in the studied communal villages increased at least until the 1970s, and decreased thereafter. This is in contrast to the Thaba Nchu area as a whole.

Table 9 Population of the four research sites, 2001-2011[341]

Village	Population 2001	Population 2011	Decrease '01-'11 (%)
Middeldeel	320	302	- 5.6%
Sediba	862	659	-23.5%
Talla	553	434	-21%
Rooifontein	1064	977	-8.2%

Finally, in order to judge the pressure on the natural resource base, one should not only consider population size, but also livestock numbers. I have shown in Chapter 4.4.5 hat in the first livestock culls in the 1940s, the number of sheep was more significantly reduced than that of cattle. This changed in the 1950s, when cattle were basically the only animals left for culling. Unfortunately, the archival documents did not contain more figures on livestock numbers for the 1960s, 1970s, 1980s and 1990s. Censuses that had been conducted by Bophuthatswana officials could not be traced in the North West provincial archives as the documents of the former Department of Agriculture and Agricor had not yet been catalogued when I conducted the archival research for this dissertation. The most recent data on livestock numbers in Sediba village are from 2011 and were collected by my colleague Sebastian Rasch from the Department of Economics. According to his survey, 547 cattle (in 1959: 785) and 501 sheep (in 1959: 262) were grazing in Sediba in 2011. Whereas recent cattle numbers are considerably lower than those before and during the 'betterment', the sizes of sheep herds have again increased significantly. It is important to note, however, that rangelands are also bigger now, since former fields were abandoned and are used as grazing land. To sum up, grazing pressure today is considerably lower than before and during 'betterment'.

7.3.4 Thaba Nchu: A Malthusian or Boserupian Case?

In the beginning of Chapter 7.3 I asked whether the neo-Malthusian 'threat' that the geographers depicted for Thaba Nchu in 1995 was realistic. The answer has to be 'no'. First, the analysis of ethnographic data on reproduction behaviour indicates that the general fertility rate declined throughout the second half of the twentieth century. Therefore, to suggest that 'drastic

---

[341] Data are based on the national census data and provided by Frith (2013b) and Frith (2013a).

behavioural changes' (De Vielliers and Booysen 1998:53) in terms of reproduction are necessary in order to prevent resource degradation is misleading. Of course, the Thaba Nchu region had and still has a high population pressure, and the population has shown rapid increases during certain periods. However, the analysis illustrates that the population pressure is a problem that is mainly rooted in apartheid politics, and not a result of the high fertility of Thaba Nchu women.[342] Sharp and rapid population increases in the Thaba Nchu area followed the passage of stricter laws regulating the residence of black people. Also with regard to the 'threat' of post-apartheid immigration, the geographers' assumptions were wrong. The rural areas, in fact, have experienced a clear population decrease in the last two decades. Additionally, the grazing pressure on the rangelands has reduced. Thus, De Vielliers and Booysen's prognosis must be rejected.

Another question that emerges is whether the Thaba Nchu case is a good example of an intensification of land use, as suggested by Boserup. In other words: was the system resilient enough to deal with population increase? The archival sources indicate that after the formation of 'native reserves', land use was intensified through an increase in livestock numbers and the size of cultivated fields. Government regulation of land use prevented a further intensification by determining an economic maximum that households were allowed to own in terms of agricultural capital. Officially, an increase in agricultural efficiency was encouraged, which is reflected for instance in the establishment of irrigation schemes. But the limits on intensification that were enforced by the state did prohibit the development of a successful class of small-scale farmers. Households responded with labour migration and income diversification and today depend only to a small degree on agricultural incomes. Thus, over the long term, the region is characterised by an extensification of land use rather than intensification, and therefore does not reflect the Boserupian ideal. However, to argue that the system was not 'resilient' enough to deal with the population increase would be misleading. Rather, the conclusion should be that the hindrance

---

[342] A comparison with other cases shows that the total fertility rate of the sampled women in Thaba Nchu, at 3.3 children per woman, is not especially high. The Wampar in Papua New Guinea have a considerably higher rate, with 7.3 children per woman (Schulze 1997b:146). In Pueblo Nuevo, Mexico, the total fertility rate is about 10.3 (Pauli 2000:222).

of intensification decreased the system's resilience, and thus the conditions described in Boserup's model were never to be fulfilled.

7.4 Preliminary Summary

The transformative capacities of rural communities, i.e. the ability to access assets from a broader wider socio-political arena, have increased with the end of apartheid. This is not least reflected in the successful claim for the restitution of land rights in the former Sediba reserve. In the following chapter, I will illustrate that access to state transfers is another measure that can be understood as a state's attempt to increase people's social resilience. With regard to land reform, the figures for the provincial level suggest that in the Free State, the programme is a reform on paper rather than in practice. The process is slow: only a few individuals benefit from the reform, some of whom could not carry the financial burden in the long term.

The state's role with regard to environmental governance changed after 1994. Land use is still regulated to a certain extent, but not violently as in the past. Instead, projects are implemented on a voluntary basis. In terms of cultivation, the state focusses on food security, which means that all people should be enabled to have 'economic and physical access to food' (Esterhuyse 2012 95). However, food security projects in Thaba Nchu have up to now experienced only minimal success, have had hardly any impact on rural livelihoods and, paradoxically, excluded the poorest members of communities. A closer look at these projects suggests that officials assume agricultural development can only be achieved by expensive technological inputs. The management practices of commercial farmers are the benchmark for land use planning in communal areas. In other words, the way of implementing agricultural development has changed in recent years, yet the underlying ideology is still the same as in the past.

The situation is similar with regard to policies governing communal rangelands, which focus on low stocking rates, fencing of rangelands and rotational grazing management (Vetter 2013). Despite growing evidence that South African rangelands are complex, non-linear systems, conventional approaches still dominate the policies made in the agricultural departments (Allsopp 2013; Salomon et al. 2013). In the studied communities of Thaba Nchu, the state was not actively involved in rangeland management in 2011. In Middeldeel, the sheep and wool project was introduced with a re-erection of grazing camps and watering points in order to facilitate rotational grazing. However, the management plan did not work in the long term: on the one hand, this was due to lack of control of the infrastructure by rangers, and on the other hand because of informal arrangements in which some actors of the community claimed exclusive access to certain grazing camps. Moreover, people wanted to watch their animals during the day, which was only possible when livestock grazed in areas close to the community; this was a reason also identified in other regions as a hindrance to rotational grazing (see for instance Salomon et al. 2013). As in the past, the bureaucratic institutions implemented by officials are not in accordance with the institutional arrangements and social practices on the local level.

The dominance of conventional approaches and explanations is connected to the persistence of certain narratives. The TLRRP report compiled in order to inform government departments about the recent land situation in Thaba Nchu depicted the whole area as degraded and overpopulated, despite the fact that no data were collected to support such judgements. Both narratives, i.e. 'degradation' and 'overpopulation', were not novel. However, the rationale that high fertility and immigration threaten natural resources was not stated explicitly beforehand and was clearly inspired by debates that emerged in the context of the World Bank. Thus, the argument of a powerful actor of the international development arena was translated into the Thaba Nchu context. By doing so, rural communities, households and women in particular were not only depicted as incapable of dealing with natural resources, but were also diagnosed as suffering from a social 'organisational dysfunction' (Fairhead and Leach 1996:273).

8. MAKING A LIVING IN A POST-AGRARIAN SOCIETY

In the previous chapters, I have illustrated that agricultural production has continually decreased in rural Thaba Nchu since the mid-twentieth century. While parts of communities' subsistence requirements were still met by cultivation and animal husbandry, more and more households were pressed into labour migration in order to earn additional income. When Thaba Nchu became part of Bophuthatswana, agro-pastoral production had become a mere supplementary income strategy for many households. Communities were described as non-agricultural, yet a proportion of arable lands was still used for crop production at that time. With the end of apartheid, cultivation of fields ceased and villages experienced a significant population decline due to families migrating from rural to urban areas. In short, the Thaba Nchu area underwent a process of 'de-agrarianisation' (Bryceson 1997a) throughout the past 70 years.

Against this background, I want to elucidate in this chapter how the social structures of rural communities and patterns of labour migration have changed since the end of apartheid. Moreover, I will explore how the people of Thaba Nchu make a living in a post-agrarian setting. The main focus of the analysis will be on the household (Sets. *lelapa*),[343] which is the central unit of consumption and co-residence. By analysing the livelihood strategies of a stratified sample of households, I will illustrate how people of different wealth strata try to improve their livelihood security and increase their resilience. In this regard, it is important to ask what role land use plays in household economies. Finally, the analysis will focus on the question of how intra-communal cooperation and support networks have changed since people were resettled and agriculture lost its importance.

8.1 Demographic Patterns in the Research Sites

Rural households have changed considerably in post-apartheid South Africa (Madhavan and Schatz 2007). Democratisation, freedom of movement, unemployment and HIV/AIDS are only some of the variables that affect household structures (ibid. 86) which in turn have impacts on the social organisation of rural communities. In Thaba Nchu too, villages and

---

[343] The Setswana word *lelapa* has several connotations and refers not only to the household, but also to the homestead itself and the wider family group (see also Steyn 2012:123).

households have undergone profound transformations over the past two decades, as indicated by the ethnographic census results. The ages of the people recorded ranged from 8 months to 86 years, with an average of 30.7 years. Typically, in South African population statistics all people who are 15 years of age and older are regarded as adults (Lahiff 2000:148). Thus, they are counted as people of 'working age' (15 to 59 years). However, in my ethnographic census sample all children younger than 20 were either not yet attending or still attending primary, secondary or tertiary[344] education. Hence, I decided to regard people who were 20 years of age and older at the time of data collection as constituting the working-age population.

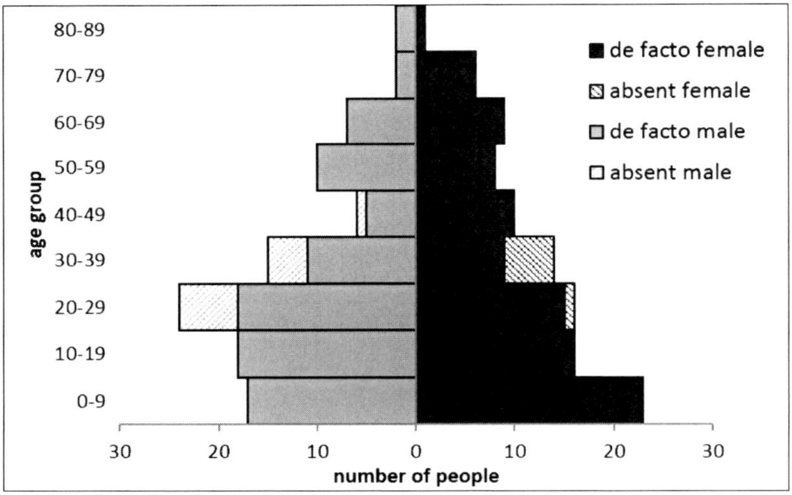

Figure 22 Age pyramid of the de facto and de jure population (n=204)

I first want to look at the *de facto* population of the research sites – i.e. people who were present in the village at time of data collection – and compare it with the municipality level. According to the ethnographic census 39.6% of the sampled population were children; 46% were between 20 and 59 years old and 14.4% of the sample were people aged 60 years or older. About 51.9% of the *de facto* population was female. Thus, women

[344] Only one individual was a college student though; all the other children had not yet finished secondary school.

slightly outnumbered men (48.1%). When looking at the Mangaung municipality level, i.e. including other parts of Thaba Nchu as well as Botshabelo and Bloemfontein in the statistics, the picture is different (Statistics South Africa 2013). The national census[345] conducted in 2011 showed that children and youth under 20 years of age make up 36.3% of the Mangaung population. About 55% of the population is between 20 and 59 years old and only 8% are over 60 years of age. Thus, compared to the municipality level the numbers of children and particularly old-aged people are considerably higher in rural Thaba Nchu.

Figure 22 illustrates the age pyramid for the ethnographic census sample. Strikingly, the figure does not display the formerly typical pyramid form, which indicates that several factors shaped the population structures. Children, except for females between 0 and 9 years of age, do not clearly outnumber the adult population. A major reason for this trend is presumably the above-mentioned decline in the fertility rate (Chapter 7.3.2). Moreover, data from other regions indicate child mortality increased since the early 1990s (Madhavan and Schatz 2007:88). However, it is difficult to estimate whether this is also true for the sample as only very few cases of child mortality were reported. It is also difficult to explain why, in children younger than 10 years of age, girls outnumber boys; but it could be related to the fact that infant mortality is higher in boys than in girls (Pongou 2013). The working-age population, particularly those between 30 and 50 years old, are under-represented in the communities. One reason for this is labour migration, which will be addressed in more detail below. It is also likely to be a consequence of higher mortality rates amongst adults caused by HIV/AIDS (see also Madhavan and Schatz 2007:88). The older generation is growing in number, with a higher proportion of women. This might indicate that some male labour migrants never came back to their homes after they retired. Additionally, women have a longer life expectancy (Leibbrandt et al. 2010:10).

8.2 Changing Patterns of Labour Migration

As a next step, I will differentiate between people who *de facto* live in the village and those who have temporarily left for work (Figure 22). When looking at the former group, the proportion of women is higher than that

---

[345] The national census also refers to the *de facto* population (Statistics South Africa 2012b:1).

of men, particularly in the group of 40- to 49-year-olds. Yet considering the whole working-age population, women are only slightly overrepresented (50.5%). The image is somewhat different for the absent population. While about 12.5% of all working-age women live outside the household, 20% of the men between 20 and 59 years of age were absent for longer periods of time. Migration plays a bigger role for younger people. 25% of males aged between 20 and 29 years of age, and 26.6% between 30 and 39 years of age, had left their homes. In the latter age group, the proportion of women who had migrated is even higher, at 35.7%.

In most cases, migrants left their home households in order to work in an urban area. These migrants supported their home households in cash or in kind (Chapter 8.4). Furthermore, of the 17 absent people, two were studying in Pretoria, three were unemployed but looking for work, and in one case, the household head did not know whether the migrant had a job or not. Additionally, it may happen that people leave their homes temporarily in order to take care of other relatives. The length of the period of absence from the home household varies, and in some cases a migrant may never return. Most of the 17 migrants visited their home households regularly at weekends, months' ends or during holidays.

When comparing the recent gender distribution with historical population figures, it appears that the numerical predominance of women in the homelands – typical during the twentieth century – has declined. One reason for this is certainly the rising unemployment figures in South Africa since 1994 (Banerjee et al. 2008). High unemployment rates make it more difficult for men to find jobs in town, yet at the national scale the proportion of men migrating from rural areas has not significantly decreased. What has changed, however, is the proportion of women who are leaving rural areas in search of work in town (Posel 2003:8). These changes in national migration patterns are also visible in the villages in Thaba Nchu. What are the causes of this higher proportion of female migrants? Posel (2003) suggests that changing marital behaviour might be a motivation for contemporary women to migrate more often than in the past (ibid. 15). Indeed, at the national level, marriage rates are declining (Seekings 2014:79). The census questionnaire did not contain a question on marriage dates. However, since I recorded the marital status and birth date of each person, I can draw some conclusions about marriage behaviour today and compare this with older data from the mid-1960s.

According to Schüler's (1965) household survey data, the majority of the married women had their weddings at the age of 22 or 23. More than 82% of the wives had been married before they reached 30 years of age. For the male household heads, Schüler stated that about 56% of them had their first marriage between the ages of 20 and 29 years (ibid. 153–54). Today, people tend to marry considerably later. Of the 19 women aged between 20 and 30 years, 13 never had been married (68.4%). The remaining women had married in a 'white wedding' in a church, or had a 'traditional wedding'.[346] With two exceptions, all married women were at least in their late twenties. Schaumburg (2013) has shown for women living in Pretoria's township, Mamelodi, that they perceive marriage as one of several options. Thaba Nchu women's reasons for deciding against an early marriage are manifold; here I only want to give two brief examples. One young woman remarked that she wanted to sustain her personal freedom, which would not be possible within a marriage as she would have to cook for and serve her husband. 'My family are my two daughters; that's enough for me', she explained (Kedikilwe Olifant, 24 March 2011). Another woman in her twenties stressed that men were not reliable and serious about relationships, which was the main reason why she preferred to remain unmarried. In the future, she hoped to find a lucrative job that would enable her to buy a house and a herd of cattle for herself and her two daughters (Jeanette Rancho, 19 April 2011). For both women, marriage was – at least at the time of data collection – not a preferred option. Amongst young men, too, marriage is less frequent today than in the mid-1960s. In 2011, of 26 men aged between 20 and 30 years, 92.3% had never been married. The few men that I asked about their reasons for not being married offered very similar responses to those of the two women above. In addition, the inability to raise enough money or livestock to pay *bogadi* (bridewealth) was also given as a reason.

How do these transformations in marriage behaviour influence labour migration? Based on national data, Posel (2003) has illustrated that unmarried women were significantly more likely to migrate than their married counterparts (ibid. 13). This could also be a possible explanation for the recent

---

[346] Two types of weddings can be identified: a 'civil wedding' that is held in a church – and often called a 'white wedding' – or at a public office; and a 'traditional wedding' that follows customary rituals. The latter must be registered at the Department of Home Affairs within three months after the wedding date in order to be officially recognised by the state.

changes in labour migration patterns in Thaba Nchu. However, I argue here that education also plays a critical role. The expectations of the labour market have changed in the last four decades and for less-skilled people it had become increasingly difficult to find jobs (Banerjee et al. 2008:724). When looking at the levels of education of the working-age population in the sampled villages, it becomes clear that women in the ethnographic census are on average better educated than men (Figure 23). Only 6.7% of the women never went to school or did so only for a few years. In contrast, 20% of the men aged between 20 and 59 years never finished primary school or did not even start it. Almost 50% of the women aged between 20 and 59 years had at least started secondary school, though never finished it. The differences at the highest level of education – i.e. matriculation, college, technikon or university degree – are less clear. Apart from their likely impact on labour migration patterns, higher levels of education may also be implicated in the decline in marriages (Francis 2000:92).

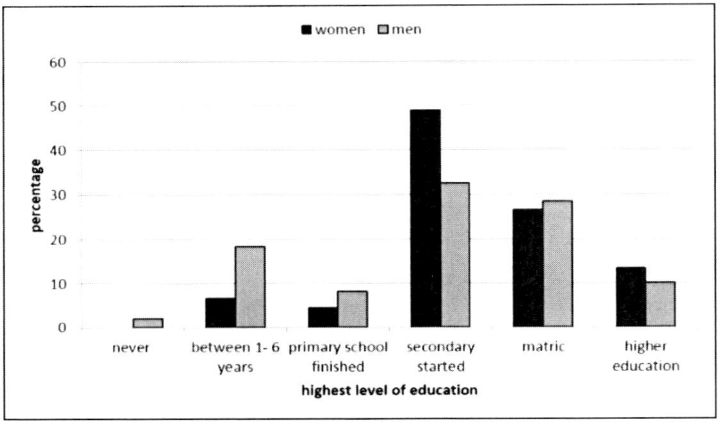

Figure 23 Level of school education of women (n=45) and men (n=49) aged 20-59 years[347]

To sum up, children and elderly people are in the majority (54%) in the villages. In mid-1980s Bophuthatswana, 18.3% of the rural population in Thaba Nchu was over 50 years of age. Today 24% of the rural residents are older than 50; thus the 'ageing' of the villages has increased over the last three decades. Migration of people between 20 and 59 years of age is

---

[347] People who were still attending school (4 men) or college (2 men; 3 women) were excluded from the diagram.

still very common in the area. However, migration is more balanced in terms of gender nowadays than in the middle of the twentieth century. Posel (2003) argues that the changes in the proportion of female migrants happened in the 1990s. Oral histories of Thaba Nchu women suggest that female migration is not something entirely new, but that it also played a role for households during Bophuthatswana times (Chapter 6.3.3). The increased proportion of old people in the rural areas is not only connected with the migration of the younger, but likely also a result of HIV/AIDS. However, as it is taboo to ask whether a person's death was caused by AIDS, it is very difficult to quantify to what extent HIV/AIDS-related deaths affect the age-structures of the villages.

8.3 Transforming Household Structures

The first matter that must be considered when approaching household economies is the specification of the unit of analysis. For decades, anthropological studies have dealt with the question of how a household should be defined (for instance Netting et al. 1984a; Rössler 1997; Berzborn 2006; Greiner 2008). The search for a useful definition seemed to be difficult because households show a high degree of cross-cultural variation and significant transformations over time. The classification and analysis of those variations and processes often remained fuzzy, as anthropologists had the tendency to 'confuse or to fail to differentiate between what households look like as structural units and what they do and how they function' (Wilk and Netting 1984:2). To overcome this methodological and theoretical problem, it is necessary to distinguish between three dimensions: (i) morphology, i.e. the household's structural composition and size; (ii) activity, i.e. things that households do; and (iii) the cultural rules, obligations and norms that organise and give meaning to the household unit (Netting et al. 1984b:xxix). Such a methodological differentiation is particularly necessary in those cases 'where households are changing or adapting their functions to new economic, political, or ecological circumstances' (ibid.).

The functions or activities of households involve production, distribution and consumption; transmission of resource rights; reproduction; and co-residence (Wilk and Netting 1984). All these activities *may* be found in a household unit, yet this is *not always* the case. Co-residence was a condition observed in many Thaba Nchu households, but not necessarily everywhere. In some extended-family households, male relatives of the head of household, e.g. unmarried sons or brothers, ate every day in the main homestead, but slept in their own huts, which were not necessarily adjacent. In these cases, co-residence was still observable to a certain degree, as the men visited daily. The situation was more complicated when one household member was absent most of the time because she or he worked or studied – and lived – somewhere else. In these cases, I asked the informants why they included the absentees when counting the members of their household. Commonly, the answer was that those absent members were unmarried children who visited the home regularly – at least several times per year – and who spent parts of their income on the household's budget, or – in the case of pupils or students – were still supported financially. Based on the observations I made, the following working definition of a household can be made for the research sites:

A household is a social unit where people are related and living together in one homestead, though individual members sometimes have a separate sleeping place or are temporarily absent in order to work or study. People that belong to one household eat and consume together, distribute, produce and reproduce, and have access to certain resources from which they can exclude non-household members, e.g. the garden, kraal, or a water tap in the yard.

8.3.1 Household Size and Development Cycle

At the time of data collection, the 45 sampled households in Sediba, Middeldeel, Talla and Rooifontein had 204 members in total. The mean *de jure* household size was 4.5 members (SD 2.5), with a minimum of one and a maximum of 13 persons (Figure 24). When considering only the *de facto* population, the average household size was slightly smaller, with 4.1 members (SD 2.2). In other rural areas of South Africa, households have a mean size of more than 6 members (see for instance Shackleton et al. 2002:132; D'Haese and van Huylenbroeck 2005:103; Berzborn 2006:144; Madhavan and Schatz 2007:89). Thus, for a former homeland, the figures in Thaba Nchu are comparatively small.

Historically, Thaba Nchu households were considerably larger. Schüler (1965) reported a mean household size of 5.9 members in the 1960s (ibid. 142); in the mid-1980s, the average size was 6.2 people (Chapter 6.3.3). Nationwide, a decline of household sizes since 1994 can be observed and cannot be explained by demographic factors alone, i.e. declining fertility and increasing mortality (Wittenberg and Collinson 2008:1–2). Young people's changing attitude toward marriage gave rise to the formation of smaller households of single men or single mothers with their children. The relaxation of settlement control has enabled these people to find a plot for a new home: aerial photographs indicate that new households emerged at the fringes of villages between 1992 and 2004 (Chapter 7.3.3). Moreover, families who emigrated from rural communities left empty plots behind which could be allocated to others.

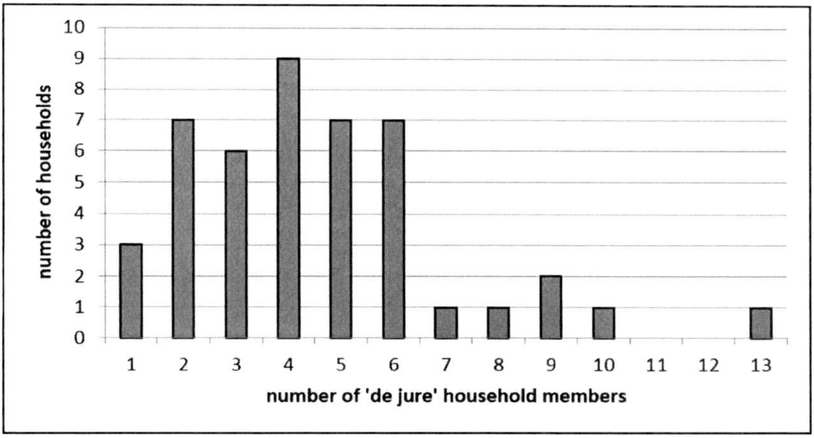

Figure 24 Frequency distribution of household size (n=45)

A new attitude toward marriage is not the only cultural transformation; other norms connected to the development cycle of households and inheritance are contested, too. Today, young men and women themselves decide whom – or whether – they want to marry. Arranged marriages were only reported by some of the older women. Commonly, marriages are preceded and followed by repeated rituals and negotiations between the bridal couple's patrilineal families. In the meetings that take place before the wedding, the involved parties discuss *bogadi,* which is normally also paid in 'civil marriages' and depends on the woman's age and whether she already has children. According to several informants, a typical amount for *bogadi* would be between five and 10 cattle or the monetary equivalent. Discussions arise around the question of whether a higher bridewealth has to be paid for women who have graduated from college or university. Whereas younger people stated that a higher price was justified, older informants disagreed.

In the premarital meetings, the families also agree on the details of the prospective wedding ceremony. Similarly, to recent trends in rural Namibia (Pauli 2009), 'white weddings' are often connected to high costs for food, drinks and other goods. Some poorer informants stated they would only marry 'traditionally', as this is cheaper. In the ideal case, postnuptial residence is virilocal. After the wedding, the new wife moves into the house of her husband's parents, where she must stay all day for a certain time period; this is determined by his patrilineal family and ranges from a couple of days to a couple of months.[348] Both the groom's and bride's families organise ancestral feasts and invite one another in order to strengthen the relationship to their affines.

Before the resettlement, a young man would have built a new hut close to his father's home and then moved in with his wife (Schüler 1965:182). The restrictive regulation of the location of the residential sites made this procedure more difficult, as normally no plot can be obtained next to the father's homestead. Thus, the married couple would move to any other plot situated in the same village. Yet young couples now do not necessarily follow the virilocal residence patterns and instead strive for independence from parental control, leaving the rural area to establish their own household in Thaba Nchu or somewhere else. Moreover, cohabitation is found in villages, though this practice is not approved by everyone.

In the past, the eldest son of the household was deemed to be the principal heir of the father's livestock (Schapera 1994:230–31). In contrast, the youngest son – who typically remained in his parent's house after his own wedding and took care of the parents – would have a claim to the fields of the deceased mother (ibid.: 235). Schüler (1965) observed that the practice of choosing the eldest son as the main heir was only adopted by some families in the 1960s. Moreover, some fathers made provision for their daughters by giving them cattle before they passed away (ibid. 288). My informants were at odds with each other on the question of who would be heir to the family's properties. A younger woman said that traditionally the firstborn son would inherit everything, i.e. the house and the kraal. Sometimes the youngest son takes care of his parents, and would then, perhaps, inherit their house. An older informant stated that the youngest son inherits the house and – according to him – even the kraal. When the first son was

[348] A relatively short time period is chosen when, for instance, a woman has a permanent job, since a stay of several months would impede her occupational duties.

born, this informant explained, the parents were still healthy and could send him to school, which enabled him to find a good job. Today, however, children would 'fight' about the inheritance, which is why everybody should make a will that considers all their children (Frans Kganare, Sediba, 07 November 2011). Other informants, too, stated that they planned to give each child an equal share after their death in order to avoid such quarrels.

8.3.2 Household Composition

Table 10 illustrates the various household types in terms of composition. As is typical for households of black South Africans (Posel 2001:655), the sampled households were often multi-generational, and in four cases up to four generations lived together under one roof. Three cases were skipped-generational households, in which one of the middle generations – e.g. the children of the household head – were missing. Almost two thirds of the sampled households consisted of singles, nuclear families[349] or nuclear families with grandchildren or great-grandchildren. In most cases, these grandchildren were the descendants of a daughter of the household head. About every tenth household also included a married son and his nuclear family. In all cases this was the youngest son of the household head. In about 13% of the sampled households, siblings of the head and their children or grandchildren were part of the household. The last four columns show households in which even the broad definition of 'nuclear family' applied in this thesis would not be appropriate. In these households, single widows, widowers or people who were never married lived together with other relatives that were not their children.

[349] When I speak of nuclear families, I also include single parents that live together with their children.

Table 10 Household typology based on composition (n=45) [350]

Household types	no	%
Singles	3	
Nuclear family or smaller (F-M-C; F-M; M-C; unmarried couples)[351]	16	
+ (great-) grandchildren	10	**64.5**
+ married son's nuclear family	1	
+ married son's nuclear family + (great-) grandchildren[352]	4	**11.1**
+ unmarried/divorced siblings	2	
+ sibling's (grand)children	3	
+ sibling's children + others[353]	1	**13.3**
+ MB's nuclear family	1	**2.2**
Singles + grandchildren	1	
Singles + sibling's children	1	
Singles + unmarried sibling + grandchildren	1	
Singles + married sibling's nuclear family	1	**8.9**
	45	**100**

The question of whether it makes sense to use the concept of 'household heads' when people are asked to identify the head by themselves is discussed by Posel (2001). One point of critique is that it is not always clear on what ideas of 'headship' such a self-definition is based. Moreover, the determination of one head assumes that households are organised hierarchically and that a single person is the sole or final decision-maker representing the interests of all members (ibid. 653). Still, Posel argues that the concept should be retained as it also provides valuable insights. Based on national survey data she shows that the determination of headship correlates with age, income and decision-making power; however, differences

[350] Abbreviations used in the table and footnotes: F: father; M: mother; C: child; H: husband; B: brother; D: daughter; S: son; Z: sister.

[351] In one case, the child of the nuclear family was adopted.

[352] In one case, the grandchild was the grandchild of the brother-in-law of the household head (HBDD).

[353] In this case the household consisted of the head and her child; the head's father, who was very old and depended on care; and the head's brothers' children; as well as a cousin of the head (FZS).

can be observed between female- and male-headed households. Thus, it is important to take into account the gender of the head. In the case that it is a woman, one can make inferences about the household's economic vulnerability, as female-headed households often have less access to economic resources than those headed by men (ibid. 669).

Figure 25 Female-headed household (photograph: C. Naumann)

The proportion of female-headed households increased after the end of apartheid. Whereas in the mid-1980s, 38% of rural households were headed by women (Chapter 6.3.3), in 2011 51.1% of the sample were female-headed. This high proportion is also found in other areas in Southern Africa (van Driel 1994). Except for one case – a young man who worked in a mine near Johannesburg – the heads of all 45 households were living in the villages. Thus, the often cited 'absent head of household' (Posel 2001:656) was not common in my sample. This also means that the female-headed households were *de jure* led by a woman, and not only *de facto*. Peters (1983) argues that households headed by unmarried women are in a pre-marriage stage of the household development cycle, and would be supported by their future husbands and patrilineal family. Thus, they could not be regarded as independent households. However, in my sample I had several cases where young, unmarried women were economically independent and did not receive financial, material or physical support from their partners.

The average age of female heads was with 59.9 years (SD: 15.3) higher than that of the male heads (mean: 52.5; SD: 16.6). Such age differences between female and male household heads are not uncommon. Similar results were reported by other researchers* (for instance Posel 2001:658; Berzborn 2006:145). The reason for the age difference is that widows take over the household after their husband's death. Almost 70% of the female heads in my sample were widows. The remainder had never been married, though some of them had a partner.

In summary, similarly to demographic changes on the village level, households' structures have transformed significantly after 1994. One major change is a decline in the average household size, which was caused by various factors including a decline in birth rates, increase of mortality, the freedom to move and changing marriage patterns. The biggest group of households consist of a nuclear family, single parents with children, grandparents with grandchildren, couples without children or single individuals. Slightly more than 50% of the households are headed by women. This indicates that female headship has gained increasing importance in the last two to three decades, yet it is important to stress that most of these heads are widows. Therefore, these households have not always been female-headed, but are at a specific stage in their development cycle.

8.4 Household Economy

After discussing the structure of households, I will turn to the activities in which households are engaged in the following section. The focus of the analysis centres upon stratification and wealth distribution, strategies of income generation, and livelihood diversification. However, some preliminary considerations on the question of income pooling and distribution within households are necessary in order to illustrate that a household is not always a social unit whose members decide unanimously over the use of resources.

8.4.1 Whose Money? Preliminary Considerations on Households Economies

> Jeanette Rancho (24), a single mother, lived with one of her two daughters in the house of her grandmother, who was the head of the household. Apart from Jeanette and her daughter, an unmarried son of the grandmother was living there, along with a married son and his nuclear family as well as several other grandchildren and one great-granddaughter. 13 people stayed in the household, whose only regular income were social transfers from the government; i.e. an old age pension, four child-support grants and one grant for a foster child. In total, these came to less than R3,000 per month. The grants were not equally pooled in the household, as Jeanette complained. When the grants had been paid out, the married uncle, his wife and children would visit the wife's family in another village. When they came back, they had spent most of their own money, and lived off the grandmother's pension for the rest of the month. Jeanette further complained that her uncle would sometimes buy a coke, but would only give his children some of it to drink. Jeanette, on the other hand, invests parts of her child support grants in food for the whole household; the remainder she uses to pay the school and kindergarten fees for her daughters and to buy clothes. Jeanette was ambitious about her future. In contrast to her uncles, she applied for jobs in Thaba Nchu or Bloemfontein and participated in voluntary courses held by the Department of Agriculture as she hoped this training would improve her curriculum vitae. Once in a while she had short-term jobs, for instance as an election support worker at the national elections. Last time she received about R1,000 for three days' work in the polling station. Jeanette used her salary for herself and spent it on a new mobile phone. She insisted that it is money that she has earned with her own work and thus has the right to decide what to do with it.

This vignette illustrates various aspects that could often be observed in Thaba Nchu households: members do not always contribute the same amount to the household budget; nuclear families within extended family households have their own fund; and wages are not shared by the earner. Households – often referred to as the smallest economic unit in anthropology (Schapera and Comaroff 1984:39; Berzborn 2004:7) – do not always manage their money together; nor does a household head have ultimate control over the members' incomes. Thus the 'smallest economic unit' does not always act and make decisions as a unit. That various household incomes do not necessarily coalesce into a mutual, general fund has been argued by Wilk (1994) and others (Acheson 1996; Berzborn 2006). Individual members or groups of individuals sometimes have personal funds. Historical studies have shown that personal budgets became more prominent with the increase of labour migration as absent young men tried to maintain control over their financial resources (Beinart 1982:96). In Thaba Nchu, the issue was first mentioned by the Native Commissioner in 1933, when he complained in a letter to the Sediba reserve board that younger people migrated to town and earned good salaries, but failed to support their parents back in the rural areas.[354] In the census interviews, several household heads stated that they did not know how much money their sons or daughters were earning, yet it must be emphasised that with one exception all parents whose children were migrants or commuters either received an old-age pension or had jobs themselves, and thus were financially secure. In some cases, the absent labourers gave a fixed amount every month to the household's fund; in other instances, wage earners bought food which they brought home at the end of the month. However, as the vignette above illustrates, not only employees but also people who live in the household permanently and receive a state grant have personal funds.

Without doubt, there are also households where the classical 'pooling' can be observed, i.e. incomes, goods and stock belong to all household members (Wilk 1994:379). Normally, these were nuclear families. Whereas in households headed by a single mother[355] the decision over the spending of money was vested in the head alone, the situation was more complex in the case of married couples. Several scholars have shown that household expenditure decisions are not made unanimously by married couples, but are in fact negotiated. Members have differential bargaining power in such discussions (Posel 2001; Eder 2006). This is of course also true for extended family households.

I did not systematically study the aspects of personal funds, the negotiation of expenditures and intra-household power relations. Based on my qualitative data gathered through interviews, conversations and observations,

[354] SAB NTS 8142 2/341 part 2, Native Affairs Commissioner, Thaba Nchu, to the Sediba Native Reserve Board, 25 July 1933.
[355] No nuclear families with single fathers were found in the survey sample.

some generalisations about different household types in terms of income pooling are possible. Besides the above mentioned nuclear families that in many cases combine their incomes, one finds households that are described in the literature as 'reciprocal' (Wilk 1994; Berzborn 2006:202–03). The household of Jeanette Rancho, described in the vignette, could be an example of this. Several members have smaller incomes, e.g. government grants or 'piece jobs' (see below), and use these mainly for themselves or – as in the example in the vignette – for their nuclear family. In addition, the members of the households have to spend parts of their income for certain goods and expenses that serve the household as a whole. For instance, in a male-headed household that received two old-age pensions, two daughters of the head – both having good jobs – paid parts of the food and the electricity vouchers for the whole household. Finally, so-called redistributive households were also found in Thaba Nchu. In this type, household members have personal funds of which they give part to a general budget (see also Berzborn 2006:198–202). In my ethnographic census, this was the case in some households where children of the head were working for wages. Generally, two arrangements were found: either the children gave money to a common household fund irregularly, or they paid a monthly, steady amount, though this was not reported very often. To sum up, only in a few cases are household budgets administered by the household head alone or by all members together. As they are not necessarily pooled together, the concept of 'household income' must be interpreted with caution.

8.4.2 Socio-Economic Stratification and Wealth

When conducting the village mapping, I realised that households in rural Thaba Nchu have a higher diversity with regard to wealth than I expected in the beginning. At first sight, this was visible in the material wealth and possessions of the households. Whereas most people depended on public transport, some households had cars, others even tractors. Housing ranged from corrugated iron shacks or little clay houses with old and broken furniture to large brick houses with tiled sleeping, dining and living rooms equipped with expensive couches, beds, tables, chairs and flat-screen TVs. But commodities are not necessarily a good indicator of wealth as they do not necessarily reflect the actual financial situation of the household (Rössler 1997:324; Berzborn 2006:161).

The stratification observed is not a recent phenomenon. An old woman summed it up: 'we have always differed and we will always differ'. Indeed, the historical data and interviews revealed that significant economic differences also existed in former times. In the first half of the twentieth century, these differences manifested mainly in agricultural capital, for instance the sizes of herds, number of fields owned, or possession of agricultural implements. Today too, families have varying herds and differential gardening capital. However, wealth is also displayed in several other ways. To understand the complexities of the cultural meaning of wealth and stratification, a ranking of households was appropriate.

Although the method is often applied in studies of rural economies (see for instance Mearns et al. 1992; Grandin 1994; Pretty et al. 1995; Schwinge 2001; Berzborn 2006; Bollig 2006; Tönsjost 2013), its use is also contested. Ferguson (1992) argues that the method's underlying assumption is that wealth is measurable on a linear scale. He uses the example of rural Lesotho and illustrates that the cultural meaning of wealth is very complex as different kind of wealth holdings exist, i.e. land, livestock, cash and consumer goods, housing, wealth in people. Whereas some of these holdings are exchangeable – cash, for instance, is the most convertible property in the Lesotho case, and can be transformed into basically everything except land (ibid. 66) – others are not. This makes a comparison, and thus a linear ranking, difficult. Ferguson concludes that a study of stratification and economic inequalities must first consider the 'topography of wealth' and then concern itself with 'situating economic actors within this already well mapped cultural-economic space' (ibid. 69).

In the first half of the twentieth century a mapping of the wealth topography in rural Thaba Nchu would probably have led to very similar results to those of the Ferguson Lesotho case. However, the cultural meaning of wealth has changed. Some prior wealth indicators – e.g. fields – no longer exist. Others, e.g. cattle and money, are exchangeable. Whereas in Lesotho the sale of cattle was strongly disapproved of and would only occur when people desperately needed money (Ferguson 1992:63), Thaba Nchu livestock owners – except for those who have very few animals – also sell cattle when they are not in need of cash. Other wealth indicators, namely wage labour and state grants, have gained importance as major sources of income and as a sign of economic well-being.

To address Ferguson's critique and to avoid the confusion of certain wealth domains, the 13 rankings were combined with questions that dealt with local concepts of wealth, which I discussed briefly before the ranking task. While the informants conducted the wealth rankings, I encouraged them to talk about their decisions. After the wealth groups had been identified, the informants were asked to explain which indicators differentiated the various groups. With regard to Ferguson's criticism, the interviews revealed that the concept of wealth does indeed comprise several domains, some of which are difficult to compare. The results are therefore not used as single indicators of a household's economic position and well-being, but are correlated with other data and thereby put into context. Moreover, when wealth rankings are accompanied by further questions it is possible to understand the informant's decisions and thereby the meaning ascribed to different forms of wealth. For instance, whereas older men clearly regarded livestock as very important for the definition of wealth, younger people stressed that, besides livestock holdings, a high level of good quality education is an indicator. Such aspects could certainly also be explored by using other methods, yet from my experience, the sorting of cards helped the informants to explain their concepts of an abstract term and illustrate the fine nuances that determine socio-economic stratification in their communities.

When indicating signs of wealth, informants most often mentioned livestock, wage labour and big houses. Most people regarded only those employees who had 'good jobs' – generally meaning all types of government jobs, e.g. police officers, teachers, soldiers or nurses – as rich or wealthy. In Rooifontein, running a private taxi enterprise was also regarded as a good job. For rich households, wealth had continuity – or, as a 36-year-old man put it, 'they are not in danger of becoming poor'. Older household heads had held good jobs in the past and were now receiving a private pension. Their children often had a high level of education and were employed. Thus, these households successfully accumulated not only economic, but also cultural capital. Indeed, the seven households from the census sample with at least one member who had graduated from college or university belonged, with one exception, to the rich and average-wealth group. One of the rich households had four members who held a degree from a tertiary education facility.

Wealth manifested not only in livestock, jobs, degrees and houses, but also in consumer goods, e.g. furniture and food. One male informant from Rooifontein mentioned two additional aspects of wealth: land ownership and the ability to employ poorer people for so-called 'piece jobs'. Casual jobs for women typically include washing the laundry or cleaning houses; common male jobs, on the other hand, entail house construction, repairs of any kind or preparing gardens for cultivation. By allocating piece jobs, affluent households are able to transform economic capital (i.e. money) into social capital. Piece jobs are not randomly offered to other people; for instance, a woman of a wealthy household that needs to employ someone for doing laundry would ask a woman with whom she already maintains social relations. The ties between an employer and a piece jobber are not only reproduced through exchanging labour force and money, but also through gifts (e.g. second-hand clothes) or mutual visits. Therefore, I argue that these forms of employment have actually taken over the role of *mahisa* cattle that constituted patron-client relationships in the past. The employer has the economic capital and is in the more powerful position, yet piece jobbers also benefit from these relations, as they earn some money and receive additional goods. By investing time in social relations, the poor can thus transform social capital into economic capital.

Piece jobbers often belong to poor households. Informants described them as having nothing: no animals, no income and neither state nor private pensions. Some of them perhaps receive a child support grant, but nothing else. These households struggle a lot to 'put food on the table', as a young woman in Sediba put it; she added that they have 'no helping hand'. An older man explained that the poorest have no parents, no husbands and no children who could support them. There was a high consensus amongst the informants of each village on which households belonged to the wealthiest and the poorest categories.

For households situated between the rich and the poor, the consensus was less pronounced, but here too certain similarities were found. For the middle groups, it is more difficult to cope, but they can still survive, be it through 'helping hands' or social transfers. Nine of the 13 informants stated that state pensioners belong to this group. This is different from the findings of Berzborn (2006) in the Richtersveld, where pensioners belonged to the poorer households as they often had many children and grandchildren that did not generate any income.[356] However, average household sizes were considerably smaller in Thaba Nchu than in the Richtersveld.

---

[356] However, Berzborn (2006) stresses that pensioners were not regarded as poor *per se* (ibid. 166–167).

In order to prove whether the indicators of wealth are also reflected in the sample, the findings of the wealth ranking were compared with other data I collected with the ethnographic census.[357] Table 11 illustrates the correlation of the wealth scores with several variables.[358] The first block shows the variables 'livestock ownership', 'cattle herd size' and 'garden'. Whereas the first two variables correlate significantly with the wealth scores, the picture is different when looking at gardening. For rural Lesotho, Ferguson (1992) has shown that wealth was connected to land, i.e. cultivated fields. In former Thaba Nchu, too, a man's status was influenced by his household's achievements in agricultural production, though success in livestock production was always more important (Schüler 1965:365). Today, households are still involved in the cultivation of vegetables and maize on a small scale. However, gardens are planted by both poorer and wealthier households, and thus the mere existence of a garden is not a good indicator of wealth.

Significant correlations were found with between the wealth score and the variables 'car ownership' and 'house type'. Houses made out of concrete blocks or bricks from the long-established South African company Corobrik are clear indicators of wealth. In the 1970s the latter were only purchasable in Bloemfontein, and people who wanted such material had to pay for the expensive transport of the bricks. Normally, older Corobrick houses belong to former labour migrants. Similarly, to other regions, for instance Northwest Namibia (Greiner 2008:164–65), wage labourers in Thaba Nchu

[357] Although 45 households were part of the ethnographic census sample, I considered only 42 in the final analysis. In two cases, the households that were randomly selected for the survey regarded themselves as part of another household for which the wealth rank was unknown or different. In one case, a woman lived together with her children in the house of her uncle and took care of his livestock. The uncle himself was a former labour migrant and lived in Thaba Nchu. Apart from child support grants, the woman had no income and held no rights to the house or the livestock, and thus was not actually rich. However, the wealth ranking informants placed her in the wealthiest category, apparently because they had her uncle in mind. I excluded this special case as it would distort the correlations.

[358] For the correlation of the wealth score with 'cattle herd size', 'household size' and 'age of household head' I used Pearson's correlation coefficient. For all other variables, I made a point-biserial correlation (Field 2005:131–33). This coefficient can be used to correlate a metric variable, i.e. the wealth score, with a dichotomous, nominal variable. Therefore, the variables should be read as follows: e.g. 'livestock ownership' means that the household owns grazing livestock.

brought new ideas of housing – as well as the financial means to buy them – and thereby shaped the appearance of rural villages.

Table 11 Correlation of wealth scores with land use types, material goods, house-hold size and head, and income types (n=42)

Variables	Correlation with wealth scores
Livestock ownership	0.449**
Cattle herd size	0.446**
Gardening	0.011
Car	0.559**
House type (concrete blocks or 'Corobrik')	0.630**
Household Size	0.111
Gender Household Head	0.047
Age of Household Head	0.418**
Income from wages	0.301
State pension	0.164
Private pension	0.382*
Confidence level: * p<0.05; ** p<0.01	

Whether the household's position in the development cycle, its composition or the gender of the head are indicators for wealth are aspects that have been discussed by several scholars (for instance Posel 2001; Schwinge 2001; Berzborn 2006; Tönsjost 2013). For the Thaba Nchu sample, the data suggest no correlation between wealth and 'household size' or 'gender of household head'. However, 'age of household head' is correlated with wealth. These findings are interesting as they are in contrast with some of the results of the above-mentioned authors. With regard to the variable 'age of household head' my case rather corresponds to the observations made by Tönsjost (2013) in Namibia, where the households of older heads belong to the wealthier groups (ibid. 112–13). In contrast, rich households in the Richtersveld are often headed by younger people who are engaged in wage labour (Berzborn 2006:167).

It is surprising that the gender of the household head does not correlate with wealth. Several studies show that female-headed households are found first and foremost in the poorer wealth classes (Schwinge 2001:127; Berzborn 2006:167). Households headed by women often belong to the lower-income classes and are economically vulnerable (Posel 2001:651).

What accounts for the difference in Thaba Nchu? I suggest that my findings might reflect a development that was initiated with the increasing access to old-age pensions described in more detail in Chapter 8.4.3. The comparatively large monthly transfers[359] changed patterns of wealth and stratification for female-headed households in particular. The analysis of household structure has shown that female heads are often older widows. Whereas in the past, these cases were typically the poorest households, the old-age pensions guarantee a degree of financial security. Thus, social transfers increased the resilience of households headed by older women.

No significant relationship exists between the wealth scores and the income from wages. It is important to note that 'income from wages' was dichotomised here in the use of the categories 'wage income' and 'no wage income'. That wealth and income do not correlate might be surprising at first sight, yet the discussion of local wealth concepts has already revealed that the prestige of wage jobs varies. Households whose members are engaged in low-paid jobs, i.e. construction worker and cleaner, are not necessarily wealthy, as their incomes might not be higher than those of the private-pension receivers. Moreover, wage incomes, particularly in low-paid jobs, are insecure, as loss of employment is possible. Private pensioners are regarded as wealthy, as the correlation coefficient indicates. They normally have a long employment history, during which they were able to accumulate pension entitlements. Thus, their wealth is not a result of the pension *per se*, but of a former salary. The variables 'state pension' and 'wealth' do not correlate positively; but neither do they correlate negatively – as they do for instance in the Richtersveld (Berzborn 2006: 166).

Household Incomes: An Estimation

Households in rural Thaba Nchu normally depend on several types of incomes. How these types are combined will be elucidated in Chapter 8.4.3. Here, I would like to focus on the quantitative aspects and illustrate the stratification of incomes. Similarly, to studies that have investigated household economies in other parts of South Africa (see for instance Lahiff 2000; Berzborn 2006), it was difficult to collect precise income data in Thaba Nchu. Only in the case of social transfer payments do I have reliable data.

---

[359] In Namibia, for instance, the old-age pensions consisted of only 500 Namibian Dollars in 2011 (Menestrey Schwieger 2017:136) which equals the same amount in Rand.

The amount received in the grants is no secret, and all informants were willing to admit whether they received such payments or not. The situation with the incomes gained from wage labour was different. Some people were not willing to state exact amounts, and others understated their salaries. As indicated above, the wages earned by absent migrants were often unknown to their respective household heads. To fill the data gaps, wherever possible I estimated the wages. First, I was able to do this for those jobs that appeared several times in the survey. For instance, two people were working as police officers in Thaba Nchu, but only one of them was willing to provide information about earnings. For the other person, I assumed a similar income. Second, I asked people about minimum incomes for certain occupations to get at least an idea of the amount earned. Third, I researched online job advertisements, in which wage ranges were sometimes provided. By these means I was able to calculate at least rough salaries for most wage-earners.

The correlation of each household's wealth score and monthly income per capita is significant ($r = 0.605^{**}$; $n=37$).[360] Table 12 illustrates the average monthly income per capita per household. The mean income per household member was R981.47 per month (SD 967.71), but there was great variation between individual cases. The minimum income was R100 per household member, the maximum R4,087.50. In total, 42.5% of the households in the sample had a monthly per capita income of R500 or less, and thus fell below the South African poverty line (Leibbrandt et al. 2010:17).[361]

As indicated above, it is not always clear how much money labour migrants give to their families at home. Remittances are sometimes sent irregularly and may be in the form of food, clothes or other material goods. A precise investigation of remittances and their importance for the household economy requires a consideration of the sides of both the receiver and the giver, and thus a multi-sited ethnography is necessary (Greiner 2008). To overcome the methodological problem of uncertainties regarding remittances, Lahiff (2000) assumed for his case study in Venda that 50% of a migrant's

---

[360] Besides three cases that were excluded because no wealth rank data were available (see above), I did not use the data from five other households as their incomes were uncertain.

[361] There is no official determination of a poverty line in South Africa. An OECD study uses a monthly income of R515 per capita as an indicator. This is also used by South Africa's National Planning Commission (http://www.npconline.co.za/).

wages are contributed to the fund of the home household (ibid. 159). This appears a plausible value for the cases where the household head is a migrant. But in my sample, in most households the children or grandchildren of the head migrated. Interview data suggest that in these cases less than 50% of the salaries were contributed. Still, households that earn wages have an economic security that others do not, even though the employees might not share their whole incomes.

Table 12 Monthly income per household member in Rand (n=40) [362]

Monthly income per household member in Rand	Frequency	%	Accumulated %
< 250	9	22.5	22.5
251-500	8	20	42.5
501-750	8	20	62.5
751-1000	3	7.5	70
1001-1250	1	2.5	72.5
1251-1500	4	10	82.5
1501-1750	3	7.5	90
> 1750	4	10	100

In summary, households in rural Thaba Nchu are highly economically stratified. Whereas in the past the stratification was manifested in cattle holdings and agricultural capital, now cash income, along with cattle holdings, is another factor that determines a household's wealth. The local conceptualisation of wealth has changed. In social-ecological system terms this could be interpreted as a decoupling of wealth and land leading towards a stronger coupling of wealth and money. This decoupling is also reflected in the discontinuation of former patron-client relationships, i.e. the lending of cattle. Another transformation was initiated by people's increasing access to social grants, and particularly to the old-age pensions. Households that in the past were the most vulnerable, i.e. the widow-

[362] Five households whose incomes were uncertain were excluded. The line under the second row indicates the poverty line. Berzborn (2006) points out that the calculation of income per capita is certainly imprecise as not all household members need the same amount of money. For instance, a household had to spend more money for school children than for toddlers, as for the former school fees and school uniforms have to be paid for.

headed households, are now economically secure. Certainly, this economic well-being is strongly connected to the household's cycle. A household headed by a pensioner is not regarded as poor, as long as the elders are still alive. Thus, this form of security is only temporary.

8.4.3 Diversification of Rural Livelihoods

The concept of livelihood diversification can be defined as 'the process by which rural families construct a diverse portfolio of activities and social support capabilities in order to survive and to improve their standards of living' (Ellis 1998:4). This phenomenon, which gained scholars' attention in the 1990s, is observable in many regions in the rural South. Some interpreted it as a process of de-agrarianisation which is having negative effects on the sustainability of rural livelihoods (Bryceson and Jamal 1997; Meagher and Mustapha 1997; Bryceson 1999). Others argued that income diversification is a strategy to minimise the ecological and economic risks faced by households in uncertain environments and to increase resilience (Ellis 1998; Francis 2000; Slater 2002; Berzborn 2006; Schnegg 2009; van Leynseele and Hebinck 2011; Bollig and Schnegg 2013).

It is undeniable that 'diversification is the norm' (Barrett et al. 2001:315). An issue that is less clear is whether diversification is actually a unanimously decided household strategy, or driven by individual decisions – for instance those of women or youth that want to obtain more power over financial resources (Bryceson 1999:183–84). The observations I made with regard to the pooling of incomes suggests that both alternatives are found in rural Thaba Nchu. Clearly, simple generalisations about livelihood diversification are difficult and even misleading as it is a 'heterogeneous social and economic process, obeying a myriad of pressures and possibilities in the rural economy' (Ellis 1998:29). To understand the reasons for and impacts of diversification processes it is necessary to elucidate the local circumstances. In the following section, I will therefore introduce the main approaches to income generation in contemporary rural Thaba Nchu, and analyse how these various ways are combined and whether patterns are observable.

Wage Labour, Remittances and Self-Employment

In his study of northwestern Namibia, Schnegg (2009) differentiates between wage labourers and people who are employed by the government. This makes sense in his case as the former normally work as herders on commercial farms, and thus have significantly lower incomes than the latter (ibid. 234-235). In my sample, incomes are also highly diverse. Some jobs are commonly low-paid, e.g. work in the construction sector as a brick layer. However, it is not only government employees who receive high salaries, but also some people working in the private sector. For instance, a lawyer or a supervisor in a major supermarket chain earned more than people employed as watchmen or cleaners in a governmental institution. Hence, a simple differentiation of wage labourers and governmental employees would be misleading.

In the ethnographic survey 28 households had at least one member that was working. Of all people between 20 and 59 years of age who were not disabled or attending a school or college (n=92), 39 people (42.4%) had a job in 2011. However, four of them did not have a permanent position. When considering only the secure regular jobs, 38% of the working-age population were employed. Eleven of the employed people were migrants and sent remittances in cash or in kind. Thus, permanent labour migration was less important for the sampled households than were the employment facilities that could be found in the Thaba Nchu-Botshabelo-Bloemfontein region. Wages of the employees ranged from R1,000 – the wages of e.g. a general worker in a private telecommunication company, or a construction worker – up to R12,500 – the net wage of the principal of a local primary school.

Officially, the unemployment rate in the rural Thaba Nchu area was indicated at about 50% in 2008 (Mangaung Local Municipality 2008:7). Assuming that in my sample not all non-working people between 20 and 59 years were actually looking for a job, the ethnographic census figure corresponds roughly with the high unemployment rate reported officially. Certainly, unemployment is not a new problem in the region. Labour shortages that were typical in the 1940s and 1950s 'had given way to open and visible unemployment' (Seekings and Nattrass 2005:166) by the mid-1980s in all homelands. However, national data indicate that the unemployment rate has increased tremendously since the end of apartheid (Seekings and Nattrass 2005:317; Banerjee et al. 2008). A major change for Sediba residents was the loss of South African Railways, nowadays Transnet, as a major employer in the area. After 1994 many corporations in the public sector underwent significant transformations, resulting in job losses. Transnet too, reduced its work force by more than 27,000 people within two years (Vermeulen and Wiesner 2000:389). In my sample, nobody had worked in railway construction, wagon design or related fields in recent times. Moreover, the agricultural sector lost its importance as an occupational sphere – a trend that was already visible during Bophuthatswana times. In the sample, two men were paid to herd the animals of a relative. However, as they were both over 70 years old and received a pension, their wages were not their main income sources. One man did casual labour on farms but had no permanent employment. Another group not represented in my sample is that of the self-employed, such as shop owners and taxi drivers. Whereas the latter were only found in Rooifontein, one or more shops exist in each village.

Social Grants

> Early in the morning I already observed several women walking to the 'Thusong Service Centre' where the grants are paid out today. One woman carried a huge plastic box filled with fried fish that she wanted to sell at the pay point. Kedikilwe Olifant picked me up at ten o'clock and we left to the service centre. She was well-dressed, almost as if she was on her way to the Sunday service in church. When we reached the pay point a long queue had already developed. Around the fence of the service centre many people were sitting and selling self-made food. Some traders had come from town with their pick-ups and offered vegetables, fruits and other goods. We walked to the entry gate of the fence where the cars of the SASSA (South African Social Service Agency) parked. The gate was safeguarded by a man with a machine gun. He checked whether the people entering had a chip card, which proved their entitlement to a social grant. Since I had no card, he did not want me to enter the premises. Kedikilwe tried to explain who I was and that I wanted to observe the payday procedures. Eventually, I was allowed to enter and make observations. Grant receivers – or 'customers' as they were called by the supervisor – had to produce their cards, which were inserted in a chip-card reader. Their identity had then to be proven using their fingerprints. The card reader was connected to a portable ATM that dispensed the exact amount money due in each social transfer. After handing out all grants to the Sediba residents, which took only about one hour, the cars of SASSA drove in a convoy to the next village on their route; the village of Talla (field notes, Sediba, 4 July 2011).

Social benefits have a long history in South Africa, though a highly racialised one. In 1928, old-age pensions were introduced for 'whites' and 'coloureds' aged 65 years and older. 'Indians' and 'natives' became eligible about 15 years later, yet black South Africans – like the other 'non-white' groups – received considerably lower amounts. Moreover, until 1948 rural elders only qualified if they were landless (Devereux 2007:541–45). Despite their entitlement to a pension, many black South Africans in the rural areas did not obtain any old-age support due to lack of information and difficulties in accessing payment points. After the end of apartheid, South Africa has expanded its government transfers considerably (Neves et al. 2009) and the 'social grant system (...) ranks among the most impressive in the so-called "developing" world' (Marais 2010:3). Today, in addition to old-age pensions the government provides several other benefits, such as the child support grant, disability grant, and foster care grant.[363] The grants are 'means-tested', i.e. citizens are only entitled when they have an income below a certain threshold. The means test varies from grant to grant. In 2014, for instance, the monthly income threshold for the old-age pension was about R4,100 for a single adult and R8,320 for a married couple (South African Government Services 2014d). The child grant was paid to single parents that had an income of less than R2,900 per month, or to married couples when they had less than R5,800 (South African Government Services 2014a).

Recently, several studies have stressed the usefulness of grants for poverty alleviation (Legido-Quigley 2003; Meth 2007; Neves et al. 2009), and their positive effects on child nutrition (Aguëro et al. 2006). With regard to poverty alleviation, Neves et al. (2009) conclude that 'grants are crucial to the survival of the poor as they provide for their basic needs and consumption' (ibid. 68). Moreover, their findings suggest that the social transfer payments lead to investments and savings and allow borrowing, and thus support informal social security and reciprocity (ibid.). In the Thaba Nchu sample, government transfers were an important income for many households. In total 33 households in the sample received a grant. Not all of the benefit recipients belonged to the poorer groups in the villages. Even of the households identified as rich in the wealth ranking (n=10), 50% received benefits. 21 of 45 households stated that social grants were their main

[363] The child support grant and the disability grant had also been paid in Bophuthatswana (Lund 1990:71).

source of income. In 13 of these not a single household member had a regular job, and thus the state support was their only certain monthly income. The following grant types were found in the sample:

(1) *Old-Age Pension*: 21 people in 19 households received a state pension in 2011. In one case, the pension was paid from Lesotho, which was the home country of the household head. To receive an old-age pension from the South African state the applicant must be 60 years of age or older and a citizen or permanent resident (South African Government Services 2014d). In 2011, the pension was R1,140 for 60-year-olds and R1,160 for people who were 70 years or older.

(2) *Child Support Grant:* for 42 children in 24 households a child support grant was paid. This transfer is paid to the primary caregiver of the child, who may be a parent, grandparent or any other person who can prove they are responsible for the child. Grants are paid until the completion of the 18th year of life (South African Government Services 2014a). In 2011, the child support grant was R260 per month.

(3) *Disability Grant:* a disability grant is paid to people aged between 18 and 59 years of age who are physically or mentally disabled and cannot work for a period longer than six months. The government differentiates between a temporary disability grant – locally often called 'sick pension' – which lasts between 6 and 12 months, and a permanent one that lasts more than 12 months (South African Government Services 2014b). The disability grant amounts to R1,140 per month, and thus corresponds to the old-age pension. In the sample, one person received a disability grant while two received a 'sick pension' grant. Another informant was still waiting for approval of his application.

(4) *Foster Child Grant:* the foster grant is paid to people who take care of an orphaned, abandoned or neglected child. As with the child support grant, the child must be younger than 18 years (South African Government Services 2014c). However, the foster grant is significantly higher than the child support, and amounted in 2011 R760 per month. For three children in two households of the sample a foster child grant was paid by the government.

Table 13 shows the frequency of grants per household. Slightly more than one quarter receive no state support at all, and more than 55% of the sample households have one or two grants at their disposal. A correlation of wealth scores – of those households where such data existed – and the number of grants per household showed no significance (r= -0.056; n=42).

Table 13 Number of social grants per household (n=45)

Number of grants per household	Frequency	%
0	12	26.67
1	14	31.11
2	11	24.44
3	3	6.67
4	3	6.67
5	0	0
6	2	4.44
Σ	**45**	**100**

Besides the old-age pension paid by the government, some former government employees – e.g. teachers or South African Railway workers – receive a private pension. Four men of the sample were entitled to a private pension ranging from R800 to R5,800 per month. The entitlement to a private pension is, as indicated above, a sign of wealth.

Gardening and Animal Husbandry

Land use – I refer here only to gardening and animal husbandry – is the most common economic activity in the sample. Of 45 households, 35 had either a garden (22.2%) or livestock (6.7%) or both (46.7%). However, the reported monthly incomes from these types of production are considerably smaller than for the other economic activities. Not a single household mentioned 'land use' as a main income source.[364] In Chapter 8.5 I will shed light on the various kinds of land-based activities and elaborate on animal husbandry, horticulture and other forms of land use in more detail. In the

[364] Certainly, these data must be interpreted with caution, since a monetary evaluation of land use in communal areas often underestimates what is gained *de facto* through direct use (Shackleton et al. 2001:582).

following I will address only the marketing of agricultural products, and show which households are involved in land use activities.

Whereas 90% of the wealthy and average-wealth households (n=20) were engaged in agricultural production, only 68% of the poor and poorest (n=22) were involved.[365] Of those who did no gardening or animal husbandry, only two belonged to the wealthy and average-wealth group respectively. The former was the household of an old widow who lived together with two sons and two grandchildren. Together they had several incomes and received one pension and two child support grants. Additionally, one son was absent and working in Thaba Nchu and the other son did casual piece jobs. Moreover, they had inherited a large amount of money in the past, thus were financially secure. The household that belonged to the average-wealth group was also headed by an old widow. She lived together with one son – who was absent and working in Bloemfontein – and three grandchildren. Until recently they had owned cattle; however, all these had died because of drought. Both households had only a restricted labour force as the grandmothers were old, their sons were working and their grandchildren attended school. Therefore, an investment in land use – be it a garden or animals – made no sense at that point in time.

Similarly, to other regions (see for instance Shackleton et al. 2001), vegetables and crops from the gardens were primarily consumed by the household members, but some were given for free to family members, friends or neighbours. In a few cases crops served also as animal fodder. The most common plants cultivated were: pumpkin (planted by 87% of the gardeners); beans and spinach (both 48.3%); maize (38.7%); beetroot (25.8%) and carrots and potatoes (less than 20%). Moreover, several households in the research area have cultivated peach or apricot trees. The assessment of the precise incomes gained from the sale of agricultural products with a survey is difficult, as incomes vary annually and seasonally (Ellis 1998:9). Often, the cash earned with last year's harvest is only vaguely recalled. 12 interviewees stated that each year they sell some vegetables and earn between R70 and R500 per annum. The main animal products sold are milk and wool. Prices for the latter vary greatly depending on their quality. After shearing, wool is typically packed in big bags for which farmers obtain prices ranging from R50 to more than R300. In Middeldeel, where the Department of Agriculture implemented a sheep and wool project in around

[365] I excluded the three households for which no wealth rank data were available.

2006, prices were considerably higher than in the other research sites. Better types of sheep, better shearing facilities and additional training in wool production resulted in higher quality wool. One Middeldeel household that owned about 80 sheep earned up to R3,000 per year from wool.

The marketing of animals is another activity that provides households with cash. 83% of the livestock owners in the ethnographic census stated that they sold animals to traders, at auctions or to other village residents. The remaining four people had very small herds of only two or three cattle, and no interest in sales. Traders are often white farmers who buy even skinny animals, feed them and sell them again. Access to public auctions is difficult. Middeldeel has facilities for sheep auctions that were part of the implemented wool project. Cattle auctions are not conducted in any of the four research sites. Some people from Sediba bring their animals to Excelsior, a small town about 15 km east of Sediba, yet farmers regard this as a great effort as prices are not always good. Thus, many prefer to sell to the traders, though they pay a meagre price. For a full-grown cow prices paid in 2011 ranged between R3,000 and R4,000, but good animals can fetch prices up to R7,000. The number of animals sold per household varies greatly and also depends on the size of the herd. Whereas owners of larger cattle herds sell several animals per year, households that have only a handful rather keep them for ritual purposes and only sell them if there is a financial need. Sheep are mainly purchased by other residents, who slaughter them for ancestral rituals or other feasts. Prices normally ranged from R750 up to R900 per sheep.

Piece Jobs, Petty Trade and World Vision Stipends

In addition to wage labour, social grants and agricultural production, income was also gained from petty trade and piece jobs. The former refers to the small-scale selling of edibles or goods. The importance of petty trade as a means of income has been described for various regions in the global South (for instance Arizpe 1977; Hogg 1986; Little 2000). As it does not require a high capital input it is a popular source of income, particularly for women (Bryceson 2002:732). For households engaged in petty trade it was commonly not their main source of income. Once or twice per month these households bought sweets, snacks, paraffin or 'airtime' vouchers for mobile phones at the wholesale market in Thaba Nchu and sold the goods informally at home. Another form of petty trade was the production of home-made maize beer by older women. The men involved in petty trade engaged in more capital-extensive activities. For instance, one man offered to transport goods with his tractor, and rented out chairs and tables for functions. Another one offered his car for casual transports to town. Piece jobs are an important source of income for the poorest people. They differ from petty trade in that for piece jobs no capital is needed at all, but only one's own labour force.

In the sample, 20 people living in 19 households (42.2%) were involved in petty trade or piece jobs. Normally it was only one household member who was mainly responsible for petty trade. In only one case a young couple who sold vouchers for 'airtime' and electricity said that they both did the trading. Women are more often involved in petty trade and men in piece jobs. Of ten households whose members were doing piece jobs, seven had been identified as among the poorest in the wealth ranking. The situation was slightly different in the case of petty trade, which was engaged in by about eight households, with only one being in the poorest group. The findings indicate that petty trade was not an option for the pauperised households of the sample as it requires at least some capital to make small investments. The mean income of traders and jobbers was about R125.65 per month, though a standard deviation of 259.56 indicates high variation.

Another type of small-scale income should be mentioned separately: three women assisted the Christian NGO World Vision, which is active in the Thaba Nchu area. World Vision sponsors individual children and their families, supports education programmes and distributes basic-needs goods, such as blankets and food parcels. Each of the research sites had several so-called 'home visitors' that helped to identify children in need. Although principally regarded as volunteers, home visitors receive a stipend and are thus also beneficiaries of the NGO's activities (Kotze et al. 2012:36). In 2011, these assistants received a stipend of about R500 per month.

Basing Livelihoods on Several Incomes

In the following section, I will elucidate how households combine the various sources of income and whether there exist certain types of strategies (see also Schnegg 2009). Table 14 illustrates how many income types are combined by the surveyed households. Less than ten percent of them derived their incomes from only one source. In one case, the household lived on income from piece jobs; one other lived on the salary of one member, and in two cases the households depended on social benefits. No household made a living from land use alone. Almost 70% of the sample had two or three different income sources at the time of data collection. When comparing the number of strategies with the wealth rank, the data indicate that the rich households have on average the highest number of income sources (mean: 3; SD 0.92), followed by the average households (mean: 2.9; SD 0.97) and the poorest (mean: 2.7; SD 0,79). The lowest level of diversification is found in households classified as poor, i.e. the third group in the wealth rankings (mean: 2.2; SD: 0.91). Thus, my data support to a certain extent the findings of Hogg (1986), who argued, for the case of pastoralists in northern Kenya, that it is only the very wealthy and the very poor that diversify (ibid. 321). However, Thaba Nchu also differs from Hogg's case in that livelihood diversification is not exclusively found in these groups.

Table 14 Number of income sources and distribution (n=44) [366]

Number of Income Sources	Frequency	%
1	4	9.1
2	17	38.6
3	14	31.8
4	9	20.5
Σ	45	100

In the next step, I consider four different strategies: wage labour and remittances; state grants; agricultural production (here I examine households that have a garden or livestock or both); and piece jobs and petty trade. Table 15 shows the five most common types of combinations. All other types only appeared once or twice in the sample. I am interested in the first three rows in particular, as these combination types were found in slightly more than 50% of the households in the ethnographic survey. Type A indicates households that received a state grant and were involved in agricultural production. All of them cultivated a small garden, and five owned both cattle and sheep. One household had sheep only. Type B consists of cases in which at least one household member had a salary, which was combined with state grants and agricultural production. Similarly, to type A, all households of the second group cultivated a small garden. In this

[366] Here I again excluded the case of the woman that depended on her uncle (see above).

group too, five households owned livestock; all except one had cattle only. Thus, the herds of type-B households are less diversified than those of type A. In group C, people are combining four different strategies, and receive salaries and government transfers, but are also involved in agricultural production and smaller jobs or trades. Except in one case, all households of this type had a garden. Additionally, five of them also owned livestock.

Table 15 Combination and frequency of the most common income strategies (n=44) [367]

Type	Wage labour & Remittances	State grants	Agricultural production	Piece jobs & petty trade	Cases	%
A		+	+		8	18.2
B	+	+	+		8	18.2
C	+	+	+	+	7	15.9
D	+		+		3	6.8
E	+	+			3	6.8
				∑	**29**	**65.9**

The diversification of livelihoods is dependent on several factors, such as access to land and means of production, jobs, the household's position in the development cycle, and the available labour force (Francis 2000:48–50; Berzborn 2006:176–83). Almost all households in the studied villages have access to land for gardening and animal husbandry. Gardens can be cultivated on the home plot and the communal grazing land is open to all residents that hold a PTO. The question of access to means of agricultural production will be addressed in Chapter 8.5. Wage labour has been discussed above and is certainly a restrictive factor in the area, particularly for people who have no matriculation. At first sight, there seems to be a connection between type of diversification and available labour force. The households of type A are on average smaller (mean: 4.375; SD: 3.74) than those of type B (mean: 4.875; SD: 1.89). Households of type C have the largest average size (mean: 6; SD: 2.77), yet these differences are not statistically significant (ANOVA: $p > 0.05$). The results are similar when

[367] The variable 'agricultural production' was dichotomised, i.e. coded with 'true' when the household either had a garden, livestock or both; and 'false' when the household had no animals and no garden.

looking at the relationship between average number of adults per household – i.e. the income-generators – and diversification type. Here, too, statistical tests indicate no significance (ANOVA: $p > 0.05$). Finally, the mean age of the household heads differs between these three groups, yet not statistically significantly (ANOVA: $p > 0.05$).

For now, the ethnographic survey can only prove that there exist certain variations in the ways in which households combine their livelihood strategies. More than 90% have at least two different income sources to rely on. Thus, the Thaba Nchu case shows similarities with other regions where high economic, social and environmental risks require a diversification of incomes (Hogg 1986; Francis 2000; Berzborn 2006; Bollig 2006; Schnegg 2009). The most common are wage labour, social grants and agricultural production, yet their importance in terms of monetary income varied. The question that remains is whether diversification is a voluntary strategy for protection against risk or an involuntary response to crisis. In Thaba Nchu, the motivation for diversification seems to be different for the poorest and for the wealthy. For the former it is first and foremost a necessity for survival, as they have only very small regular incomes, if any at all. They tend to choose diversification strategies that do not require a high capital input, such as piece jobs. Agricultural production is also important for the poor, but in contrast to the richer households only about two thirds of them were involved in animal husbandry, gardening or both. Again, I suggest, this is mainly due to a lack of capital. From a social resilience perspective diversification is a coping strategy for the poorest households, i.e. a response to poverty that they face daily, but it does not enable them to increase their resilience in the long term. This is different for the wealthier households, who invest in capital-intensive forms of income diversification, such as petty trade. For them, diversification is not a necessity but rather a strategy to accumulate wealth. Thus, in terms of social resilience, for the rich income diversification can be interpreted as an adaptive capacity that guarantees 'security for future well-being' (Keck and Sakdapolrak 2013:10).

8.5 Land-Based Activities

Whereas some scholars argue that Sub-Saharan Africa experienced a long-term de-agrarianisation process and an increasing dependence on non-farm income sources (Bryceson and Jamal 1997; Barrett et al. 2001), others claim that rural households are still involved in land-based activities the importance of which is often underestimated (Cousins 1999; High and Shackleton 2000; Shackleton et al. 2000; Shackleton et al. 2010; Twine 2013; Vetter 2013). In particular the direct use of land-based resources, as well as non-use values (e.g. the symbolic meaning that is attached to certain resources) are neglected when studies merely focus on cash outcomes gained from agricultural production. Whereas the marketing of agricultural products is often 'relatively poorly developed' (Shackleton et al. 2000:59), studies suggest that most rural households rely at least on two types of resources, i.e. livestock, cultivation or wild resources, for direct use. Thus, a holistic approach to rural livelihoods and their sustainability should pay attention to these non-market-oriented uses of natural resources (ibid. 58–59). In the following, I will provide a detailed picture of land-based activities found in rural Thaba Nchu. Whereas the cash income from agricultural production was briefly discussed in Chapter 8.4.3, here the focus will be placed on more general aspects.

8.5.1 Animal Husbandry

> FK: 'Here at the rural places […] we have grown up with livestock. We know how to use livestock and we got a life from the livestock because we used to get milk, meat, selling them. Selling them is only just to get clothes to wear. Sometimes we get money, like in the days before we used to send children to institutions, those institutions that you call high schools and secondaries […]. We people, we used to pay *lebola* with livestock. We do our custom with livestock, burials with livestock. Why do we bury our people with livestock, especially cattle? It's because in days before, long, long ago we did not have a coffin. Before the coffin was the skin of the cattle. […] When we are busy with a funeral you must have cattle in between, a livestock, a cattle. And after we have used the skin for wrapping the body […] the meat of that cow it will be used by the people who are gathering together at the funeral. And then from there, we used to plough with cattle, with livestock. At that time travelling [was done] with livestock.'
>
> CN: 'So actually cattle have a lot of purposes?'
>
> FK: 'A lot of course! A lot, a lot! […] that time we did not have coal to make a fire, you got the dung from the kraal of cattle for cooking, boiling water, making a tea. […] Livestock has many,

> many things to do.' (interview Frans Kganare, Sediba, 7 June 2011)

Even though the old farmer in the interview referred to the past, animal husbandry still has a multipurpose character in South African communal areas today (Cousins 1996:181). Livestock not only has manifold uses, such as draught power and provision of milk, meat, dung and other animal products, but also high economic value, as it can be turned into cash; as one informant put it, 'cattle is our bank account'. Moreover, people are socially tied through livestock, e.g. by cattle loans, bridewealth, labour contracts and payments, and at the same time are enabled to 'maintain meaningful connections with their ancestors' (Ainslie 2013:36; 37) through ritual slaughters at ancestral feasts or funerals.

Many of these uses and meanings of livestock are also relevant in Thaba Nchu. In addition to the above-mentioned sale of animals and livestock products, three quarters of the livestock owners (n=24) in the ethnographic census stated that ritual use of animals was a main motivation for keeping livestock. At the funeral of an adult villager, a cow or ox and several sheep are slaughtered, whereas for ancestral feasts only sheep are used. Considerably fewer people (33.3%) said that they kept animals because they or their sons might use them for bridewealth. Less than 30% of the interviewees stated that they used the animals for consumption. Except for one household all of them owned small stock. The slaughtering of cattle for consumption only, i.e. with no ritual purpose, is not common. Only when a cow died would people eat the meat and share it with family members, friends and neighbours. Non-livestock owners also benefit from cattle through direct use of livestock goods. Dung, for instance, is used as fuel, building material for mud houses and fertilizer for gardens. Moreover, milk is sometimes given for free to people who do not own cattle.

A household's herd is normally kept in a kraal on the plot, though some people with larger herds used empty plots in the village. A herd does not necessarily include only the animals of one owner, but also cattle or small stock of relatives. In some cases, I observed that neighbours combined their herds and shared the work and provisions, e.g. one neighbour was responsible for leaving the animals out in the morning and bringing them back in the evening, whereas the other neighbour provided additional fodder for them. Some livestock farmers formed loose groups in which they contributed cash to buy medicine and vaccinations. This was particularly necessary for the owners of small herds, as they only used small amounts of veterinary drugs. In general, however, cooperation between land users seemed to be rather weak in most villages and was restricted to individual households that knew each other very well.

In the majority of the households (67%) no herder was employed. Those who pay someone to take care of the animals pay a monthly wage ranging from R50 up to R1,000, yet even when a herder is paid this does not necessarily mean that they herd the animals throughout the day. Typically, in the morning herders would drive the animals away from the village, which can take up to a couple of hours. Around midday the herders normally come back to the homestead. Often the cattle will turn back toward the village themselves in the afternoon. Herders meet them halfway and lead them back to the kraal. Rooifontein is an exception in that regard; I observed there that several herders remained throughout the day in the *veld,* a practice that is mainly a result of the village's vicinity to Thaba Nchu's townships and the livestock owners' fear that animals might be stolen. Theft of animals is perceived as a severe problem, and not only when cattle, sheep and goats are grazing. Livestock – even entire herds – are also stolen at night from households' kraals:

> On a morning in October, 77-year old Andrew Thepe discovered that 29 sheep had been stolen from his kraal at night. People were convinced that the thieves had used *muthi* in order to make Andrew and his wife sleep so deeply that they would not wake up at night. The whole issue had stressed him out so that he had felt sick for a couple of days. The news spread fast to other villages and rumours emerged about possible culprits. Village residents, people from town, Basotho herders, the police, everybody was suspected. Yet about two weeks after the incidence, policemen found some of the animals near the Rustfontein dam, which is west of the township Botshabelo. Some days later Andrew had 18 sheep in his kraal again. He said he knew who was involved in the theft but did not want to reveal any names (field notes, Sediba, 20 October 2011 and 4 November 2011).

A recent phenomenon is that of absentee owners, i.e. households that do not live in the rural area but keep livestock there and employ herders. In each village between 5 and 10 absentee owners kept cattle in 2011. Often, the herders employed were not from the village but migrants from Lesotho.

Although some people stated that absentee ownership would not have been possible in the past, it is known from historical sources that people residing in townships gave their animals to relatives in rural areas who claimed that they were their own livestock (Chapter 5.4.3). During my research only a few informants were concerned about absentee ownership. It remains to be seen whether the numbers of livestock of absentee households will increase in the future and whether means are developed for the prevention of this.

Transactions Involving Livestock

There are several forms of animal transaction, namely inheritance, purchase, gift, payment, bridewealth and exchange. All forms can be combined in order to increase the size of one's own herd, yet transactions are not only used to build up a herd: purchase, gift and exchange, in particular, also play crucial roles in the performance of ancestral rituals. For instance, when people have a funeral and are in need of cattle and sheep,[368] they have to purchase animals, obtain them from family members, or exchange them. Since in Thaba Nchu the death of a woman requires the slaughtering of a cow, while a male animal is slaughtered for a man, the exchange of cattle of different sexes sometimes becomes necessary.

Table 16 illustrates the transactions of animals between July and October 2011 that were undertaken by a sample of livestock-owning households (n=12). Four of them undertook no transactions at all during that time period. Looking first at cattle only, one can see that animals were only sold and bought, not slaughtered or given as gifts. Except for one ox, all of the cattle were sold to traders. Asked about the reasons for sale, informants gave different answers. One man sold two of his cows because they were too old. A young couple sold a heifer because they needed the money. The other eight animals were sold by a poorer household who owned 11 cattle in total. The woman heading the household had a sister-in-law (HZ) who was very sick and suffered from diabetes and severe arthritis. She was in urgent need of good medical care as one of her legs had to be amputated. The owner of the cattle sold the animals mainly to provide her affinal relative with money. As the in-law had assisted her financially when the household head's husband and her daughter had died, she felt obliged to

[368] The number of animals slaughtered also depends on the age of the deceased. While for a child only sheep would be used, the death of an old man or woman requires the slaughtering of cattle.

reciprocate now. In the sample only one heifer was bought during the time of monitoring, which was used to build up the household's herd.

Table 16 Transactions of cattle and sheep of 12 livestock owners between July and October 2011[369]

	calve	cow	ox	∑	lamb	ewe	wether	∑
sold	2	8	1	**11**		4	1	**5**
slaughtered				**0**	4	3		**7**
gift				**0**		2		**2**
bought		1		**1**		1		**1**

Whereas cattle were only sold and bought in the sample, sheep were transacted in various ways. Of the five animals sold two were bought by village residents that had ancestral rituals and thus needed meat. Three were sold by a female-headed household. Besides cattle that stayed on a farm, the households owned only these three sheep. The animals were sold because they had destroyed the home garden several times and the household head was tired of the work connected with the small stock. In total, seven animals were slaughtered: four lambs were used for the household's own consumption, but only because they were sick; three ewes were used for ancestral rituals. Two sheep were given away as a gift by a wealthy woman to help her brother, who had lost his wife and used the meat for the funeral. One sheep was bought in Thaba Nchu and also used for an ancestral feast. The data clearly illustrate the important role that sheep play in the ritual practices of Thaba Nchu inhabitants.

Herd Size, Diversification and Reproduction

Cattle found in the Thaba Nchu area are typically mixed-breeds of Nguni, Brahman, Bonsmara, Afrikaner and Brown Swiss. The latter were introduced in the 1940s to increase the milk production of cows. The previous chapters have illustrated that herds were not only improved by the introduction of certain breeds, but also significantly diminished during 'betterment'. The effects of culling are still visible of today, as Figure 26 illustrates for cattle herds. Although there are owners of larger herds in the villages, the vast majority of the resident households (80%) had fewer than

[369] Neither bull nor rams have been sold, slaughtered, given as gifts or bought; thus, they are not included in the table.

ten cattle; almost 50% had only between one and five cattle. That the figures for Talla for the first (1-5 cattle) and second group (6-10) are different compared to the other villages is probably due to methodological circumstances. The assistant in Talla who did the livestock census faced difficulties during the assessment as many herd owners did not want to reveal precise animal numbers. She used categories that I had proposed before, i.e. the first category was 1-4 cattle and the second 5-10 animals. Thus, the categories used in the livestock census and those presented in Figure 26 are overlapping.

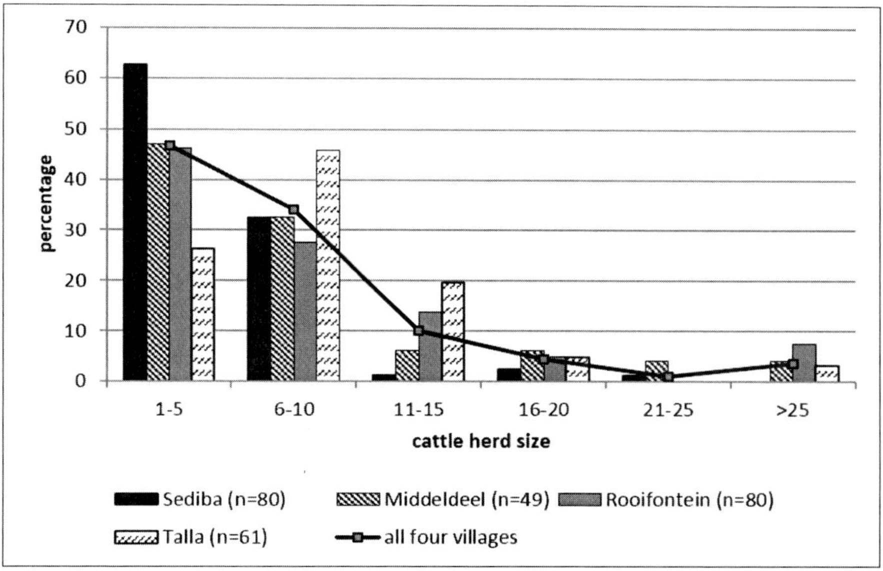

Figure 26 Cattle herd sizes in the research sites (n=270)

Sediba is the only village where the number of owners of large cattle herds is almost non-existent, with only 3.75% of the households keeping more than 15 cows. It is difficult to say why this is the case. Cattle owners stated that keeping larger herds would cause envy among residents (see also Rasch et al. 2016). In the other villages, between 8.2% and 14.3% of livestock owners kept more than 15 cattle. In some of these cases, I collected additional data about the livestock-owning households from the ethnographic census or other interviews. It turned out that these households had an income from wage labour recently or in the past. Salaries enabled them to purchase some animals and to forgo sales in times of financial need, thereby preserving their herds. For these households, cattle had a savings function.

Table 17 Frequency of herd diversification of livestock owning households in Mid-deldeel, Talla, and Rooifontein

	Middeldeel (n=51) Number (%)	**Talla (n=61)** Number (%)	**Rooifontein (n=87)** Number (%)
Cattle	36 (70.6%)	47 (77%)	64 (73.6%)
Sheep	2 (3.9%)		2 (2.3%)
Goats			4 (4.6%)
Cattle & Sheep	10 (19.6%)	13 (21.3%)	13 (14.9%)
Cattle & Goats	1 (2%)		
Cattle, Sheep & Goats	2 (3.9%)	1 (1.7%)	4 (4.6%)
Sum	51 (100%)	61 (100%)	87 (100%)

Although cattle are the dominant type of livestock, some households diversify their herds. Table 17 illustrates the herd diversification of livestock owners in Middeldeel, Talla and Rooifontein.[370] Between 70% and 77% of the households concentrate on cattle only; and between 15% and 21% additionally keep sheep. Only a minority have more than two grazing animal types in the kraal. Particularly in Rooifontein, a diversification with goats is noticeable. Over the past 100 years, goats were never very common in the Thaba Nchu grasslands, but a few households owned small herds. In

[370] Unfortunately, I did no livestock census in Sediba. The economic survey done by my colleagues concentrated on cattle owners. Thus, I do not know the exact number of people who kept only sheep or goats or both. In the village mapping that was done at the beginning of the second research stay, in seven cases it was recorded that the household owned sheep, and one household had goats and sheep. However, as indicated in the methods chapter sometimes these data were provided by the assistant or a neighbour. Hence, here I decided to consider only Middeldeel, Talla and Rooifontein.

the 1940s goats were, together with donkeys, the first animal types to be culled. Recently some farmers had invested in goats and built up a herd. I assume that the fact that goats are predominantly found in Rooifontein is due to the village's vicinity to town, as possible customers live there. Whereas Setswana-speaking people in rural Thaba Nchu hardly use goats for rituals at all, amaXhosa prefer the animals, as one goat owner stressed. An army soldier, he had invested part of his money in goats because the small ruminants achieve up to R1,300 each; i.e. higher prices than sheep. This is not the case in all South African provinces, he explained, but in the central Free State goats are scarce. Whereas the small ruminants are perceived as a worthwhile investment by some households, donkeys are nowhere evident in the four villages.

Which households diversify, then? To answer this question, I compared the cattle herd sizes of people who farmed with cattle only with those of people who owned both cattle and sheep. The cattle herd sizes were significantly different (permutation-based t-test $p < 0.001$) with non-diversified herds consisting on average of 6.96 animals compared to 10.08 cattle in diversified herds. Thus, the data indicate that in particular it is the wealthy households with larger cattle herds that diversify. As in the case of incomes, their herd diversification is a means of accumulating capital. Moreover, a diversified herd increases the household's resilience to price fluctuations and livestock loss through animal diseases.

As there are no grazing camps that allow a separation of male and female animals, the reproduction of the herds is uncontrolled. Although cattle can reproduce every year, the majority of the farmers stated that their cows would calve only every second year. One farmer said 'only if the cow loves you she gives a calf every year'. Sheep and goats reproduce faster and can give birth to lambs or kids twice a year, and may also have twins (cf. Berzborn 2006:246). Yet the possibility of two births per year also depends on the availability of fodder. Moreover, the mortality of lambs is high. In the four months of monitoring, in 7 sheep-owning households between 30 and 40 sheep were born.[371] In the same period nine lambs died, in addition to the four that were slaughtered because they were sick (see above). The low rate of reproduction and survival has several reasons. Firstly, the *veld* is of low quality in winter, which was the period of monitoring. Grasses in the

---

[371] In one month, an owner of a large herd could not remember precisely how many sheep were born.

central Free State grow between December and April, thus animals eat dry grasses from May onwards. After five to six months of dry fodder all the animals are in bad condition if they do not receive additional food and licks (Herman Fouché, personal communication). Whereas commercial farmers supply their herds with these supplements and thereby maintain a good animal quality, the situation is different in the communal area. Although all of the 12 monitored households provided additional fodder for at least one or two months, and 10 even provided salt licks, the amounts fed were in most cases very small. Still, the data of the agricultural economists indicate that there is a positive correlation between the amount of money spend on medicine and supplementary fodder and the birth rate of cattle (Sebastian Rasch, personal communication). Another factor contributing to the low productivity of the herds is a general lack of bulls. Of the 11 cattle owners, who had been monitored, only one owned a bull. The successful insemination of cows is thus left to chance, which is not increased by the fact that grazing camps are not fenced off.

Besides grazing animals, some households in the research areas diversified their herds by keeping pigs. According to an official of the GLEN Agricultural College, pork achieves about R14 to R15 per kilogram (interview, Bloemfontein, 15 September 2010), yet the pig breeders I talked to do not always earn this amount. In the sample, no household was engaged with pig-keeping. In contrast, 40% of the ethnographic survey sample owned chickens and use them and the eggs for own consumption or sale. Cooperative projects dealing with pigs or poultry had been funded by the Bophuthatswana government in several Thaba Nchu villages, but all such projects had run down in the research sites over the last twenty years.

8.5.2 Gardening

In the ethnographic survey 68.8% of the sample cultivated a garden. The size of gardens ranged from 0.1 ha to only a few square metres, which is much smaller than cultivated areas considered by scholars that stress the importance of agriculture as an income generator. For instance, Shackleton et al. (2000) cited several studies in which households cultivated plots or fields ranging from 0.9 ha to 4.5 ha (ibid. 56, table 8). However, they remarked that former 'betterment' regions typically have smaller residential plots than those that were not rehabilitated (ibid. 55). Two or three households in Sediba also cultivated empty plots (about 0.2 ha) for animal fodder, e.g. lucerne, fodder rape or maize. In contrast to pastoral farming, which is normally a consistent, albeit fluctuating, long-term activity, gardening is practiced less regularly. Sometimes gardens lie fallow for several years or people only cultivate when they get seeds for free. These are distributed by several actors, e.g. the Department of Agriculture and Rural Development or World Vision. However, the majority of the gardening households (80%) stated that they normally purchased their seeds. When possible, e.g. in the case of pumpkins, seeds from the previous harvest are kept aside and used the following year.

In my sample, it was typically one household member that was responsible for the garden work and made decisions about cultivation and harvest, but there were also exceptions. Four married couples stated that they made decisions, and often also worked, together. Three wealthy households employed workers for the gardening, particularly for the strenuous turning of the soil. In contrast to the findings of Esterhuyse (2012), who stated that husbands and boys were reluctant to be involved in cultivation (ibid. 99), in my sample slightly more than 50% of the gardeners were male household heads. The observations made in Sediba village support these findings. However, I noticed, as Esterhuyse did, that most children and youth showed little interest in cultivation activities.

Of the 19 households that took part in the monthly monitoring of land use activities, 12 cultivated their gardens between July and October 2011. Except for one case, the households that had no gardens at all belonged to the poor or poorest wealth groups. This indicates that poorer households do not always have the financial means for cultivation, e.g. seeds and tools. Most monitored households used their gardens only in the wet season, but a few also planted vegetables in winter, e.g. peas or beans. In 2011, households started to prepare their gardens by turning the soil between July and September. Done manually by many people, the labour-intensive procedure is normally not finished within one or two days but rather continues for a couple of weeks. After the turning was finished some cultivators applied fertiliser. All of them used organic material based on cow or sheep dung. After the first extensive rainfalls – which was around September and October in 2011 – people planted vegetables. Then, the time-intensive care started, as the gardens had to be cleaned and watered regularly and protected from hungry animals. Moreover, particularly in the first weeks and months after the cultivation plants were very sensitive to heavy precipitation, as young plants can easily be washed away and a season's vegetables lost. Most of the monitored households used no herbicides in their gardens, but there are people who apply industrial products. Pesticides used include Bulldog, Turbo, Metasystox and Roundup. Common mineral fertilizers are Ammonium nitrate and Superphosphate (Kundhlande et al. 2004:239).

8.5.3 Other Types of Land Use

Apart from livestock and garden-related activities, people in the villages use the land for other purposes. Except for one case, all 19 households that were monitored between July and October regularly collected land-based resources. Thus, when considering not only animal husbandry or gardening but all kinds of land-based activities, almost all households use the land in one way or another. Table 18 gives an overview over the most common types of natural resource use found in the research sites.

Table 18 Households collecting wild resources and dung between July and October (n=19)

Resource type	% of households involved
Cow dung	73%
Firewood	57% collected 21% purchased[372]
Wild food plants (*merogo*)	57%
Medicinal plants	42%
Hunting or Fishing	10%

Almost three quarters of the monitored households collected regularly dung between July and October. Whereas livestock owners generally use dung from their own kraal, other people either have to ask for access to kraals, or collect the dried excrement in the rangelands close to the village. The collection of firewood is a more time-consuming activity, as shrubs and trees are normally only found in the mountains. Informants said that they collected only dead wood, but some also admitted that they cut shrubs when they did not find enough dry material. Although it is illegal, on three occasions I observed that people crossed fences to neighbouring private land and cut wood there. Besides the wild wood resources, each village has a small site with *Eucalyptus* spp. which were cultivated in the course of 'betterment'. About every fifth year the trees are cut and given to the villagers. In Sediba people get the wood for free, but in other places people

[372] Some households both collected and purchased wood.

have to pay a fee to the traditional council in Thaba Nchu. The transport of the trees to the village, normally done by one of the tractor owners, is not free. Thus, the cut firewood is not affordable for the poorest. Households that did not collect dung or firewood in the four monitored months almost all belonged to the rich or average wealth groups. This does not mean to say that they never would do so – in fact, when there is an ancestral ritual or funeral households need large amounts of dung and wood because food for the guests has to be cooked. However, the data indicate that in contrast to poorer households, wealthier ones do not have to rely on these resources in their everyday lives as they use paraffin or electricity.

Wild food plants whose green leaves are eaten are called *merogo* (Sg. *morogo*) or *merogo ya naga* (vegetables of the *veld*). The term also refers to the green leaves of some cultivated plants, e.g. pumpkin and beetroot. Normally the leaves are prepared, cooked and spiced like spinach. Many of these plants grow in the gardens and are thus sometimes described in the literature as weeds. However, for rural populations in southern Africa these plants have a considerable 'social, nutritional or financial value' (High and Shackleton 2000:142). The informants of the ecological knowledge survey (n=34) knew on average 5.3 spinach-like food plants (min: 3; max: 10). The most common *merogo,* which were named by more than 75% of the informants, are *Amaranthus* spp. (*thepe*), *Chenopodium album* (*seruwe*), *sepatlapatla* (*Sisymbrium thellungii* according to Thornton-Barnett 2013) and *leswabe* (Asteraceae family, species not identified). Normally these were collected in the household's own garden, but I also observed cases where people asked for access to other gardens and even private farms adjoining the communal areas. When questioned about the names of spinach-like food plants, a male informant answered '*basadi ba itse merogo*' ('women know *merogo*'), indicating that it's normally women and girls who collect these plants. More than half of the households collected *merogo* several times within the four months of monitoring. Households that did not gather wild food plants at all belonged to both the wealthier and poorer spectrum of the village population.

Besides the vegetables of the *veld*, people also eat fruits, roots or flowers of other wild plants. Subsumed under the term *dijo tsa naga* (food of the *veld*), between 1 and 10 species (average: 4.8) could be named by informants in the ecological knowledge survey.[373] Most frequently mentioned were *Cyperus usitatus* (*monakaladi*), *Searsia dentata* (*lebelebele*), *Graziana krebsiana* (*sweswe*) and *lekgaba* (species not identified). Some case studies have shown that the sale of plants and wood products makes a significant contribution to the incomes of Southern African rural households, particularly to women (Shackleton et al. 2001:583). Although I also met two women who sold brooms made out of *Fingerhuthia africana*, this is less often done in Thaba Nchu than in the cases cited by Shackleton et al. (ibid.). This is presumably related to the fact that the research sites are in the relatively close vicinity of urban centres. Studies indicate that less isolated villages also rely less on natural resources than rather remote regions (ibid. 51). 42% of the households collected medicinal plants once or several times during the monitoring. Although the people have access to state healthcare, the data show that natural remedies are still widely used. The importance of medicinal plants is also reflected in the data collected in the ecological knowledge survey (n=34). On average, the informants listed 7.3 names of plants (min: 0; max: 22)[374]. An overview of the ten most frequently mentioned species, their names and their uses is provided in Appendix 11.4.

Whereas the collection of dung, wood and plants was predominantly a female task, hunting and fishing was only done by boys and young men. In summertime, fish are caught in the largely ephemeral rivers or in the Grobler dam in Sediba village. Hunting, which according to a former ranger was not allowed under the Bophuthatswana government, is done throughout the year. Young men meet in the morning or at night and go hunting with their dogs. Others, who do not own a dog, place self-made traps in the rangelands which they check every second day. All wild animals that cross the area are hunted: for instance, rabbits, jackals, porcupines, guinea fowl and smaller antelopes, such as steenbok or impala; though the two-latter

---

[373] In the monitoring, I did not include a question for *dijo tsa naga* as this would have required asking the whole household. For instance, the wild berries or *Searsia dentata* are loved by children, who eat them when playing outside. Thus, unlike *merogo*, *dijo tsa naga* are not always consumed together with the whole household.

[374] One informant stated that he did not know any medicinal plants. However, he seemed to be a bit irritated by the free listing tasks generally, so I assume that he just gave the answer to hasten the end of the interview.

species are only found on rare occasions. Asked about their motivations for hunting, a group of young men stated that it's firstly for the meat, which they share with each other. The purchase of meat is too expensive and hunting is a good alternative. Secondly, they perceived it as a kind of sport, as they had to follow their dogs when they ran after the animals. One man said it is like a competition as every dog owner wants his dog to win the hunt. Thirdly, the skins of the animals were used for the making of clothes. Some of the boys and young men were part of dancing groups, which exist in most villages. When they have a public performance in the village, they wear dresses that resemble 'traditional' Batswana clothes and are made out of these skins. The best hunting grounds, the young men stated, were not the hills but the former fields where predominantly rabbits could be found. Sometimes they would hunt down up to seven rabbits within a day; however, on other days they came home empty-handed.

Figure 27 Young hunters (photograph: N. Dreber)

In summary, most rural households are still involved in land use activities. However, there are striking differences between wealthier and poorer households. The former diversify their herds and some of them, as illustrated in the previous chapter, intensify vegetable cultivation by applying tested seeds and fertilizers. Herd diversification is particularly conducive to improving a household's resilience against dry spells, livestock diseases and price fluctuations. By building up larger herds consisting of different types of animals, these households accumulate capital, which strengthens their ability to cope with future financial crisis. Poorer households, on the other hand, cannot afford herd diversification and sometimes not even the cultivation of a garden. Much more than the wealthy, they depend on natural resources, especially dry dung, as an energy source for cooking and heating. That most households in the Thaba Nchu area are in one way or another engaged in land use activities indicates that livelihoods and land are not completely decoupled. However, the human-environment relationship has transformed considerably and the coupling has lessened in intensity. The importance of food production has clearly decreased. Although gardening adds to many households' livelihoods, its significance for consumption is in most cases rather ephemeral throughout the year. Instead, livelihood security is achieved through social transfers and wage labour.

8.6 Declining Support and Cooperation?

Besides livelihood diversification, mutual cooperation and support networks are further means to minimise risk and uncertainty (Bollig 2006; Berzborn and Schnegg 2007). Before the resettlement in the mid-1960s, one or more agnatic family groups lived together in smaller settlements. I have shown in Chapter 3.3 that family groups constituted important units of social organisation. These groups were involved in marriage negotiations, rituals and conflict resolution, and members cooperated in agro-pastoral production and 'helped one another with gifts of food, livestock and other commodities' (Schapera 1994:16). Mutual support within the patrilineal family group, particularly helping the poorest members of the group, was an obligation; or, as an older informant put it, 'people [...] used to fill *pitse* [pots] for people who have nothing; your children would never sleep without food at that time' (interview Lydia Pobe, Sediba, 4 September 2011). Today, in comparison, informants depicted other people in their communities as 'selfish' and 'uncooperative'.

It would be wrong to argue that mutual support is non-existent. Households have broad social networks through which they gain emotional, material, financial and advisory support from relatives, friend or neighbours (Esterhuyse 2012:101). I did not collect data for a proper network analysis, but the ethnographic survey assessed regular support in cash and kind. More than 30% of the households received money, clothes, shoes and food from non-household members living in the same village, in other Thaba Nchu communities or further away. In the majority of these cases, the supporter was a relative of the household head or another member. Less frequent was support from friends. Two retired domestic workers still received financial help from their former employees or the employee's children. Thaba Nchu households are not only receivers but also support others. 15% stated they would assist others with cash or goods. Interestingly, the majority of them also received help.

The interviews, along with observations made in the field, suggest that both generalised and balanced reciprocal relations are maintained by households. In the past, the former type of reciprocity was mainly found in the patrilineal family group. Although people still maintain close social ties to their relatives today, the obligation to support the poorest members of one's family group seems to have lost importance. When a single mother has no money to buy food for her children at the month's end, she will not necessarily be supported by all agnatic relatives, but mainly by those with whom she has a close relation anyway. Thus, poor households are still assisted by certain relatives in times of need, yet the rendering of assistance to all members of the family group *per se* is not normatively demanded. Still, a differentiation should be made here between close relatives – e.g. parents, children, siblings and grandchildren – and others. Among the former, assistance in times of need is a social norm that should be complied with.

Today, the wider family group still plays a role with regard to other functions, for instance the performance of ancestral rituals. Thus, it is rather the day-to-day assistance between members where the social unit's importance has decreased. I suggest that the supportive function of the family group was transformed after the resettlement of households: with the destruction of the territorial units, the localised kinship group became scattered in Trust villages, and some relatives had to leave the area for good. New relations had to be built with neighbours and other non-related people, which replaced some of the interrupted relations to patrilineal kinsmen. This may be the reason why older people perceive that support has declined: reciprocity within the patrilineal family group was rather generalised, whereas relations to friends and neighbours need more attention and are instead based on balanced reciprocity. Still, better social network data are needed to support this assumption. It is clear, however, that support networks, i.e.

to have 'helping hands', as one woman expressed it, increase the social resilience of rural households.

In the historical chapters I have shown that institutionalised forms of co-operation were especially important in the agricultural production. The formation of *letsema* was necessary in environments where the harvest of several hundred hectares sometimes had to be finished within days; yet the *letsema* practice was increasingly contested by wealthier farmers who the possessed agricultural capital and wanted a financial compensation for their assistance. Today, older informants stated that institutions such as *letsema* would not work anymore as everyone expected money in return for help. Indeed the 'monetisation' of cooperation (see also Berzborn 2006:352–53) was observed on several occasions in the field. For instance, when the public taps in Sediba were out of order people were forced to get water at the provisional tanks erected by the municipality, from the lorries that came once or twice a week, or at one of those households that had a private hand pump. For older people, the transport of several 20-litre canisters was too strenuous. When they had no helpers available in their own household they normally had to pay someone for assistance. One Sediba resident offered to transport canisters with his horse cart for R2 per 20-litre canister, or R2.5 for 25-litre. Buying water from a household with a hand pump cost about the same, but then the transport still had to be paid. Even when asking the neighbour's children, one woman complained, she had to pay them, as if she refused to give money they would never help her again. A situation of need, such as very limited access to water, was thus used by some inhabitants to make money.

Households of a community still cooperate in times of sorrow: 'in death […] they really associate […] they heartily do their work, really sincerely […] that time they really unite, until everything is over' (interview Lydia Pobe, Sediba, 4 September 2011). Mourning households get support from many people in the village, not only their family members. In the days before the funeral, people gather and help with preparations. Men slaughter cattle and sheep, collect and cut wood, and dig the grave, whereas women cook for the bereaved family and their helpers, wash the dishes, and prepare the food for the funeral. When the deceased's family are poor, other people come and bring some food so that visitors who offer their sympathy will be served some juice or snacks. Informants stressed that mutual assistance is important in such cases; not only because the bereaved need help to organise the funeral but also because it was expected by the community.

'When you do not help at funerals', one older man explained, 'nobody will help when you are dead' (informal interview with Frans Kganare, Sediba, 27 April 2011).

Cooperation in times of death was observed in all four research sites, yet interviews indicate that there might be some differences with regard to cooperation amongst the villages. In Middeldeel, for instance, the cooperation of pastoral farmers is said to be better than in other villages (interview with extension officer of the Department of Agriculture, Sediba, 17 September 2011). This is supported by the fact that each week Middeldeel holds community meetings, which was not the case in Sediba, for instance. Middeldeel residents collaborated also on other occasions: a couple of years before I went to Thaba Nchu a young woman from a wealthy household realised during a funeral that the village's cemetery was in desperate need of a proper clean-up. She organised a get-together that followed the rules of a *letsema*. After announcing her plan to clean the cemetery, the woman bought ingredients for traditional beer and a meal in order to 'compensate' about 20 helpers that came to work at the graveyard. Whether such cooperation is indeed only the result of a certain 'spirit' that Middeldeel people have – as my assistant always suggested – is difficult to say. When considering the origin of the village, one could assume that the joint resettlement history that Middeldeel families experienced (Chapter 5.4.1) had a positive effect on the cooperation of resident households (see Naumann 2014), but this is impossible to prove. What is clear, however, is that cooperation is not completely monetised as many statements by elders might suggest.

Savings and Credit Associations

A specific form of financial cooperation must be addressed separately, namely the saving and credit associations. Often subsumed under the name *stokvel*,[375] this form of cooperation is widespread in South Africa (Schulze 1997a; Verhoef 2001; Bähre 2002) and also plays an important role in the research area. A *stokvel* consists of a group of people that voluntarily agrees to 'regularly contribute money to a common pool and circulate the pool among the group' (Verhoef 2001:263). The first *stokvel* groups

---

[375] According to Schulze (1997), the term *stokvel* derives from the 'stock fairs' held by English-speaking farmers in the nineteenth century, which were also attended by black labourers who used them as social and economic gatherings (ibid: 20).

emerged in urban areas, where they have been common since the late 1920s. Many of them were established by women as their membership in these associations enabled them to earn supplemental income (Schulze 1997a:20; Verhoef 2001:265). Today, various forms exist, with different purposes and regulations. In Sediba village three main types of *stokvel* associations were found: (1) savings clubs, (2) burial societies and (3) borrowing associations. Some people were involved in several of these societies.

> When asked about forms of cooperation, Rose Sebitloane told me that she is part of a saving society that has more than thirty members. She called it a kind of 'do-it-yourself' society, as the members helped themselves financially. Each member contributes R50 per month, and funds are saved in a bank account in Thaba Nchu. Once a year, mostly at the end of October, the money is withdrawn and each member gets an equal share. Members use the money for expensive purchases such as microwaves, ovens or the like. The monthly saving of money for such goods was easier when one had an obligation to contribute a certain amount to the pool (field notes, Sediba, 21.07.2011).

Some saving clubs use the money for specific goods, for instance to buy food or pots. Burial societies are associations that provide financial and organisational help for the funerals of deceased members. Costs for a decent funeral are constantly increasing. Today, two older women of Sediba explained, it is expected that funeral expenses will include buying an expensive coffin, renting a tent for the service, and offering an impressive meal with two types of meat and several types of vegetables for up to several hundred guests. At a funeral whose preparations I attended, the renting of a bus for R2,000 was discussed in order to transport the mourners from the homestead to the graveyard 2 or 3km away. Families face high expenditures for funerals, which they often cannot finance without assistance. Bähre (2002) argues that burial societies are 'demutualising' (ibid. 19), as insurance companies offer policies that cover the costs of the funerals and also provide administrative assistance of the ceremony. I have no quantitative data about the importance of such insurance for rural households in Thaba Nchu, but many people who can afford it do insure themselves and their families against such events.

The third type of *stokvel* is the above-mentioned borrowing society. Variations exist here too, but a general practice is that each member contributes either a larger amount at the beginning of the year or a smaller monthly amount. The fund is then used for credits or loans that can be borrowed by both members and non-members at a fixed interest rate. These vary greatly from society to society and range from 20% up to 50%.[376] As with the saving societies, the profits are shared among all members of the association at certain intervals. The lending of money is a highly lucrative business – for individuals, too, in the case that they can afford to lend. One woman interviewed in the ethnographic survey even mentioned credit-giving as one source of income, since she received monthly interest. I gained an impression of the overall dimensions of capital flow within the village for the first time when observing people's activities on the payday for the social grants:

> After Kedikilwe Olifant collected the grants for her children she left the premises of the 'Thusong Service Centre'. As soon as she stepped through the gate, her creditors approached her. The pay point was an ideal place to collect outstanding debts. Everybody knew that the borrower had money now, and thus could not use an excuse to postpone the payback. Money was passed from hand to hand. Borrowers paid their creditors, who sometimes also had to pay their own debts. Kedikilwe Olifant was in debt to one woman who had sold her a suitcase for R100. Another woman requested R200 that Kedikilwe still owed her. Ten minutes after receiving two child support grants amounting R520, she was left with R220. Obviously, this was not enough for the whole month and Kedikilwe had no choice but to obtain another loan from one of the societies (field notes, Sediba, 4 July 2011).

The vignette illustrates the advantages and disadvantages of these money-lending associations. On the one hand, people with no access to formal financial markets are provided with economic capital (Verhoef 2001). Moreover, members of these societies are able to save larger amounts of money for expenditures that they cannot afford alone. In other words, membership in such societies makes a transformation of social capital into

[376] High interest rates of 50% were also reported by Bähre (2002) who did a research on financial support among amaXhosa migrants in Cape Town (ibid. 29).

economic capital possible. On the other hand, money-lending can also create vicious cycles of debt for the poorest, particularly when interest rates are extremely high.

8.7 Preliminary Summary

Today, rural communities in Thaba Nchu can be characterised as post-agrarian societies. Since the working-age population seeks employment in urban centres, villages are the home of children and elderly people. The patterns of labour migration have considerably changed since 1994, with a higher number of women who now leave rural areas. The investment in, and accumulation of, cultural capital in the form of grades has become more important in the past decades, which is visible in many ways. The period of education is extended, and most young people attend school until the age of 20. From the perspective of youth, wealth – besides that of money, livestock holdings and commodities – is reflected in a person's level of education, and many people regard a college or university degree as the key to economic success. However, access to tertiary education is confined to wealthy families who can transform their economic capital into cultural capital.

Livelihood diversification 'is the norm' (Barrett et al. 2001:315) in Thaba Nchu villages. Wage labour, remittances, social grants, agricultural production, petty trade and piece jobs are the most important income pillars of rural households, but social transfers in particular provide a stable income for most people. Wage labour, especially permanent jobs in government departments and the private sector, generates more cash than the other strategies; however, not everyone is able to find a job. The recent transformations of the South African economy have resulted in a decoupling – at least in part – of Thaba Nchu communities from labour markets, which is reflected in high numbers of unemployed people. The comparatively high old-age pensions, on the other hand, offer a regular income that enables households without employment to survive. To increase their resilience, households engage in piece jobs and – in the case of the better-off households – invest in more favourable income spheres such as petty trade. Diversification is not a new phenomenon, yet the social transfers that became increasingly accessible after 1994 changed the composition of 'livelihood portfolios' (Boyd and Folke 2012:10). Certainly, these transfers are an important means of poverty alleviation, and increase the resilience of rural households. However, as they also create a stronger dependency on state support, it is questionable whether this form of resilience is indeed sustainable in the long term. Moreover, for all households the state funding is only accessible as long the person entitled to transfers is still present.

To increase their resilience, households pursue different strategies: wealthy households accumulate different types of capital, invest in cultural capital and diversify their livelihood portfolios and livestock herds. Several engage in petty trade and sell commodities in their homes. Poorer households diversify too, yet they choose other means. By offering their labour force for piece jobs, they earn some money without making larger investments themselves. Often, they maintain close social ties with their employers, thereby increasing their chance of obtaining access to further piece jobs. Last but not least, for all households, be they wealthy or not, the maintenance of social networks is crucial to resilience, as these networks provide emotional, economic and advisory support.

All households are in one way or another involved in land-based activities, which supports the arguments of scholars who stress the importance of land use for rural communities. However, the importance of recent land-based activities cannot be assessed without considering the historical perspective. I argue here that there is a decoupling of rural livelihoods from land and farming (ibid. 183). Today, households do not produce a significant proportion of their food or make a living from agriculture. The cultivation of fields that was the main basis for the household's food security in the past has almost completely vanished in the entire communal area. The loss of this livelihood base, one could argue, is today replaced by social transfers.

A similar decoupling of social and ecological components of the system can be observed with regard to institutions. Patron-client relationships that were previously established through the lending of *mahisa* cattle are based on money today. Institutionalised forms of cooperation constituted the foundation for agricultural production in the past, but today only a few households in each community join together and cooperate in terms of animal husbandry; in the case of gardening, such collaboration is even less frequently observed. Still, larger groups of people now participate in savings and credit associations where they are able to gain access to economic capital. Thus, it would be wrong to argue that cooperative behaviour is not found on the community level at all; rather, it is found in other domains than land use.

9. CONCLUSION

This book has offered a historically and ethnographically informed case study of environmental governance, institutional and land use change, and livelihood strategies in rural communities in the South African Free State province. Based on rich archival material, I have reconstructed how state authorities rationalised and legitimised the regulation of human-environment relations in rural Thaba Nchu. It has been shown that the motivation for intervention was not so much based on ecological concerns, but on a political master plan that aimed at racial segregation and the creation of a labour reserve. With regard to political motivation, Thaba Nchu was a typical case of 'betterment'; however, Thaba Nchu was special in terms of the point in time of intervention, as it was one of the few districts that were identified as soil erosion hotspots in the early 1930s. Moreover, Thaba Nchu was also one of the districts where extensive livestock culls had been enforced shortly after the implementation of 'betterment'. In many other areas, culling had only started after the Second World War (Beinart 1984:73). After implementing measures for soil conservation and livestock improvement, the second phase of 'betterment' focussed on a stabilisation of rural areas.

South Africa's 'betterment' approach was not unique. Throughout the world similar social engineering schemes had been implemented, of which many have failed over the long term. In his seminal book *Seeing Like a State* James Scott (1998) identifies four elements that are inherent in many of these schemes and that can also be identified in the present case. First, nature and societies had to be 'simplified': i.e. settlements were surveyed and mapped, inventories were made, land subdivided into 'discrete categories' (Neumann 2005:121) and households eventually ordered in blueprint-based grid-patterned villages. This standardisation was necessary in order to make rural populations more 'legible' (Scott 1998:2) and thus more easily controllable. Second, engineering schemes reflect a high-modernist ideology drawing on 'Western' science and technology. However, as this book has argued, the methods and technologies applied were not always sound and appropriate. In Thaba Nchu most scientific assessments of ecological and social conditions remained superficial. Moreover, the introduced technologies and materials were often as cheap as possible. Third, the high-modernist thinking implemented in these schemes becomes particularly problematic if an authoritarian state is willing to enforce the ideology by any means, which was clearly the case in South Africa. And fourth, connected to that, is the issue of the existence of a civil society that 'lacks the capacity to resist the [government's] plans' (ibid.:5).

Despite the similarities to Scott's analysis the South African case also differs from other social-engineering schemes as it was less development-oriented than other projects, such as the *ujamaa* villagisation programme in Tanzania in the 1970s. Even though conservation and development were the official justification for state intervention in South Africa, the underlying rationale can rather be found in the state's pursuit of racial segregation and apartheid. Land was appropriated and settlement patterns destroyed in order to create enough 'space' for 'surplus people' who had been evicted. The improvement of 'backward' land-use methods was said to be necessary in order to make agricultural production more efficient, which in turn would allow more people. In addition to that, the South African state aimed to maintain a reservoir for cheap labour power in the reserves. This was necessary in order to protect the South African capitalist economy, particularly the mining and the agricultural sector. Both sectors where constantly in search for cheap workers. A restriction of access to land – as enforced by the 'betterment' – forced all rural families to engage at least temporarily in labour migration. Hence, the rationale for 'betterment' – at least in its later phase – emerged from political motives rather than ecological concerns.

Control – over people and resources – was the motivation for intervention, but what conditions made it possible? Jacobs (2003) points out that communal tenure was a prerequisite for state involvement. Indeed, coercive measures such as the culling of livestock were not enacted on the farms owned by Barolong, let alone land owned by white farmers. Moreover, the structures of indirect rule provided the political conditions for an enforcement of drastic interventions on the local level (ibid. 205). Without the cooperation of the Barolong elite, 'betterment' would have been less 'successful' in Thaba Nchu. Finally, I have argued here that the invention of a degradation narrative was yet another condition of this control because, in the words of Fairhead and Leach (1996), it constructed 'inhabitants as incapable resource stewards, while simultaneously instilling the imperative to intervene and improve the situation on their behalf' (ibid. 292).

Ethnographic data collected during field research indicate that Thaba Nchu communities have increasingly transformed into post-agrarian societies in the second half of the twentieth century. Despite the agro-pastoral activities in which most rural households had been involved in the past, wage labour became a more important remuneration strategy. Compared to other regions in southern Africa (see for instance Murray 1981; Sharp and Spiegel 1985), past patterns of labour migration had been different in the studied rural communities: many people worked in the railway industry and the domestic sector in nearby Bloemfontein. Historical labour migration thus resembled the conditions Berzborn (2006) found in the Richtersveld, where migrants were able to keep stronger ties with their relatives and engage in land use activities due to the proximity of their homes to their workplaces in the diamond mines. In contrast to the Richtersveld, where wage employment in the mines still played a crucial role for rural household economies after the end of apartheid, the job opportunities for people of rural Thaba Nchu drastically declined after 1994.

In the introductory chapter of the thesis, I proposed several theoretical questions and hypotheses concerned with social-ecological systems, coupling and decoupling processes, institutional development and social resilience. In the following sections of my conclusion, I will summarise the main findings.

9.1 Reframing the Social-Ecological System Approach

In a framework for complex adaptive rangeland systems, Walker and Abel (2002) visualise the interconnections of different scales and how these scales constitute a nested hierarchy. One of the variables the authors consider is termed 'colonial history'. However, in their diagram, the system's socio-political history is only considered as important for the understanding of processes on the national scale, but not on the regional and local scale (ibid. 304–05). I do not want to imply that the authors are ignorant of historical aspects and their role in local processes. However, their framework for complex adaptive rangelands systems is symptomatic of approaches toward social-ecological systems, as it does imply that past policies, debates and events are of secondary importance for the understanding of recent processes and conditions in such systems. As they often neglect historical context, studies of social-ecological systems have been criticised as 'ahistorical' by social scientists. Moreover, many studies are 'apolitical' too, as they tend to ignore stratification within agrarian societies, different motives and interests of actor groups and 'the crucial role of sovereign authorities' (Swanstrom 2008:16) in regulating human-environment relationships. However, a consideration of historically evolved power inequalities, competing interests of social actors and contestations over resource use and management are indispensable for explaining recent configurations of social-ecological systems (Beymer-Farris et al. 2012). Therefore, this study has included perspectives of political ecology and environmental history that pay attention to some of the 'blind spots' of social-ecological system approaches.

Ann Kinzig (2012), who has published extensively on social-ecological systems and resilience, states in a recent contribution that case studies using a social-ecological system approach should be concerned with the system's embeddedness in dynamics at broader regional, national and global scales (ibid. 317–19). I have tried to illustrate that such scale linkages are not a recent phenomenon induced by progressive globalisation; past events in the Thaba Nchu social-ecological system were connected to other scales, too. After the settlement of Barolong in the area, the system was never entirely 'isolated' or 'closed'. Throughout the region's history explored here, Thaba Nchu households and communities were linked to the broader economic system, be it through involvement as transport raiders in the nineteenth century or by sending labour migrants to South African Railways or other companies throughout the country during the twentieth century.

On the other hand, both communities and individual households were directly and indirectly affected by national policies and bureaucratic institutions that reorganised land management, settlement patterns and social and political organisation. Officials who passed these laws were in turn influenced by global events – for instance, the American Dust Bowl in the 1930s and political upheavals in other parts of Africa in the 1950s. This study especially highlighted the linkages between the local, regional and national levels and focussed on the state as an agent that regulated land management. Others have shown that the state is not necessarily the only actor that has previously or recently intervened in rural Africa: churches and missionaries (Grove 1989; Showers 1989; Comaroff and Comaroff 1997), non-governmental organisations and international donors (Neumann 1997; Ferguson 2003), and global private investors (Peters 2013), have played – and continue to play – crucial roles in shaping discourses about rural Africa, regulating land use and appropriating natural resources.

Thaba Nchu was no case of a clear-cut 'them' – i.e. apartheid officials – against 'us' – the local population. Rather, the study has shown that at the local level, too, various groups pursued their own interests and had different levels of bargaining power in enforcing them. The land-owning Barolong elite supported the introduction of 'betterment' in the Thaba Nchu reserve. Since they used the rangelands in the Thaba Nchu 'native reserve' for their own stock, they had an interest in reducing the grazing pressure. The elite's critical role in implementing 'betterment' in the Sediba area reflected their intentions to gain more control over the renegade population in the northern reserve. However, in the Sediba reserve, different actor groups had also diverging interests. Households with livestock and access to arable lands feared that 'betterment' would undermine their livelihoods and subsequently lead to a decline of their resilience. On the other hand, households with less economic capital – those without land and livestock – hoped to benefit from 'betterment' by getting full access to natural resources which would enable them to diversify their incomes. Translating the case of Thaba Nchu into a more general insight into social-ecological systems and resilience (see also below), the study suggests that indeed – as others have pointed out (see Cannon 2008; Keck and Sakdapolrak 2013) – some actor groups have an interest in excluding others from improving their resilience.

9.2 Coupling and Gradual Decoupling

A central assumption of the social-ecological system approach is that the system's components are intricately coupled. This assumption implies that a contrary state is possible too: components of a social-ecological system can be decoupled, or in other words, feedback between different components can be disrupted. Some scholars have provided case studies for decoupling processes, yet they tend to romanticise past social-ecological conditions (e.g. Li and Li 2012) or present clear-cut examples of displaced communities that have been decoupled from their specific environments (e.g. Hoole and Berkes 2010). Particularly with regard to the latter aspect, the book has shown that decoupling does not necessarily affect all parts of a system.

In rural Thaba Nchu, people still live in their environment and engage in livestock husbandry and gardening. Thus, they are not entirely 'detached' from the ecological system, yet a decoupling of rural livelihoods from natural resources took place during the twentieth century. Even though rural households still use natural resources today, the importance of wage labour and, more recently, social transfers for economic survival and wellbeing is undeniable. However, it is not only the significance of salaries and grants that indicates a decoupling, but also a decline of feedback between land and livelihoods. It is unlikely that a severe drought, as happened in the early 1930s, would have the same devastating effects (i.e. starvation) today. Strikingly, the drought of the 1930s is most salient in the memory of elders, despite the fact that dry spells occurred during the second half of the twentieth century too. I suppose that the effects of the later droughts, such as livestock loss and harvest failure, were buffered by incomes from wages.

The decoupling of livelihoods from land is a particular characteristic of post-agrarian societies (Bryceson 1997b). The Thaba Nchu case illustrates that the South African state played a critical role in forcing a decoupling process. However, as other studies have suggested, in some former homelands livelihoods are still comparatively closely coupled with natural resources (Shackleton et al. 2000). Why, then, is the situation different in Thaba Nchu? First, the communities studied are not as 'deep rural' (Shackleton et al. 2001:583) as other communities that depend more on the natural resource base. A study of three South African villages by Shackleton et al. (2002) showed that households in very remote places without easy access to alternative markets were consuming more natural resources than households in a setting similar to Thaba Nchu. Second, residential plots laid out during 'betterment' were often much bigger in areas in which gardening was still a central livelihood base. Third, natural woodland, which constituted one of the most valuable resource bases for communities in which livelihoods were still closely coupled with land (see for instance Shackleton et al. 2002), was less important in the Thaba Nchu area, which is situated in the grassland biome. Still, the survey data suggested that differences exist between poor and wealthy households. For instance, those households that did not regularly use dung or additional wood resources for cooking and heating belonged to the wealthier groups. Thus, the data indicated that decoupling is often not an 'on' or 'off' state, but rather a gradual process.

In the mid-1960s, several households had to move from the Sediba reserve and other rural communities to urban townships. For these families, resettlement not only meant a disruption of ties to relatives and neighbours, but also a decoupling from their specific natural resource base. Other households in the Sediba reserve were relocated only a few meters or kilometres away, yet found themselves in a new social, economic and ecological setting. The family group as a residential unit was spatially disrupted as the new residential plots were to be occupied by only one household. Influential men (e.g. headmen) could arrange the new settlement patterns in a favourable way by placing their relatives on neighbouring plots. However, less powerful, marginalised and uncooperative people had no choice of where to stay. The relocation of newcomers from white farms on the former homesteads of long-established family groups resulted in conflicts. Moreover, the settlement of strangers on ancestral homes created additional antagonisms in a culture in which the worship of ancestors is a crucial part of life. Finally, the reorganisation of villages directly affected the ecological system. Settlements were removed from mountainous, rocky areas to other sites that had been used as grazing lands. By defining clear-cut land categories – i.e. arable, residential and grazing land – a new coupling was created between land and the type of use.

The Role of Institutions

It was pointed out in the theoretical framework that social institutions play a critical role for coupling within social-ecological systems, as they structure humans' engagement with their environment. Hence, a discontinuation of these institutions would indicate a decoupling of system components or at least a change of coupling. In Thaba Nchu, institutionalised forms of cooperation in the agricultural sector ceased to exist. Ethnographic data indicate that reciprocal harvesting groups (*letsema*) were still common in pre-'betterment' times. However, to suggest that it was first and foremost the state's intervention that undermined the formation of working groups would be misleading. Even before the top-down-implemented restriction of field sizes was enforced, institutional arrangements of agricultural cooperation had changed. Wealthy households negotiated existing norms and found more profitable arrangements by ploughing and harvesting other households' fields for money or a part of the harvest. Thus, the replacement of reciprocal forms of cooperation by monetary arrangements was a development initiated within the social-ecological system.

Additionally, the lending of *mahisa* cattle that constituted patron-client relationships in the past ceased to exist. In contrast to *letsema*, the discontinuation of cattle-lending institutions was mainly a consequence of state intervention. Livestock numbers of basically all households were reduced to a minimum by forced culls, which made it impossible to maintain cattle-lending networks with other households. Today, wealthy households pay other people for services, e.g. herding, laundry or cleaning. The employment of people for piece jobs resembles former patron-client relationships, as wealthier households allocate jobs not randomly, but to specific households. Employers and piece jobbers often maintain additional social relations that are reproduced by the exchange of gifts and visits. Thus, *mahisa* cattle have been replaced by money and, accordingly, patron-client relationships are no longer directly coupled to the ecological system.

The South African state played a critical role in establishing institutions that couple components of the social-ecological system. In the communities studied, formal rules that aimed at a regulation of land use were introduced in the early twentieth century. These reserve regulations were the first attempt in formalising land types and land use patterns, yet they were extended in the following decades through additional regulations. Some bureaucratic institutions, such as the Land Act of 1913, were in contrast with the social practices on the local level and thus did not directly take effect. Others, such as forced livestock culling, undermined socially-embedded institutions and resulted in their discontinuation. One common feature of state-introduced institutions was that they were – and are still today – based on scientific knowledge which legitimised the intervention from the state's point of view. However, the allegedly objective measures on which institutions were based were adopted very flexibly, as the example of the carrying capacity shows. In such cases, institutions did not create a real coupling of social and ecological components of a system, but rather an ideological coupling: the carrying capacity was first and foremost adapted to political needs and not to ecological ones.

State-implemented institutions had a multipurpose character. Officially, they sought improvement of land use methods and the conservation of natural resources, but they were in fact used to increase control over local populations. The coercive enforcement of conservation by the state is not specific to South Africa. In the context of international environmental agreements, Peluso (1993) has shown that some nation-states – she takes the examples of Kenya and Indonesia – use the 'ideology, legitimacy, and technology of conservation as a means of increasing [...] their control over valuable resources and recalcitrant populations' (ibid. 199). The appropriation of a conservation ideology as a political legitimation for intervention is therefore not confined to colonial or authoritarian states that strive for racial segregation.

In the long term, many of the top-down institutions that regulated land use patterns were not enduring. After the end of apartheid, rangeland infrastructure in the form of fences and water points broke down and rotational grazing management was discontinued. In most of the communities visited in 2011, the regulation of rangeland use was reduced to a minimum, i.e. exclusion of grazing livestock that did not belong to the community. A main reason for the collapse of rangeland management institutions was the abolition of the rangers who enforced these regulations on the local level. Other institutions, such as the definition of residential areas, were maintained: people did not return to their original settlement sites after the end of apartheid as happened in other areas, such as Hobeni in Eastern Cape (see for instance Fay 2012).

9.3 Social Resilience

Over the last few decades, resilience has emerged as a buzzword in several disciplines. Scholars that investigate social-ecological systems have defined resilience as the system's capacity to cope with shocks and to maintain and develop major functions, structures and feedback either by recovering or reorganising. However, the question of whether or not an entire social-ecological system is resilient cannot be answered with 'yes' or 'no' (Kinzig 2012:316). Instead, the researcher has to define which structures and variables of the system are of interest (i.e. 'resilience of what to what?' Carpenter et al. 2001) in order to make conclusions about the degree of resilience, i.e. whether a system is highly resilient or less so (Kinzig 2012:316).

In this book, I have focussed on the concept of social resilience, which is defined as the ability of a household or community to cope with, adapt to and transform in ways that secure future wellbeing. 'Coping capacities' refer to the ability to deal with immediate environmental or social threats and shocks in order to maintain wellbeing in the short term. The term 'adaptive capacity' is the ability of social actors to learn from past experiences and transformations and adapt their strategies in order to prevent and cope with future threats and shocks. Thus, 'adaptive capacity' is related to what other scholars call 'buffering mechanisms' (e.g. Bollig 2006), which refer to institutions that 'lower the vulnerability of communities and enhance their potential for quick recovery' (ibid. 269). Finally, actors have 'transformative capacities' when they are able to gain access to assistance from governmental and non-governmental organisations, take part in decision-making processes and shape institutions that improve future wellbeing. To operationalise social resilience, I have identified surrogates that indicate resilience at the household level (i.e. livelihood diversification and the accumulation of different types of capital) and at the community level (i.e. maintenance of networks for support and cooperation).

Social Resilience from a Historic Perspective

Common drought-coping strategies of pastoral societies have been described, for instance, by McCabe (1990) for Kenyan Turkana herders and by Bollig (2006) for Kenyan Pokot and Himba in Namibia. These strategies were applied during a period of environmental crisis and included *inter alia* an increase of slaughter, changes in food habits, sale of livestock, spatial mobility and the generation of income from low-paying jobs. The oral histories collected in Thaba Nchu indicate that some of these coping strategies were also applied in the area in the twentieth century. During the severe drought in the early 1930s, households changed their diets by eating the meat of slaughtered animals and protein-rich locusts. Moreover, people worked temporarily for food only and thus diversified their livelihood strategies by engaging in unpopular, poorly paid jobs. In contrast, spatial herd mobility, as found amongst Pokot and Himba pastoralists (Bollig 2006:210–31), was no option in the small Sediba 'native reserve' surrounded by privately-owned farms. Today, people no longer depend on self-produced food, and hence the importance of the drought-coping strategies applied in the 1930s has declined: instead of slaughtering animals during dry spells, farmers either do nothing or sell some of their livestock and buy additional fodder for the remaining herd.

In the past, Thaba Nchu households found several ways to buffer their vulnerability to environmental shocks. By maintaining the production of drought-resistant grains, such as sorghum, households tried to prevent food crises in the first half of the twentieth century. Livestock-owning households that could afford it have diversified their herds by keeping cattle, sheep, horses and, more recently, goats. In a similar way to the lending of *mahisa* cattle, the herd-diversification strategy enables farmers to cope better with animal diseases. At the same time, diversification reduces a herd owner's vulnerability to fluctuations of livestock prices. Not only herds, but livelihoods in general have diversified in the past century. Men, and some women as well, left the reserves to search for permanent work that would make a significant contribution to their households' livelihood. Moreover, parents sent their children to school, where they were equipped with capacities that improved their access to the labour market. The long-term diversification of livelihoods as a strategy to minimise future risks is also found amongst Himba and Pokot pastoralists (see Bollig 2006). However, the Thaba Nchu case and South Africa as a whole differ in one aspect: the very nature of South African capitalism required the constant circulation of migrants. Minimum wages were paid in order to tie the migrants to rural areas, where they could earn additional income from agriculture (Wolpe 1972). On the other hand, the restriction of resource access and increasing population pressure ensured that households would not be able to produce self-sufficiently. Therefore, livelihood diversification was not only a strategy pursued by rural households, but a political aim of the South African government.

The case of Thaba Nchu illustrated that the apartheid state systematically undermined the 'transformative capacities' of rural households and communities. The access to and amount of governmental transfer payments was racially biased and the ability to participate in political bodies curtailed. The Sediba reserve board, which was established in the first decade of the twentieth century, included representatives of each village in the 'native reserve' and held meetings in the vicinity of the villages. Even though board members did not have the final say in any decision, they could voice their concerns. In contrast, the Barolong tribal authority established in place of the reserve boards restricted the small means of participation that rural people had previously held. First, paid chiefs were released from their accountability to the *kgotla*, the assembly of all men. Hence, they were 'answerable to the Department [of Bantu Administration and Development] rather than their subjects' (Lodge 1984:266). Second, as it was an administrative body situated far away, rural people perceived the tribal authority as an exclusive assembly that was only concerned with the issues of the Thaba Nchu elite. The political establishment of the Barolong was restricted in its decision-making power, too. However, the elite could pursue their own interests to a certain extent when they cooperated with government officials. Furthermore, the Barolong elite and other influential men (e.g. headmen and large land owners) were those who were able to cope with and adapt to other threats, as they possessed comparatively large amounts of economic and social capital. Thus, apartheid policies did not affect the social resilience of all Thaba Nchu people to the same degree.

The 'betterment' period was the most critical socio-political event to have an impact on the social resilience of communities and households. The resettlement was particularly damaging, as the displacement of people is often connected with a breakdown of social resilience (Adger 2000:357). In Thaba Nchu, this is reflected, for instance, in the destruction of territorially organised family groups and the connected weakening of social network structures; the loss of assets in form of material wealth, land and livestock; and the destruction of 'landesque capital' (Brookfield 2001; Widgren 2007), i.e. durable innovations that generations of land users had made regarding land, including labour investments, technologies and skills. The decline of social resilience during 'betterment' made all communities and households vulnerable to further threats. At the village level, increasing conflict and distrust caused tension and hindered cooperation. At the household level, the forced culling of livestock negatively affected the social resilience of rural households, as small herd sizes made it difficult to buffer livestock losses and maintain cattle-lending networks. The restriction of land sizes made households vulnerable to drought and bad harvests. Moreover, the restriction of resource access in general made households more vulnerable to unemployment, since job losses could not be easily buffered by people who only used half 'economic units' or less.

Social Resilience in a Post-Agrarian Society

With the end of apartheid, the capability of people for political participation in Thaba Nchu and South Africa generally improved. Free elections, free movement, the right to restitution of past dispossession and improved access to social welfare payments are just some of the examples that demonstrate an improvement of people's 'position of power in the society' (Keck and Etzold 2013:87). However, a real transformation of the system's structures is yet to come: the land reform process is very slow, and in the Free State province it is rather a reform on paper than in practice. People in rural areas are still not taken seriously by government officials and are put off with community meetings that lead nowhere, and development approaches remain dominated by narratives that depict rural communities as incapable to govern natural resources.

The social welfare payments are perhaps the state's most popular measure for addressing poverty. A critical question that emerges in this regard is whether the high dependency of rural households on state welfare is an indicator of low resilience. Ferguson (2013) argues that social transfers are a 'rightful share' (ibid. 170) of the poor in a state in which unemployment is a normal condition and formal full employment will not be achieved, even over the long term (ibid. 178). Since receivers of state transfers are also sources of income for other dependents (e.g. children and grandchildren), he argues that dependence on the state 'is not the name of the problem; it is the name of the solution' (ibid. 172). Beyond a doubt, social grants have predominantly positive effects on rural households and communities, as they enhance the distribution of resources. For instance, people are able to make small investments, buy products in town and trade them in their communities. In particular, households that include at least one pensioner are no longer locally perceived as destitute. However, these households are only entitled to social grants as long as the grandparents are still alive. The death of a pensioner marks not only an emotional crisis for the bereaved, but often an economic crisis as well. Moreover, grants mean that there is no direct benefit to households without grandparents, children or disabled people. Thus, state transfers cannot be the only solution for increasing social resilience; yet it is perhaps the easiest one in terms of implementation.

In post-agrarian rural Thaba Nchu, resilient livelihoods are characterised by a diversification of non-farm incomes. Today, a majority of rural households have at least three different income sources, which makes livelihoods less vulnerable to hazards. Yet the analysis of households from different wealth strata has illustrated that resilience is, as Cannon (2008) suggested, unequally distributed within communities. Wealthy households earn income from various sources: wage labour, private or state pensions, sale of animals and petty trade. Profits from the former two sources, i.e. salaries and pensions, are used for investments in the latter. Livestock herds are diversified, commodities purchased in town and sold in the village and capital-intensive services rendered to other residents, such as renting out chairs for functions. By diversifying non-farm and farm incomes, wealthy households not only build robust livelihoods, but also accumulate economic capital that can be transformed into social and cultural capital in terms of higher education. To summarise, for wealthy households, livelihood diversification and the accumulation of different types of capital is a strategy to buffer against future crises.

The situation is different with regard to poor households, who can also diversify their livelihood strategies but only have access to less profitable domains. Although they can earn some money from insecure piece jobs, they are excluded from the formal labour market and receive only small social transfer payments (e.g. a child support grant), if at all. Not all of them own livestock, and those who do keep merely a few animals as backup in case there is a funeral. They do not have enough economic capital for herd diversification and depend on free available natural resources, such as dung or firewood, as energy sources. The poor are not necessarily poor with regard to social capital: they maintain close social networks, through which they receive access to material, financial and emotional support as well as piece jobs. However, they often lack cultural capital in the form of higher education certificates and thus do not possess the same variety in types of capital as wealthy families. The most vulnerable households, from a local perspective, are those with only minimal economic, social and cultural capital, e.g. the unemployed who have no 'helping hand'.

Social capital is important for strengthening the adaptive capacities of both households and communities (see for instance Berzborn 2006; Bollig 2006; Keck and Sakdapolrak 2013; Leslie and McCabe 2013). Rural households in the Thaba Nchu area maintain networks of mutual support with relatives, neighbours and friends. Ethnographic data have indicated that generalised reciprocity, which was first and foremost found within the patrilineal family group, declined after the destruction of the settlements in the mid-1960s. Support networks between friends, neighbours and relatives that do not belong to the nuclear family are instead based on balanced reciprocity. Institutionalised forms of cooperation are less common than in the past. Individual pastoralists pool money to purchase medicine, and some collaborate in terms of herding animals. The strenuous turning of garden soil in late winter is not done in working groups, but individually. In terms of land use, households act individually, and people complain about a monetisation of cooperation. This change in reciprocity and cooperative behaviour is both a result of dynamics that have taken place within the social system, i.e. bargaining over norms, as well as forces from outside the system, i.e. the destruction of settlement patterns and social units by the government.

Still, when a family member passes away, unlimited support for the bereaved is a social norm. Thus, in times of deep emotional crisis, institutionalised cooperation still exists. This is for several reasons: death is part of life and therefore concerns all families; the recent rise of the mortality rate increases pressure on communities and households, as occasions of death occur more often; and the social norm of organising a large and special funeral cannot be achieved without additional support in terms of labour, food and goods.

As they are surrogates for social resilience, the condition of support networks and cooperation indicates that communities are socially resilient, yet the level of resilience has declined and is lower today than in the past. One crucial condition for social networks and cooperation as well as for institutional adaptation is the existence of trust. The reconstruction of Thaba Nchu's environmental history has shown that trust in officials has constantly been undermined in various ways, including the arbitrary decisions of native commissioners, unfulfilled promises of 'betterment' for the poor and multiple cancellations of usufruct rights to land. This list could be easily extended. Even though the ANC is the ruling party today, trust in government officials is still substantially damaged. This is reflected not least in the failure of development projects and the general fatigue regarding election participation that I observed in 2011. Trust amongst community residents has been undermined as well, particularly by the resettlement of strangers on former homesteads and the betrayal of those who protested against 'betterment'. Therefore, to improve the social resilience in Thaba Nchu's rural communities, trust needs to be regained.

10. BIBLIOGRAPHY

10.1 References

Abel, T., and J. R. Stepp. 2003. A New Ecosystems Ecology for Anthropology. *Conservation Ecology* 7(3):12. http://www.consecol.org/vol7/iss3/art12/ (accessed October 24, 2014).

Acheson, J. M. 1996. Household Organization and Budget Structures in a Purepecha Pueblo. *American Ethnologist* 23(2):331–51.

Adger, W. N. 2000. Social and Ecological Resilience: Are They Related? *Progress in Human Geography* 24(2):347–64.

Agricor. 1986. Annual Report 1985/1986. Mmabatho.

Aguëro, J. M., M. R. Carter, and I. Woolard. 2006. The Impact of Unconditional Cash Transfers on Nutrition: The South African Child Support Grant. SALDRU Working Paper 06/08. Cape Town: SALDRU, University of Cape Town.

Ainslie, A. 2013. The Sociocultural Contexts and Meanings Associated with Livestock Keeping in Rural South Africa. *African Journal of Range & Forage Science* 30(1&2):35–38.

Allen, C. R., L. H. Gunderson, and C. S. Holling. 2010. Commentary on Part One Articles. In *Foundations of Ecological Resilience*. Lance H. Gunderson, Craig R. Allen, and C. S. Holling, eds. Pp. 3–18. Washington, DC: Island Press.

Allman, J., S. Geiger, and N. Musisi. 2002. Women in African Colonial Histories: An Introduction. In *Women in African Colonial Histories*. Jean Allman, Susan Geiger, and Nakanyike Musisi, eds. Pp. 1–15. Bloomington & Indianapolis: Indiana University Press.

Allsopp, N. 2013. Adaptive Management for Complex Communal Rangelands in South Africa. *African Journal of Range & Forage Science* 30(1&2):65–69.

Anderies, J. M., M. A. Janssen, and E. Ostrom. 2004. A Framework to Analyze the Robustness of Social-Ecological Systems from an Institutional Perspective. *Ecology and Society* 9(1):18. http://www.ecologyandsociety.org/vol9/iss1/art18/ (accessed October 24, 2014).

Anderson, D. 2002. *Eroding the Commons: The Politics of Ecology in Baringo, Kenya 1890s-1963.* Athens: Ohio University Press.

Anderson, D., and R. Grove, eds. 1987. *Conservation in Africa: People, Policies and Practice.* Cambridge: Cambridge University Press.

Archer, S. 2003. Technology and Ecology in the Karoo: A Century of Windmills, Wire and Changing Farming Practices. In *South Africa's Environmental History: Cases and Comparisons*. Stephen Dovers, Ruth Edgecombe, and Bill Guest, eds. Pp. 112–38. Athens: Ohio University Press.

Arizpe, L. 1977. Women in the Informal Labor Sector: The Case of Mexico City. *Signs* 3(1):25–37.

Bähre, E. 2002. Money and Violence: Financial Mutuals Among the Xhosa in Cape Town, South Africa. PhD thesis, University of Amsterdam, Amsterdam.

Banerjee, A., S. Galiani, J. Levinsohn, Z. McLaren, and I. Woolard. 2008. Why Has Unemployment Risen in the New South Africa? *Economics of Transition* 16(4):715–40.

Barker, C. H. 1998. Thaba Nchu Land Reform Research Project: Land Audit. Vol. 2, Report 3. Bloemfontein.

Barrett, C. B., T. Reardon, and P. Webb. 2001. Nonfarm Income Diversification and Household Livelihood Strategies in Rural Africa: Concepts, Dynamics, and Policy Implications. *Food Policy* 26:315–31.

Becker, E. 2012. Social-Ecological Systems as Epistemic Objects. In *Human-Nature Interactions in the Anthropocene: Potentials of Social-Ecological Systems Analysis*. Marion Glaser, Gesche Krause, Beate M. W. Ratter, and Martin Welp, eds. Pp. 37–59. New York & London: Routledge.

Behnke Jr., R. H., I. Scoones, and C. Kerven, eds. 1993. *Range Ecology at Disequilibrium: New Models of Natural Variability and Pastoral Adaptation in African Savannas.* London: Overseas Development Institute.

Beinart, W. 1982. *The Political Economy of Pondoland, 1860-1930.* Cambridge: Cambridge University Press.

Beinart, W. 1984. Soil Erosion, Conservationism and Ideas about Development: A Southern African Exploration, 1900-1960. *Journal of Southern African Studies* 11(1):52–83.

Beinart, W. 1989. Introduction: The Politics of Colonial Conservation. *Journal of Southern African Studies* 15(2):143–62.

Beinart, W. 1996. Environmental Destruction in Southern Africa: Soil Erosion, Animals, and Pastures over the Longer Term. In *Time-scales and Environmental Change*. Thackwray S. Driver, and Graham P. Chapman, eds. Pp. 149–68. London & New York: Routledge.

Beinart, W. 2008. *The Rise of Conservation in South Africa: Settlers, Livestock, and the Environment 1770-1950.* Oxford: Oxford University Press.

Beinart, W., and C. Bundy. 1980. State Intervention and Rural Resistance: The Transkei, 1900-1965. In *Peasants in Africa: Historical and Contemporary Perspectives*. Martin A. Klein, ed. Pp. 271–315. Beverly Hills & London: Sage.

Beinart, W., and P. Delius. 2014. The Historical Context and Legacy of the Natives Land Act of 1913. *Journal of Southern African Studies* 40(4):667–88.

Beinart, W., and L. Hughes. 2007. *Environment and Empire*. Oxford: Oxford University Press.

Beinart, W., and J. McGregor, eds. 2003. *Social History & African Environments.* Oxford: James Currey.

Berkes, F., J. Colding, and C. Folke, eds. 2008. *Navigating Social-Ecological Systems: Building Resilience for Complexity and Change.* Cambridge: Cambridge University Press.

Bernard, H. R. 2006. *Research Methods in Anthropology: Qualitative and Quantitative Approaches.* 4th ed. Lanham: AltaMira Press.

Bernstein, H. 1996. South Africa's Agrarian Question: Extreme and Exceptional? *Journal of Peasant Studies* 23(2/3):1–52.

Berzborn, S. 2004. Haushaltsökonomie, soziale Netzwerke und Identität: Risikominimierende Strategien von Pastoralisten und Lohnarbeitern im Richtersveld, Südafrika. PhD-thesis, Universität zu Köln, Köln.

Berzborn, S. 2006. *Ziegen und Diamanten: Sicherung des Lebensunterhaltes im ländlichen Südafrika*. Münster: Lit Verlag.

Berzborn, S., and M. Schnegg. 2007. Vulnerability, Social Networks, and Resilience: Rural Livelihoods in the Richtersveld, South Africa. In *Social Networks: Potentials and Constraints, Indications from South Africa*. Beate Lohnert, ed. Pp. 115–47. Saarbrücken: Verlag für Entwicklungspolitik.

Beuster, D. J. 1983. Research Priorities in Bophuthatswana Agriculture. Mafikeng.

Beymer-Farris, B. A., T. J. Bassett, and I. Bryceson. 2012. Promises and Pitfalls of Adaptive Management in Resilience Thinking: The Lens of Political Ecology. In *Resilience and the Cultural Landscape: Understanding and Managing Change in Human-Shaped Environments*. Tobias Plieninger and Claudia Bieling, eds. Pp. 283–99. Cambridge: Cambridge University Press.

Biersack, A. 2006. Reimagining Political Ecology: Culture/Power/History/Nature. In *Reimagining Political Ecology*. Aletta Biersack, ed. Pp. 3–40. Durham: Duke University Press.

Biovision Foundation. 2012. Reproductive Problems: Epivag. http://www.infonet-biovision.org/default/ct/657/animalDiseases (accessed July 29, 2013).

Blaikie, P. 2008. Epilogue: Towards a Future for Political Ecology that Works. *Geoforum* 39(2):765–72.

Bollig, M. 2006. *Risk Management in a Hazardous Environment: A Comparative Study of Two Pastoral Societies.* New York: Springer.

Bollig, M. 2012. Social-Ecological Change and the Changing Structure of Risk, Risk Management and Resilience in a Pastoral Community in Northwestern Namibia. In *Negotiating Disasters: Politics, Representation, Meanings*. Ute Luig, ed. Pp. 31–58. Frankfurt am Main: Peter Lang.

Bollig, M. 2013. Social-Ecological Change and Institutional Development in a Pastoral Community in North-western Namibia. In *Pastoralism in Africa: Past, Present and Future*. Michael Bollig, Michael Schnegg, and Hans-Peter Wotzka, eds. Pp. 316–40. New York, Oxford: Berghahn Books.

Bollig, M., and M. Schnegg. 2013. Specialisation and Diversification among African Pastoral Societies. In *Pastoralism in Africa: Past, Present and Future*. Michael Bollig, Michael Schnegg, and Hans-Peter Wotzka, eds. Pp. 1–28. New York, Oxford: Berghahn Books.

Bollig, M., and A. Schulte. 1999. Environmental Change and Pastoral Perceptions: Degradation and Indigenous Knowledge in Two African Pastoral Communities. *Human Ecology* 27(3):493–514.

Boserup, E. 1965. *The Conditions of Agricultural Growth: The Economics of Agrarian Change under Population Pressure.* Chicago: Aldine Publishing Company.

Bourdieu, P. 1983. Ökonomisches Kapital, kulturelles Kapital, soziales Kapital. In *Soziale Ungleichheiten*. Reinhard Kreckel, ed. Pp. 183–99. Göttingen: Schwartz.

Boyd, E., and C. Folke, eds. 2012. *Adapting Institutions: Governance, Complexity and Social-Ecological Resilience.* Cambridge: Cambridge University Press.

Bozzoli, B. 1991. *Women of Phokeng: Consciousness, Life Strategy, and Migrancy in South Africa, 1900-1983.* Johannesburg: Raven Press.

Bradford, H. 1996. Women, Gender and Colonialism: Rethinking the History of the British Cape Colony and its Frontier Zones, c. 1806-70. *Journal of African History* 37:351–70.

Brand, F. S., and K. Jax. 2007. Focusing of the Meaning(s) of Resilience: Resilience as a Descriptive Concept and a Boundary Object. *Ecology and Society* 12(2):23. http://www.ecologyandsociety.org/vol12/iss1/art23/ (accessed October 24, 2014).

Breutz, P. L. 1989. *A History of the Batswana and Origin of Bophuthatswana: A Handbook of a Survey of the Tribes of the Batswana, S. Ndebele, QwaQwa and Botswana*. Ramsgate: Breutz.

Brookfield, H. C. 2001. *Exploring Agrodiversity.* New York: Columbia University Press.

Bryceson, D. F. 1997a. De-agrarianisation: Blessing or Blight? In *Farewell to Farms: De-agrarianisation and Employment in Africa*. Deborah F. Bryceson and Vali Jamal, eds. Pp. 237–56. Aldershot: Ashgate.

Bryceson, D. F. 1997b. De-agrarianisation in Sub-Saharan Africa: Acknowledging the Inevitable. In *Farewell to Farms: De-agrarianisation and Employment in Africa*. Deborah F. Bryceson and Vali Jamal, eds. Pp. 3–20. Aldershot: Ashgate.

Bryceson, D. F. 1999. African Rural Labour, Income Diversification & Livelihood Approaches: A Long-Term Development Perspective. *Review of African Political Economy* 26(80):171–89.

Bryceson, D. F. 2002. The Scramble in Africa: Reorienting Rural Livelihoods. *World Development* 30(5):725–39.

Bryceson, D. F., and V. Jamal, eds. 1997. *Farewell to Farms: De-agrarianisation and Employment in Africa.* Aldershot: Ashgate.

Bureau for Food and Agricultural Policy. 2013. Land Reform in the Free State, the Land Audit, and the Way Forward to 2030. Pretoria.

Butler, J., R. I. Rotberg, and J. Adams. 1977. *The Black Homelands of South Africa: The Political and Economic Development of Bophuthatswana and KwaZulu*. Berkeley: University of California Press.

Butzer, K. W. 2012. Collapse, Environment, and Society. *PNAS* 109(10):3632–39.

Caldwell, J. C., and P. Caldwell. 1993. The South African Fertility Decline. *Population and Development Review* 19(2):225–62.

Cancian, F. 1976. Social Stratification. *Annual Review of Anthropology* 5:227–48.

Cannon, T. 2008. Reducing People's Vulnerability to Natural Hazards. UNU-Wider: Research Paper 2008/34. Helsinki. http://www.wider.unu.edu/publications/working-papers/research-papers/2008/en_GB/rp2008-34/ (accessed May 25, 2014).

Cannon, T., and D. Müller-Mahn. 2010. Vulnerability, Resilience and Development Discourses in Context of Climate Change. *Natural Hazards* 55:621–35.

Carpenter, S. R., B. Walker, J. M. Anderies, and N. Abel. 2001. From Metaphor to Measurement: Resilience of What to What? *Ecosystems* 4(8):765–81.

Carruthers, J. 1989. Creating a National Park, 1910 to 1926. *Journal of Southern African Studies* 15(2):188–216.

Carruthers, J. 2003. Environmental History in Southern Africa: An Overview. In *South Africa's Environmental History: Cases and Comparisons*. Stephen Dovers, Ruth Edgecombe, and Bill Guest, eds. Pp. 3–15. Athens: Ohio University Press.

Carswell, G. 2003. Soil Conservation Policies in Colonial Kigezi, Uganda: Successful Implementation & an Absence of Resistance. In *Social history & African environments*. William Beinart and JoAnn McGregor, eds. Pp. 131–54. Oxford: James Currey.

Centre for Development Support (CDS). 2003. Potential Demand for Social Housing in the Mangaung Local Municipality. http://econ.ufs.ac.za/dl/userfiles/Documents/00000/47_eng.pdf (accessed June 29, 2012).

Chapin, S. F., III, C. Folke, and G. P. Kofinas. 2009. A Framework for Understanding Change. In *Principles of Ecosystem Stewardship: Resilience-based Natural Resource Management in a Changing World*. Stuart F. Chapin, III, Gary P. Kofinas, and Carl Folke, eds. Pp. 3–28. New York: Springer.

Chapin, S. F., III, C. Folke, and G. P. Kofinas, eds. 2009. *Principles of Ecosystem Stewardship: Resilience-based Natural Resource Management in a Changing World.* New York: Springer.

Cleaver, F. 2002. Reinventing Institutions: Bricolage and the Social Embeddedness of Natural Resource Management. *The European Journal of Development Research* 14(2):11–30.

Cleaver, K. M., and G. A. Schreiber. 1994. *Reversing the Spiral: The Population, Agriculture, and Environment Nexus in Sub-Saharan Africa.* Washington, DC: The World Bank.

Comaroff, J., and J. L. Comaroff. 1991. *Of Revelation and Revolution: Christianity, Colonialism, and Consciousness in South Africa. Vol 1.* Chicago & London: University of Chicago Press.

Comaroff, J. L., and J. Comaroff. 1990. Goodly Beasts, Beastly Goods: Cattle and Commodities in a South African Context. *American Ethnologist* 17(2):195–216.

Comaroff, J. L., and J. Comaroff. 1997. *Of Revelation and Revolution: The Dialectics of Modernity on a South African Frontier. Vol 2.* Chicago & London: University of Chicago Press.

Commission on Native Laws and Customs. 1970 [1883]. *Report and Proceedings of the Government Commission on Native Laws and Customs 1883: With a Preliminary Report of the Commission.* Shannon: Irish University Press.

Commission on Restitution of Land Rights. 2013. Annual Report 2012/2013. http://www.ruraldevelopment.gov.za/publications/annual-report/file/1996 (accessed June 8, 2014).

Coplan, D. 2010. Innocent Violence: Social Exclusion, Identity, and the Press in an African Democracy. Paper presented at the Kölner Ethnologisches Kolloquium (KEK). Cologne, Germany, 20 July 2010.

Costanza, R., L. J. Graumlich, and W. Steffen, eds. 2007. *Sustainability or Collapse: An Integrated History and Future of People on Earth.* Cambridge, Mass.: MIT Press.

Cousins, B. 1996. Livestock Production and Common Property Struggles in South Africa's Agrarian Reform. *Journal of Peasant Studies* 23(2/3):166–208.

Cousins, B. 1999. Invisible Capital: The Contribution of Communal Rangelands to Rural Livelihoods in South Africa. *Development Southern Africa* 16(2):299–319.

Cousins, B. 2009a. Key Provisions of the Communal Land Rights Act Are Declared Unconstitutional. Where to Now? http://anothercountryside.wordpress.com/2009/11/10/key-provisions-of-the-communal-land-rights-act-are-declared-unconstitutional-where-to-now/?blogsub=confirming#blog_subscription-3 (accessed May 6, 2014).

Cousins, B. 2009b. Potential and Pitfalls of 'Communal' Land Tenure Reform: Experience in Africa and Implications for South Africa. Paper for the World Bank conference on 'Land Governance in Support of the MDGs: Responding to New Challenges'; Washington D.C., USA, 9-10 March 2009.

Cowley, J., and A. Lemon. 1986. Bophuthatswana: Dependent Development of a Black 'Homeland'. *Geography* 71(3):252–55.

Crane, T. A. 2010. Of Models and Meanings: Cultural Resilience in Social-Ecological Systems. *Ecology and Society* 15(4):19. http://www.ecologyandsociety.org/vol15/iss4/art19/ (accessed October 24, 2014).

Davenport, T. R. H. 1979. *South Africa: A Modern History.* reprinted. London: Macmillan Press Ltd.

Davis, S., A. Hughes, and K. Louw. 2008. The Impact of Free Basic Electricity on the Energy Choices of Low Income Households: A Case Study in South Africa. PESD Working Paper 80. Stanford. http://iis-db.stanford.edu/pubs/22331/WP_80,_Davis_Hughes_Louw,_FBE_in_South_Africa.pdf (accessed February 5, 2014).

De Vielliers, G., and H. J. Booysen. 1998. Thaba Nchu Land Reform Research Project: Natural Resources and Land Use. Vol. 2; Report 1. Bloemfontein.

De Wet, C. 1989. Betterment Planning in a Rural Village in Keiskammahoek, Ciskei. *Journal of Southern African Studies* 15(2):326–45.

De Wet, C. 1995. *Moving Together, Drifting Apart: Betterment Planning and Villagisation in a South African Homeland.* Johannesburg: Witwatersrand University Press.

Delius, P. 1996. *A Lion Amongst the Cattle: Reconstruction and Resistance in the Northern Transvaal.* Johannesburg: Raven Press.

Delius, P., and S. Schirmer. 2000. Soil Conservation in a Racially Ordered Society: South Africa 1930-1970. *Journal of Southern African Studies* 26(4):719–42.

Department of Agriculture and Natural Resources. 1991. Annual Report: For the Period 1 October 1990 to 30 September 1991. Mmabatho.

Department of Rural Development and Land Reform - Communication Services. 2012. Minister Gugile Nkwinti to Hand Financial Compensation to Sediba Reserve Claim. http://www.ruraldevelopment.gov.za/phocadownload/Media-invites/2012/mediainvite-sediba-07032012.pdf (accessed June 8, 2014).

Desmond, C. 1972. *The Discarded People: An Account of African Resettlement in South Africa.* reprinted. Harmondsworth: Penguin Books.

Devereux, S. 2007. Social Pensions in Southern Africa in the Twentieth Century. *Journal of Southern African Studies* 33(3):539–60.

DeWalt, K. M., and B. R. DeWalt. 2002. *Participant Observation: A Guide for Fieldworkers.* Walnut Creek: AltaMira Press.

D'Haese, M., and G. van Huylenbroeck. 2005. The Rise of Supermarkets and Changing Expenditure Patterns of Poor Rural Households Case Study in the Transkei Area, South Africa. *Food Policy* 30(1):97–113.

Dovers, S. 2000. On the Contribution of Environmental History to Current Debate and Policy. *Environment and History* 6(2):131–50.

Dovers, S., R. Edgecombe, and B. Guest, eds. 2003. *South Africa's Environmental History: Cases and Comparisons.* Athens: Ohio University Press.

Drummond, J. H. 1990. Rural Land Use and Agricultural Production in Dinokana Village, Bophuthatswana. *GeoJournal* 22(3):335–43.

Drummond, J. H. 1991. Reincorporating the Bantustans into South Africa: The Question of Bophuthatswana. *Geography* 76(4):369–73.

Du Toit, A. 2011. Comment on the Newly Released Green Paper on Land Reform. http://anothercountryside.wordpress.com/2011/09/26/comment-on-the-newly-released-green-paper-on-land-reform/ (accessed May 6, 2014).

Eder, J. F. 2006. Gender Relations and Household Economic Planning in the Rural Philippines. *Journal of Southeast Asian Studies* 37(3):397–413.

Ellis, F. 1998. Household Strategies and Rural Livelihood Diversification. *Journal of Development Studies* 35(1):1–38.

Engels, A., and T. Moss. 2003. Institutional Change in Environmental Contexts. In *How Institutions Change: Perspectives on Social Learning in Global and Local Environmental Contexts*. Heiko Breit, Anita Engels, Timothy Moss, and Markus Troja, eds. Pp. 355–75. Opladen: Leske & Budrich.

Ensminger, J., and J. Knight. 1997. Changing Social Norms: Common Property, Bridewealth, and Clan Exogamy. *Current Anthropology* 38(1):1–24.

Erasmus, P. A., and D. S. Krige. 1998. Thaba Nchu Land Reform Research Project: Land Administration and Management. Vol. 2; Report 2. Bloemfontein.

Esterhuyse, P. 2012. Social Capital in a Rainwater-Harvesting Project in Rural South Africa. *Irrigation and Drainage* 61(Suppl. 2):95–105.

Fairhead, J., and M. Leach. 1996. *Misreading the African Landscape: Society and Ecology in a Forest-savanna Mosaic*. Cambridge: Cambridge University Press.

Fay, D. A. 2012. "The Trust is Over! We Want to Plough!": Social Differentiation and the Reversal of Resettlement in South Africa. *Human Ecology* 40(1):59–68.

Ferguson, J. 1992. The Cultural Topography of Wealth: Commodity Paths and the Structure of Property in Rural Lesotho. *American Anthropologist* 94(1):55–73.

Ferguson, J. 2003. *The Anti-Politics Machine: 'Development', Depoliticization, and Bureaucratic Power in Lesotho.* 6th edition. Minneapolis: University of Minnesota Press.

Ferguson, J. 2006. *Global Shadows: Africa in the Neoliberal World Order.* Durham & London: Duke University Press.

Ferguson, J. 2013. How to Do Things with Land: A Distributive Perspective on Rural Livelihoods in Southern Africa. *Journal of Agrarian Change* 13(1):166–74.

Field, A. 2005. *Discovering Statistics Using SPSS.* 2nd ed. London: Sage.

Fischer, H. 1997. Zensusaufnahmen: Das Beispiel Gabsongkeg. In *Geburt und Tod: ethnodemographische Probleme, Methoden und Ergebnisse: Materialien zur Kultur der Wampar, Papua New Guinea*. Walter Schulze, Hans Fischer, and Hartmut Lang, eds. Pp. 37–91 Berlin: Reimer.

Flemming, L. 1927. *The Call of the Veld.* 2nd ed. Bloemfontein: A. C. White Printing and Publishing Co.

Folke, C. 2006. Resilience: The Emergence of a Perspective for Social-Ecological Systems Analysis. *Global Environmental Change* 16(3):253–67.

Folke, C., S. R. Carpenter, B. Walker, M. Scheffer, T. Chapin, and J. Rockström. 2010. Resilience Thinking: Integrating Resilience, Adaptability and Transformability. *Ecology and Society* 15(4):20. http://www.ecologyandsociety.org/vol15/iss4/art20/ (accessed October 24, 2014).

Francis, E. 1999. Learning from the Local: Rural Livelihoods in Ditsobotla, North West Province, South Africa. *Journal of Contemporary African Studies* 17(1):49–73.

Francis, E. 2000. *Making a Living: Changing Livelihoods in Rural Africa.* London & New York: Routledge.

Frith, A. 2013a. Census 2001. http://census2001.adrianfrith.com (accessed June 6, 2014).

Frith, A. 2013b. Census 2011. http://census2011.adrianfrith.com/place/499 (accessed June 6, 2014).

Gertenbach, W. D. 2000. Beef Production: The Basis. http://agriculture.kzntl.gov.za/publications/production_guidelines/beef_production/beef_prod_cover.htm (accessed July 25, 2013).

Glaser, M., G. Krause, B. M. W. Ratter, and M. Welp, eds. 2012a. *Human-Nature Interactions in the Anthropocene: Potentials of Social-Ecological Systems Analysis.* New York & London: Routledge.

Glaser, M., B. M. W. Ratter, G. Krause, and M. Welp. 2012b. New Approaches to the Analysis of Human-Nature Relations. In *Human-Nature Interactions in the Anthropocene: Potentials of Social-Ecological Systems Analysis*. Marion Glaser, Gesche Krause, Beate M. W. Ratter, and Martin Welp, eds. Pp. 3–12. New York & London: Routledge.

Government Communication and Information System (GCIS). n.d. What Are Thusong Service Centres? http://www.thusong.gov.za/about/what/index.htm (accessed February 4, 2014).

Grandin, B. E. 1994. Wealth Ranking. In *Tools for the Field: Methodologies Handbook for Gender Analysis in Agriculture*. Hilary Sims Feldstein and Janice Jiggins, eds. Pp. 21–35. West Hartford: Kumarian Press.

Greiner, C. 2008. *Zwischen Ziegenkraal und Township: Migrationsprozesse in Nordwestnamibia*. Berlin: Reimer.

Grove, R. 1989. Scottish Missionaries, Evangelical Discourses and the Origins of Conservation Thinking in Southern Africa 1820-1900. *Journal of Southern African Studies* 15(2):163–87.

Gunderson, L. H., C. R. Allen, and C. S. Holling, eds. 2010. *Foundations of Ecological Resilience.* Washington, DC: Island Press.

Halbach, A. J. 1988. *Südafrika und seine Homelands: Strukturen und Probleme der "Getrennten Entwicklung"*. München: Weltforum Verlag.

Hall, R., ed. 2009a. *Another Countryside? Policy Options for Land and Agrarian Reform in South Africa.* Cape Town: Institute for Poverty, Land and Agrarian Studies (PLAAS).

Hall, R. 2009b. Land Reform for What? Land Use, Production & Livelihoods. In *Another Countryside? Policy Options for Land and Agrarian Reform in South Africa*. Ruth Hall, ed. Pp. 23–60. Cape Town: Institute for Poverty, Land and Agrarian Studies (PLAAS).

Hart, T. G. B. 2011. The Significance of African Vegetables in Ensuring Food Security for South Africa's Rural Poor. *Agriculture and Human Values* 28(3):321–33.

Hayes, P. 1998. The 'Famine of the Dams': Gender, Labour & Politics in Colonial Ovamboland 1929-1930. In *Namibia under South African Rule: Mobility & Containment, 1915-46*. Patricia Hayes, Jeremy Silvester, Marion Wallace, and Wolfram Hartmann, eds. Pp. 117–46. Oxford: James Currey.

Hebinck, P., and C. Shackleton. 2011a. Livelihoods, Resources and Land Reform. In *Reforming Land and Resource Use in South Africa: Impact on Livelihoods*. Paul Hebinck and Charlie Shackleton, eds. Pp. 1–32. London & New York: Routledge.

Hebinck, P., and C. Shackleton, eds. 2011b. *Reforming Land and Resource Use in South Africa: Impact on Livelihoods.* London & New York: Routledge.

Hendricks, F. T. 1989. Loose Planning and Rapid Resettlement: The Politics of Conservation and Control in Transkei, South Africa, 1950-1970. *Journal of Southern African Studies* 15(2):306–25.

Hendricks, F. T. 1990. *The Pillars of Apartheid: Land Tenure, Rural Planning and the Chieftaincy.* Stockholm: Almqvist & Wiksell International.

Herskovits, M. J. 1926. The Cattle Complex in East Africa. *American Anthropologist* 28(1):230–72.

High, C., and C. M. Shackleton. 2000. The Comparative Value of Wild and Domestic Plants in Home Gardens of a South African Rural Village. *Agroforestry Systems* 48(2):141–56.

Hirson, B. 1977. Rural Revolt in South Africa: 1937-1951. *Societies of Southern Africa in the 19th and 20th Centuries: Collected Seminar Papers* 8(21):115–32.

Hogg, R. 1986. The New Pastoralism: Poverty and Dependency in Northern Kenya. *Africa* 56(3):319–33.

Holland, J. H. 1992. Complex Adaptive Systems. *Daedalus* 121(1):17–30.

Holling, C. S. 1973. Resilience and Stability of Ecological Systems. *Annual Review of Ecology and Systematics* 4:1–23.

Holling, C. S. 1996. Engineering Resilience Versus Ecological Resilience. In *Engineering Within Ecological Constrains*. Peter C. Schulze, ed. Pp. 31–44. Washington, DC: National Academy Press.

Holling, C. S., and L. H. Gunderson. 2002. Resilience and Adaptive Cycles. In *Panarchy: Understanding Transformations in Human and Natural Systems*. Lance H. Gunderson and C. S. Holling, eds. Pp. 25–62. Washington, DC: Island Press.

Hoole, A., and F. Berkes. 2010. Breaking Down Fences. Recoupling Social-Ecological Systems for Biodiversity Conservation in Namibia. *Geoforum* 41(2):304–17.

Howell, N. 1979. *Demography of the Dobe !Kung.* New York: Academic Press.

Hummel, D. 2012. Population Dynamics and Adaptive Capacity of Supply Systems. In *Human-Nature Interactions in the Anthropocene: Potentials of Social-Ecological Systems Analysis*. Marion Glaser, Gesche Krause, Beate M. W. Ratter, and Martin Welp, eds. Pp. 123–38. New York & London: Routledge.

Jacobs, N. J. 1999. Environment, Production and Social Difference in the Kalahari Thornveld, c1750-1830. *Journal of Southern African Studies* 25(3):347–73.

Jacobs, N. J. 2001. The Great Bophuthatswana Donkey Massacre: Discourse on the Ass and the Politics of Class and Grass. *The American Historical Review* 106(2):485–507.

Jacobs, N. J. 2003a. *Environment, Power, and Injustice: A South African History.* Cambridge: Cambridge University Press.

Jacobs, N. J. 2003b. The Colonial Ecological Revolution in South Africa: The Case of Kuruman. In *South Africa's Environmental History: Cases and Comparisons*. Stephen Dovers, Ruth Edgecombe, and Bill Guest, eds. Pp. 19–33. Athens: Ohio University Press.

Jacobs, P., E. Lahiff, and R. Hall. 2003. *Land Redistribution.* Cape Town: Programme for Land and Agrarian Studies (PLAAS).

Jeppe, W. J. O. 1980. *Bophuthatswana: Land Tenure and Development.* Cape Town: Maskew Miller.

Jones, P. S. 1999. 'To Come Together for Progress': Modernisation and Nation-Building in South Africa's Bantustan Periphery - The Case of Bophuthatswana. *Journal of Southern African Studies* 25(4):579–605.

Juul, K. 2001. Power, Pastures and Politics: Boreholes and Decentralisation of Local Resource Management in Northern Senegal. In *Politics, Property and Production in the West African Sahel: Understanding Natural Resources Management*. Tor A. Benjaminsen and Christian Lund, eds. Pp. 57–74. Uppsala: Nordiska Afrikainstitutet.

Keck, M., and B. Etzold. 2013. Resilience Refused: Wasted Potentials for Improving Food Security in Dhaka. *Erdkunde* 67(1):75–91.

Keck, M., and P. Sakdapolrak. 2013. What is Social Resilience? Lessons Learned and Ways Forward. *Erdkunde* 67(1):5–19.

Keegan, T. J. 1987. *Rural Transformations in Industrializing South Africa: The Southern Highveld to 1914.* Basingstoke & London: Macmillan Press Ltd.

Kinzig, A. P. 2012. Towards a Deeper Understanding of the Social in Resilience: The Contributions of Cultural Landscapes. In *Resilience and the Cultural Landscape: Understanding and Managing Change in Human-Shaped Environments*. Tobias Plieninger and Claudia Bieling, eds. Pp. 315–27. Cambridge: Cambridge University Press.

Knight, E. F. 1903. *South Africa after the War: A Narrative of Recent Travel.* London: Longmans, Green, and Co.

Knight, J. 1995. Models, Interpretations, and Theories: Constructing Explanations of Institutional Emergence and Change. In *Explaining Social Institutions*. Jack Knight and Itai Sened, eds. Pp. 95–119. Ann Arbor: The University of Michigan Press.

Kotze, F., P. Cloete, I. Eigelaar-Meets, C. Groenewald, and C. Poole. 2012. Evaluation Report: World Vision Networks of Hope Program. Bellville.

Kotzé, E., A. Sandhage-Hofmann, J. A. Meinel, C. C. Du Preez, and W. Amelung. 2013. Rangeland Management Impacts on the Properties of Clayey Soils Along Grazing Gradients in the Semi-Arid Grassland Biome of South Africa. *Journal of Arid Environments* 97:220–29.

Krige, J. D., and E. Krige. 1956. The Tomlinson Report and the Lovedu. *Race Relations Journal* 23(4):12–25.

Kundhlande, G., D. C. Groenewald, M. N. Baiphethi, M. F. Viljoen, J. J. Botha, L. D. Van Rendsburg, and J. J. Anderson. 2004. Socio-economic Study on Water Conservation Techniques in Semi-Arid Areas. WRC Report 1267/1/04. Pretoria.

Lahiff, E. 2000. *An Apartheid Oasis? Agriculture and Rural Livelihoods in Venda.* London & Portland, OR: Frank Cass Publishers.

Lahiff, E. 2002. Land Reform and Sustainable Livelihoods in South Africa's Eastern Cape Province. PLAAS Research Report 14. Cape Town.

Lahiff, E. 2009. With What Land Rights? Tenure Arrangements and Support. In *Another Countryside? Policy Options for Land and Agrarian Reform in South Africa*. Ruth Hall, ed. Pp. 93–117. Cape Town: Institute for Poverty, Land and Agrarian Studies (PLAAS).

Lahiff, E. 2011. Land Reform and Poverty Reduction in South Africa. In *Reforming Land and Resource Use in South Africa: Impact on Livelihoods*. Paul Hebinck and Charlie Shackleton, eds. Pp. 58–85. London & New York: Routledge.

Lang, H. 1998. The Population Development of the Rehoboth Basters. *Anthropos* 93:381–91.

Lang, H., and J. Pauli. 2002. Der ethnographische Zensus: Eine praxisorientierte Einführung. Methoden der Ethnographie 2. Hamburg.

Lawry, S. W. 1990. Tenure Policy Toward Common Property Natural Resources in Sub-Saharan Africa. *Natural Resources Journal* 30(2):403–22.

Leach, M., and R. Mearns, eds. 1996. *The Lie of the Land: Challenging Received Wisdom on the African Environment.* London: The International African Institute.

Legido-Quigley, H. 2003. The South African Old Age Pension: Exploring the Role on Poverty Alleviation in Households Affected by HIV/AIDS. Paper for the 4th International Research Conference on Social Security; Antwerp, Belgium, 5-7 May 2003.

Leibbrandt, M., I. Woolard, A. Finn, and J. Argent. 2010. Trends in South African Income Distribution and Poverty Since the Fall of Apartheid. OECD Social, Employment and Migration Working Papers 101. Paris.

Leslie, P., and J. T. McCabe. 2013. Response Diversity and Resilience in Social-Ecological Systems. *Current Anthropology* 54(2):114–43.

Levy, D. L., and P. J. Newell. 2005. Introduction: The Business of Global Environmental Governance. In *The Business of Global Environmental Governance*. David L. Levy and Peter J. Newell, eds. Pp. 1–18. Cambridge, Mass.: MIT Press.

Lewis, J. 1984. The Rise and Fall of the South African Peasantry: A Critique and Reassessment. *Journal of Southern African Studies* 11(1):1–24.

Li, W., and Y. Li. 2012. Managing Rangeland as a Complex System: How Government Interventions Decouple Social Systems from Ecological Systems. *Ecology and Society* 17(1):9. http://www.ecologyandsociety.org/vol17/iss1/art9/ (accessed October 24, 2014).

Linstädter, A., Kuhn, A., Naumann, C. Rasch, R., Sandhage-Hofmann, A., Amelung, W., Jordaan, J., Du Preez, C.C. & M. Bollig (2016). Assessing the Resilience of a Real-World Social-Ecological System: Lessons from a Multidisciplinary Evaluation of a South African Pastoral System. *Ecology and Society* 21(3):35. https://www.ecologyandsociety.org/vol21/iss3/art35/ (accessed September 19, 2017)

Little, P. D. 2000. Selling to Eat: Petty Trade and Traders in Peri-Urban Areas of Sub-Saharan Africa. Research Paper for the Broadening Access and Strengthening Input Market Systems-Collaborative Research Support Program. New York.

Lodge, T. 1984. *Black Politics in South Africa Since 1945.* London & New York: Longman.

Lodge, T. 2003. *Politics in South Africa: From Mandela to Mbeki.* Oxford: James Currey.

Lund, C. 2006. Twilight Institutions: Public Authority and Local Politics in Africa. *Development and Change* 37(4):685–705.

Lund, F. 1990. Welfare Under Pressure: Financing South African Social Welfare. *Transformation* 13:67–80.

Lye, W. F., and C. Murray. 1980. *Transformations on the Highveld: The Tswana and Southern Sotho.* Cape Town: Philip.

MacKenzie, J. M. 1997. *The Empire of Nature: Hunting, Conservation and British Imperialism*. Manchester: Manchester University Press.

Madhavan, S., and E. J. Schatz. 2007. Coping with Change: Household Structure and Composition in Rural South Africa, 1992-2003. *Scandinavian Journal of Public Health* 35(Suppl. 69):85–93.

Mamdani, M. 1996. *Citizen and Subject: Contemporary Africa and the Legacy of Late Colonialism.* Princeton: Princeton University Press.

Mamdani, M. 2012. *Define and Rule: Native as Political Identity.* Cambridge, Mass. & London: Harvard University Press.

Mangaung Local Municipality. 2007. Integrated Development Plan: 2007-2008. http://www.bloemfontein.co.za/docs/IDP%20Report%202007-08%20Final.pdf (accessed June 29, 2012).

Mangaung Local Municipality. 2010. Integrated Development Plan: 2010. http://www.mangaung.co.za/docs/Complete-Final-Approved-IDP-2010-2011.pdf (accessed July 17, 2012).

Mangaung Metropolitan Municipality. 2012. Annual Performance Report 2011/2012. Bloemfontein. http://mfma.treasury.gov.za/Documents/06.%20Annual%20Reports/2011-12/01.%20Metros/MAN%20Mangaung/MAN%20Mangaung%20Annual%20report%202011-12.pdf (accessed February 6, 2014).

Marais, H. 2010. *South Africa Pushed to the Limit: The Political Economy of Change.* Claremont: UCT Press.

Marx, C. 2004. *Geschichte Afrikas: Von 1800 bis zur Gegenwart.* Paderborn: Ferdinand Schöningh.

Marx, C. 2006. Zwangsumsiedlungen in Südafrika während der Apartheid. In *Wissenschaft - Planung - Vertreibung: Neuordnungskonzepte und Umsiedlungspolitik im 20. Jahrhundert*. Isabel Heinemann and Patrick Wagner, eds. Pp. 173–95. Stuttgart: Franz Steiner.

McAllister, P. A. 1989. Resistance to 'Betterment' in the Transkei: A Case Study from Willowvale District. *Journal of Southern African Studies* 15(2):346–68.

McCabe, J. T. 1990. Success and Failure: The Breakdown of Traditional Drought Coping Institutions Among the Pastoral Turkana of Kenya. *Journal of Asian and African Studies* 25(3-4):146–60.

McCann, J. C. 1999. *Green Land, Brown Land, Black Land: An Environmental History of Africa, 1800-1990.* Portsmouth, Oxford: Heinemann; James Currey.

McCann, J. C. 2001. Maize and Grace: Corn, and Africa's New Landscapes, 1500-1999. *Society for Comparative Study of Society and History* 43(2):246–72.

Meagher, K., and A. Mustapha. 1997. Not by Farming Alone: The Role of Non-Farm Incomes in Rural Hausaland. In *Farewell to Farms: De-agrarianisation and Employment in Africa*. Deborah F. Bryceson and Vali Jamal, eds. Pp. 63–84. Aldershot: Ashgate.

Mearns, R., D. Shombodon, G. Narangerel, U. Turul, A. Enkhamgalan, B. Myagmarzhav, A. Bayanjargal, and B. Bekhsuren. 1992. Direct and Indirect Uses of Wealth Ranking in Mongolia. *RRA Notes* 15:29–38.

Menestrey Schwieger, D.A. 2017. *The Pump Keeps on Running. On the Emergence of Water Management Institutions Between State Decentralization and Local Practices in Northern Kunene*. Zürich: Lit Verlag.

Meth, C. 2007. *Sticking to the Facts: Official and Unofficial Stories about Poverty and Unemployment in South Africa.* DPRU Working Paper 07/123. Cape Town.

Moberg, M. 1994. An Agency Model of the State: Contributions and Limitations of Institutional Economics. In *Anthropology and Institutional Economics*. James M. Acheson, ed. Pp. 213–31. Lanham: University Press of America.

Moiloa, T. 2012. Cash Bonanza for Sediba Beneficiaries. *Public Eye Online*, March 19. http://www.publiceyenews.com/site/2012/03/19/cash-bonanza-for-sediba-beneficiaries/ (accessed May 6, 2014).

Moorosi, L. E. 1999. Characterization of Small-scale Cattle Farming in Botshabelo and Thaba Nchu Districts of the Free State Province. Master thesis, University of the Free State, Bloemfontein.

Mora, C. 2014. Revisiting the Environmental and Socioeconomic Effects of Population Growth: A Fundamental but Fading Issue in Modern Scientific, Public, and Political Circles. *Ecology and Society* 19(1):38. http://dx.doi.org/10.5751/ES-06320-190138 (accessed October 24, 2014).

Moran, E. F. 1990. Ecosystem Ecology in Biology and Anthropology: A Critical Assessment. In *The Ecosystem Approach in Anthropology: From Concept to Practice*. Emilio F. Moran, ed. Pp. 3–40. Ann Arbor: University of Michigan Press.

Mucina, L., and M. C. Rutherford, eds. 2006. *The Vegetation of South Africa, Lesotho and Swaziland.* Pretoria: South African National Biodiversity Institute (SANBI).

Munck, V. de. 2009. *Research Design and Methods for Studying Cultures.* Lanham: AltaMira Press.

Murray, C. 1981. *Families Divided: The Impact of Migrant Labour in Lesotho.* Cambridge: Cambridge University Press.

Murray, C. 1984. Land, Power and Class in the Thaba 'Nchu District, Orange Free State, 1884-1983. *Review of African Political Economy* 11(29):30–48.

Murray, C. 1987. Displaced Urbanization: South Africa's Rural Slums. *African Affairs* 86(344):311–29.

Murray, C. 1992. *Black Mountain: Land, Class and Power in the Eastern Orange Free State, 1880s to 1980s.* Edinburgh: Edinburgh University Press.

Naumann, C. 2014. Stability and Transformation in a South African Landscape: Rural Livelihoods, Governmental Interventions and Agro-Economic Change in Thaba Nchu. *Journal of Southern African Studies* 40(1):41–57.

Nayak, P. K. 2014. The Chilika Lagoon Social-Ecological System: An Historical Analysis. *Ecology and Society* 19(1):1. http://dx.doi.org/10.5751/ES-05978-190101 (accessed October 24, 2014).

Netting, R. McC., R. R. Wilk, and E. J. Arnould, eds. 1984a. *Households: Comparative and Historical Studies of the Domestic Group.* Berkeley: University of California Press.

Netting, R. McC., R. R. Wilk, and E. J. Arnould. 1984b. Introduction. In *Households: Comparative and Historical Studies of the Domestic Group*. Robert McC. Netting, Richard R. Wilk, and Eric J. Arnould, eds. Pp. xiii–xxxviii. Berkeley: University of California Press.

Neumann, R. P. 1997. Primitive Ideas: Protected Area Buffer Zones and the Politics of Land in Africa. *Development and Change* 28(3):559–82.

Neumann, R. P. 2005. *Making Political Ecology.* London: Hodder Education.

Neves, D., M. Samson, I. van Niekerk, S. Hlatshwayo, and A. Du Toit. 2009. The Use and Effectiveness of Social Grants in South Africa. PLAAS & EPRI Research Report. Cape Town.

Newell, C. 1994. *Methods and Models in Demography.* reprinted. Chichester: Wiley.

North, D. C. 1990. *Institutions, Institutional Change and Economic Performance.* Cambridge: Cambridge University Press.

Odendaal, I. 2013. History of the Rand-Dollar Exchange. http://www.omwealth.co.za/Insights/MarketEconomicCommentary/History%20of%20the%20rand-dollar%20exchange (accessed June 12, 2014).

Ommer, R. E., and B. Paterson. 2014. Conclusions: Reframing the Possibilities for Natural and Social Sciences Dialogue on the Economic History of Natural Resources. *Ecology and Society* 19(1):17. http://www.ecologyandsociety.org/vol19/iss1/art17/ (accessed October 24, 2014).

Papilloud, C. 2003. *Bourdieu lesen: Einführung in eine Soziologie des Unterschieds*. Mit einem Nachwort von Loic Wacquant. Bielefeld: transcript Verlag.

Pauli, J. 2000. *Das geplante Kind: demographischer, wirtschaftlicher und sozialer Wandel in einer mexikanischen Gemeinde: Kölner ethnologische Studien* 27. Hamburg: Lit Verlag.

Pauli, J. 2009. (Re)Producing an Elite: Fertility, Marriage and Economic Change in Northwest Namibia. In *Networks, Resources and Economic Action: Ethnographic Case Studies in Honor of Hartmut Lang*. Clemens Greiner and Waltraud Kokot, eds. Pp. 303–25. Berlin: Reimer.

Peluso, N. L. 1993. Coercing Conservation? The Politics of State Resource Control. *Global Environmental Change* 3(2):199–217.

Peoples, J., and G. Bailey. 2015. *Humanity: An Introduction to Cultural Anthropology.* 10th ed. Stamford, CT: Cengage Learning.

Peters, P. E. 1983. Gender, Development Cycles and Historical Process: A Critique of Recent Research on Women in Botswana. *Journal of Southern African Studies* 10(1):100–22.

Peters, P. E. 2013. Land Appropriation, Surplus People and a Battle over Visions of Agrarian Futures in Africa. *The Journal of Peasant Studies* 40(3):537–62.

Planact. 2001. Introduction to Local Government Legislation: A Module for Councillors and Ward Committees. Training Manuals for Mangaung Local Municipality Ward Committees Vol. 1. Johannesburg.

Pongou, R. 2013. Why Is Infant Mortality Higher in Boys than in Girls? A New Hypothesis Based on Preconception Environment and Evidence from a Large Sample of Twins. *Demography* 50(2):445–6.

Pooley, E. 1999. *A Field Guide to Wild Flowers: Kwazulu-Natal and the Eastern Region*. Durban & Scottsville: Natal Flora Publications Trust.

Posel, D. R. 2001. What's in a Name? Racial Categorisations under Apartheid and their Afterlife. *Transformation* 47:50–74.

Posel, D. R. 2001. Who Are the Heads of Household, What Do They Do, and Is the Concept of Headship Useful? An Analysis of Headship in South Africa. *Development Southern Africa* 18(5):651–70.

Posel, D. R. 2003. Have Migration Patterns in post-Apartheid South Africa Changed? Paper for the Conference on African Migration in Comparative Perspective, Johannesburg, South Africa, 4-7 June 2003.

Potts, D. 2000. Environmental Myths and Narratives: Case Studies from Zimbabwe. In *Political Ecology: Science, Myth and Power*. Philip Stott and Sian Sullivan, eds. Pp. 45–65. London: Arnold.

Pretty, J. N., I. Guijt, I. Scoones, and J. Thompson. 1995. *A Trainer's Guide for Participatory Learning and Action.* London: International Institute for Environment and Development.

Price, R. E., and H. D. Brown. 2000. A Century of Locust Control in South Africa. In *Workshop on Research Priorities for Migrant Pests of Agriculture in Southern Africa*. R. A. Cheke, L. J. Rosenberg, and M. E. Kieser, eds. Pp. 37–49. Chatham, UK: Natural Resource Institute.

Rasch, S., T. Heckelei, R. Oomen, and C. Naumann. 2016. Cooperation and Collapse in a Communal Livestock Production SES Model – A Case from South Africa. *Environmental Modelling & Software* 75:402–413.

Redding, S. 2000. A Blood-stained Tax: Poll Tax and the Bambatha Rebellion in South Africa. *African Studies Review* 43(2):29–54.

Robbins, P. 2012. *Political Ecology: A Critical Introduction.* 2nd ed. Chichester: Wiley-Blackwell.

Roberts, B. R. 1966. The Ecology of Thaba Nchu: A Statistical Study of Vegetation/Habitat Relationships. PhD Thesis, University of Natal, Pietermaritzburg.

Rössler, M. 1997. *Der Lohn der Mühe: Kulturelle Dimension von 'Wert' und 'Arbeit' im Kontext ökonomischer Transformation in Süd-Sulawesi, Indonesien.* Münster: Lit Verlag.

Rotarangi, S. J., and J. Stephenson. 2014. Resilience Pivots: Stability and Identity in a Social-Ecological-cultural System. *Ecology and Society* 19(1):28. http://dx.doi.org/10.5751/ES-06262-190128 (accessed October 24, 2014).

Rowntree, K., M. Duma, V. Kakembo, and J. Thornes. 2004. Debunking the Myth of Overgrazing and Soil Erosion. *Land Degradation & Development* 15:203–14.

Salomon, M., C. Cupido, and I. Samuels. 2013. The Good Shepherd: Remedying the Fencing Syndrome. *African Journal of Range & Forage Science* 30(1&2):71–75.

Schapera, I. 1994. *A Handbook of Tswana Law and Custom: Compiled for the Bechuanaland Protectorate Administration*. Münster: Lit Verlag.

Schapera, I., and J. L. Comaroff. 1984. *The Tswana: With a Supplement Chapter by John L. Comaroff and a Supplement Bibliography by Adam Kuper.* 4th ed. London: KPI.

Schaumburg, K. 2013. *Bonyatsi: Heirat, Status und Beziehungsstrategien in einem südafrikanischen Township.* Münster: Lit Verlag.

Schensul, S., J. J. Schensul, and M. D. LeCompte. 1999. *Essential Ethnographic Methods: Ethnographer's Toolkit Vol. 2.* Walnut Creek: AltaMira Press.

Schmidt, S. 2004. Wohlfahrtspolitik in Südafrika nach dem Ende der Apartheid. In *Wohlfahrtsstaatliche Politik in jungen Demokratien*. Aurel Croissant, Gero Erdmann, and Friedbert W. Rüb, eds. Wiesbaden: VS Verlag für Sozialwissenschaften.

Schnegg, M. 2009. It's the Combination that Counts: Diversification of Pastoral Livelihoods in Northwestern Namibia. In *Networks, Resources and Economic Action: Ethnographic Case Studies in Honor of Hartmut Lang*. Clemens Greiner and Waltraud Kokot, eds. Pp. 229–47. Berlin: Reimer.

Schnegg, M., and H. Lang. 2008. Die Analyse kultureller Domänen: Eine praxisorientierte Einführung. Methoden der Ethnographie 3. Hamburg.

Schüler, G. M. K. 1965. Die BaRolong van Thaba 'Nchu, O.V.S. 'n Studie van Kultuurverandering. PhD thesis, University of Stellenbosch, Stellenbosch.

Schulze, W. G. 1997a. The Origin and Legal Nature of the Stokvel (Part 1). *South African Mercantile Law Journal* 9:18–29.

Schulze, W. 1997b. Die ethnodemographische Analyse der Zensusdaten von Gabsongkeg. In *Geburt und Tod: ethnodemographische Probleme, Methoden und Ergebnisse: Materialien zur Kultur der Wampar, Papua New Guinea*. Walter Schulze, Hans Fischer, and Hartmut Lang, eds. Pp. 92–189. Berlin: Reimer.

Schwinge, B. 2001. Haushaltsökonomie, Kooperation und institutioneller Wandel: Kleinbäuerliche Überlebensstrategien im Norden Namibias. PhD thesis, Universität zu Köln, Köln.

Schwingel, M. 2009. *Pierre Bourdieu zur Einführung.* 6th ed., revised. Hamburg: Junius.

Scoones, I. 1999. New Ecology and the Social Sciences: What Prospects for a Fruitful Engagement? *Annual Review of Anthropology* 28:479–507.

Scott, J. C. 1998. *Seeing Like a State: How Certain Schemes to Improve the Human Condition Have Failed.* New Haven & London: Yale University Press.

Seekings, J. 2014. The Social and Political Implications of Demographic Change in Post-Apartheid South Africa. *The ANNALS of the American Academy of Political and Social Science* 652:70–86.

Seekings, J., and N. Nattrass. 2005. *Class, Race, and Inequality in South Africa.* New Haven: Yale University Press.

Shackleton, C. M., S. E. Shackleton, J. Gambiza, E. Nel, K. Rowntree, P. Urquhart, C. Fabricius, and A. Ainslie. 2010. *Livelihoods and Vulnerability in the Arid and Semi-Arid Lands of Southern Africa.* New York: Nova Science Publishers.

Shackleton, C., S. E. Shackleton, and B. Cousins. 2001. The Role of Land-based Strategies in Rural Livelihoods: The Contribution of Arable Production, Animal Husbandry and Natural Resource Harvesting in Communal Areas in South Africa. *Development Southern Africa* 18(5):581–604.

Shackleton, S. E., C. M. Shackleton, T. R. Netshiluvhi, B. S. Geach, A. Ballance, and D. H. K. Fairbanks. 2002. Use Patterns and Value of Savanna Resources in Three Rural Villages in South Africa. *Economic Botany* 56(2):130–46.

Shackleton, S. E., C. C. Shackleton, and B. Cousins. 2000. The Economic Value of Land and Natural Resources to Rural Livelihoods: Case Studies from South Africa. In *At the Crossroads: Land and Agrarian Reform in South Africa into the 21st Century.* Ben Cousins, ed. Pp. 35–67. Cape Town: Programme for Land and Agrarian Studies (PLAAS).

Sharp, J. S., and A. D. Spiegel. 1985. Vulnerability to Impoverishment in South African Rural Areas: The Erosion of Kinship and Neighbourhood as Social Resources. *Africa* 55(2):133–52.

Showers, K. B. 1989. Soil Erosion in the Kingdom of Lesotho: Origins and Colonial Response, 1830s-1950s. *Journal of Southern African Studies* 15(2):263–86.

Simkins, C. 1978. Measuring and Predicting Unemployment in South Africa, 1960-1977. In *Structural Unemployment in Southern Africa*. Charles Simkins and Duncan Clarke, eds. Pp. 1–49. Pietermaritzburg: University of Natal Press.

Simkins, C. 1981. Agricultural Production in the African Reserves of South Africa, 1918-1969. *Journal of Southern African Studies* 7(2):256–83.

Simkins, C. 2001. Can South Africa Avoid a Malthusian Positive Check? *Daedalus* 130(1):123–50.

Slater, R. 2002. Differentiation and Diversification: Changing Livelihoods in Qwaqwa, South Africa, 1970-2000. *Journal of Southern African Studies* 28(3):599–614.

South African Government Services. 2014a. Child Support Grant. http://www.services.gov.za/services/content/Home/ServicesForPeople/Socialbenefits/childsupportgrant/en_ZA (accessed April 14, 2014).

South African Government Services. 2014b. Disability Grant. http://www.services.gov.za/services/content/Home/ServicesForPeople/Socialbenefits/disabilitygrant/en_ZA (accessed April 14, 2014).

South African Government Services. 2014c. Foster Child Grant. http://www.services.gov.za/services/content/Home/ServicesForPeople/Socialbenefits/fosterchildgrant/en_ZA (accessed April 14, 2014).

South African Government Services. 2014d. Old Age Pension. http://www.services.gov.za/services/content/Home/ServicesForPeople/Socialbenefits/oldagegrant/en_ZA (accessed April 14, 2014).

Stafford Smith, D. M., N. Abel, B. Walker, and S. F. Chapin, III. 2009. Drylands: Coping with Uncertainty, Thresholds, and Changes in State. In *Principles of Ecosystem Stewardship: Resilience-based Natural Resource Management in a Changing World*. Stuart F. Chapin, III, Gary P. Kofinas, and Carl Folke, eds. Pp. 171–95. New York: Springer.

Statistics South Africa. 2007. Community Survey, 2007: Revised Version. Pretoria.

Statistics South Africa. 2011. Statistics by Place: Mangaung. http://beta2.statssa.gov.za/?page_id=1021&id=mangaung-municipality (accessed February 6, 2014).

Statistics South Africa. 2012a. Census 2011: Municipal Report - Free State. Pretoria. http://www.statssa.gov.za/Census2011/Products/FS_Municipal_Report.pdf (accessed February 6, 2014).

Statistics South Africa. 2012b. Census 2011: Metadata Report No. 03-01-47. Pretoria. http://www.statssa.gov.za/Census2011/Products/Census_2011_Metadata.pdf (accessed June 7, 2014).

Statistics South Africa. 2013. Mangaung. http://beta2.statssa.gov.za/?page_id=1021&id=mangaung-municipality (accessed June 7, 2014).

Stevens, J. B. 2006. Adaptation of Irrigation Scheduling Methods in South Africa. PhD-thesis, University of Pretoria, Pretoria.

Steyn, G. 2012. The Influence of Botho on Social Space in Botswana Since Independence. *South African Journal of Art History* 27(3):112–129.

Swanstrom, T. 2008. Regional Resilience: A Critical Examination of the Ecological Framework. MacArthur Foundation Research Network on Building Resilient Regions Working Paper 2008-07. Berkeley, CA.

Thornton-Barnett, S. R. 2013. Ancestral Pharmacopoeias: A Paleoethnobotanical Assessment of Plant Use in the Western Free State, South Africa. Master thesis, Texas State University, San Marcos.

Tönsjost, S. 2013. *Umverteilung und Egalität: Kapital und Konsummuster bei Ovaherero-Pastoralisten in Namibia.* Münster: Lit Verlag.

Trollope, A. 1878. *South Africa.* London: Chapman and Hall.

Turton, A. R., R. Meissner, P. M. Mampane, and O. Seremo. 2004. A Hydropolitical History of South Africa's International River Basins. WRC Report 1220/1/04. Pretoria.

Twine, W. 2013. Multiple Strategies for Resilient Livelihoods in Communal Areas of South Africa. *African Journal of Range & Forage Science* 30(1&2):39–43.

Union of South Africa. 1955. Summary of the Report of the Commission for the Socio-Economic Development of the Bantu Areas within the Union of South Africa (Tomlinson Commission). U.G. 61/1955. Pretoria.

Union of South Africa. 1916a. Natives Land Commission (Beaumont Commission). U.G. 25/1916. Cape Town.

Union of South Africa. 1916b. Report of the Natives Land Commission (Beaumont Commission). Vol. 2. U.G. 22/1916. Cape Town.

Union of South Africa. 1932. Report of the Native Economic Commission: 1930-1932. U.G. 22/1932. Pretoria.

Unterhalter, E. 1987. *Forced Removal: The Division, Segregation and Control of People of South Africa.* London: IDAF.

van Driel, F. T. M. 1994. *Poor and Powerful: Female-headed Households and Unmarried Motherhood in Botswana*. Saarbrücken: Verlag für Entwicklungspolitik.

van Leynseele, Y., and P. Hebinck. 2011. Contested Livelihoods at the Interface: Ethnographic Explorations of Two Land Restitution Cases in Rural South Africa. In *Reforming Land and Resource Use in South Africa: Impact on Livelihoods*. Paul Hebinck and Charlie Shackleton, eds. Pp. 137–61. London & New York: Routledge.

van Onselen, C. 2005. *The Seed is Mine: The Life of Kas Maine, a South African Sharecropper, 1894-1985.* Johannesburg & Cape Town: Jonathan Ball.

van Oudtshoorn, F. 2004. *Guide to Grasses of Southern Africa.* 2nd ed. Pretoria: Briza Publications.

van Wyk, M. 2013. Makgoe Visits Struggling Schools. *The New Age*, January 10. http://thenewage.co.za/77629-1018-53-Makgoe_visits_struggling_schools (accessed 4 February 2014).

Vayda, A. P. 2009. *Explaining Human Actions and Environmental Changes.* Lanham: AltaMira Press.

Verhoef, G. 2001. Informal Financial Service Institutions for Survival: African Women and Stokvels in Urban South Africa, 1930-1998. *Enterprise & Society* 2(2):259–296.

Vermeulen, L. P., and R. Wiesner. 2000. Downsizing and the Survivor Syndrome: The South African Case. *South African Journal of Economic and Management Sciences* NS 3(3):387–402.

Vetter, S. 2013. Development and Sustainable Management of Rangeland Commons: Aligning Policy with the Realities of South Africa's Rural Landscape. *African Journal of Range & Forage Science* 30(1&2):1–9.

Voss, M. 2008. The Vulnerable Can't Speak: An Integrative Vulnerability Approach to Disaster and Climate Change Research. *Behemoth* 3:39–56.

Wales, J. M. 1985. The Relationship Between the Orange Free State and the Rolong of Thaba 'Nchu During the Presidency of J. H. Brand, 1864-1888. *Archives Year Book for South African History* 48. Pretoria.

Walker, B., and N. Abel. 2002. Resilient Rangelands: Adaptation in Complex Systems. In *Panarchy: Understanding Transformations in Human and Natural Systems*. Lance H. Gunderson and C. S. Holling, eds. Pp. 293–313. Washington, DC: Island Press.

Walker, B., and D. Salt. 2006. *Resilience Thinking: Sustaining Ecosystems and People in a Changing World.* Washington, DC: Island Press.

Water Research Commission. 2009. In-field Rainwater Harvesting on Croplands: Opportunities and Challenges. The Water Wheel March/April 2009. Pretoria.

Watson, R. L. 1980. The Subjection of a South African State: Thaba Nchu, 1880-1884. *Journal of African History* 21(3):357–73.

Watt, J. M., and M.G. Breyer-Brandwijk. 1932. *The Medicinal and Poisonous Plants of Southern Africa: An Account of Their Medicinal Uses, Chemical Composition, Pharmacological Effects and Toxicology in Man and Animal*. Edinburgh: E & S Livingstone.

Wells, J. C. 1983. Why Women Rebel: A Comparative Study of South African Women's Resistance in Bloemfontein (1913) and Johannesburg (1958). *Journal of Southern African Studies* 10(1):55–70.

Werner, O., and H. R. Bernard. 1994. Ethnographic Sampling. *CAM* 6(2):7–9.

Widgren, M. 2007. Precolonial Landesque Capital: A Global Perspective. In *Rethinking Environmental History: World System History and Global Environmental Change*. Alf Hornborg, J. R. McNeill, and Joan Martinez-Alier, eds. Pp. 61–77. Lanham: AltaMira Press.

Widgren, M. 2012. Resilience Thinking Versus Political Ecology: Understanding the Dynamics of Small-scale, Labour-intensive Farming Landscape. In *Resilience and the Cultural Landscape: Understanding and Managing Change in Human-shaped Environments*. Tobias Plieninger and Claudia Bieling, eds. Pp. 95–110. Cambridge: Cambridge University Press.

Widlok, T. 2003. Institutional Dynamics of Changing Land Care Practices in the Central Namib Desert. In *How Institutions Change: Perspectives on Social Learning in Global and Local Environmental Contexts*. Heiko Breit, Anita Engels, Timothy Moss, and Markus Troja, eds. Pp. 205–28. Opladen: Leske & Budrich.

Widlok, T., A. Aufgebauer, M. Bradtmöller, R. Dikau, T. Hoffmann, I. Kretschmer, K. Panagiotopoulos et al. 2012. Towards a Theoretical Framework for Analyzing Integrated Socio-Environmental Systems. *Quaternary International* 274:259–72.

Wilk, R. R. 1994. Inside the Economic Institution: Modelling Household Budget Structures. In *Anthropology and Institutional Economics*. James M. Acheson, ed. Pp. 365–90. Lanham: University Press of America.

Wilk, R. R., and R. McC. Netting. 1984. Households: Changing Forms and Functions. In *Households: Comparative and Historical Studies of the Domestic Group*. Robert McC. Netting, Richard R. Wilk, and Eric J. Arnould, eds. Pp. 1–28. Berkeley: University of California Press.

Wittenberg, M., and M. Collinson. 2008. Restructuring of Households in Rural South Africa: Reflections on Average Household Size in the Agincourt Sub-district 1992-2003. SALDRU Working Paper 12. Cape Town.

Wolpe, H. 1972. Capitalism and Cheap Labour-power in South Africa: From Segregation to Apartheid. *Economy and Society* 1(4):425–56.

World Bank. 1986. *Poverty and Hunger: Issues and Options for Food Security in Developing Countries.* Washington, DC: World Bank.

Worster, D. 1988. Appendix: Doing Environmental History. In *The Ends of the Earth: Perspectives on Modern Environmental History*. Donald Worster, ed. Pp. 289–307. Cambridge: Cambridge University Press.

Yawitch, J. 1982. *Betterment: The Myth of Homeland Agriculture.* Johannesburg: South African Institute of Race Relations.

10.2 Archival Sources

Published Reports of the South African Union are listed in the references.

10.2.1 Acts/Regulations (chronological)

- Orange River Colony, Ordinance No. 6 of 1907: Native Reserves Management Ordinance
- Orange River Colony, Department of Agriculture: Destruction of Locust Ordinance 1907
- Union of South Africa, Act No. 27 of 1913: Natives Land Act
- Union of South Africa, Act No. 20 of 1923: Native Urban Areas Act
- Union of South Africa, Act No. 38 of 1927: Native Administration Act
- Union of South Africa, Act No. 18 of 1936: Native Trust and Land Act
- Union of South Africa, Proclamation 31 of 1939: Control and Improvement of Livestock in Native Areas, Government Gazette
- Union of South Africa, Proclamation 116 of 1949: Limitation, Control and Improvement of Livestock and of Pastoral and Agricultural Resources in Native Areas, Government Gazette
- Union of South Africa, Act No. 68 of 1951: Bantu Authorities Act
- Union of South Africa, Act No. 67 of 1952: Abolition of Passes and Coordination of Documents Act
- Union of South Africa, Proclamation No. 136 of 1953: Conferment of Powers, Duties, Functions and Obligations upon the Barolong Tribal Authority, Thaba Nchu
- Union of South Africa, Act No. 46 of 1959: Promotion of Bantu Self-government Act
- Republic of South Africa, Act No. 21 of 1971: Bantu Homelands Constitution Act
- Republic of South Africa, Act No. 22 of 1994: Restitutions of Land Rights Act
- Republic of South Africa, Act No. 3 of 1996: Land Reform Labour Tenants Act
- Republic of South Africa, Act No. 28 of 1996: Communal Property Associations Act

- Republic of South Africa, Act No. 11 of 2044: Communal Land Rights Act
- Republic of South Africa, Act No. 15 of 2014: Restitution of Land Rights Amendment Act.

10.2.2 National Archives Repository Pretoria (SAB and TAB)

Department of Native Affairs
(later: Department of Bantu Administration and Development; later: Department of Cooperation and Development)

Native Reserve Boards in the Sediba and Thaba Nchu Native Reserve:
NTS 8141 1/341 part 2
NTS 8142 2/341 part 1-3
NTS 8144 5/342 part 1
NTS 8152 22/341
NTS 8154 32/341
NTS 8154 37/341

SANT Purchase of Released Areas:
NTS 3161 213/306

Soil Erosion and Betterment in Thaba Nchu:
BAO 5761 H128/1666/197 part 1
BAO 5761 H128/1666/214
NTS 7256 290/326
NTS 7335 127/327 part 2
NTS 7516 678/327
NTS 7526 717/327 part 1-4
NTS 7527 717/327 part 4
NTS 7981 259/337 part 1-3
NTS 7982 259/337/1
NTS 7982 259/377/2 part 1
NTS 9502 138/400/5 part 1-2
NTS 9509 138/400/19 vol. 1-4
NTS 10487 D1109
NTS 10496 D1309
NTS 10967 T 21083

Soil Conservation Board:
NTS 8899 140/362 (1) part 2

Agricultural Officer Reports
NTS 7394 335/327 part 1-2

Barolong Tribal Authority, Headmen, and Chief related matters
BAO 4831 F53/1666/1 part 1
NTS 308 10/54
NTS 308 13/54
URU 1793 1315

Protest/Unrest
NTS 7692 387/332

Homeland Formation
SAB SON 1606 D12/2/4/2/4/3 vol. 1

South African Police

SAP 481 15/20/50 (South African Police)

Department of Census and Statistics

STK 3/21/230
SAB STK 4/21/428
STK 4/B22/428/46

10.2.3 North West Provincial Archives (NWA)

Office of the Presidency

Development Planning Thaba Nchu
OPD 9/6/3 (34) part 1-2

Tribal Authority
OPD 6/7/2/5 part 3

Agriculture
Agri Info 1/26

10.2.4 Free State Archives Repository (VAB)

Government Secretary Oranje Vrystaat

Barolong Census
GS 1574

Orange River Colony

Census
ORC 40/28

11. APPENDICES

11.1 Map of Talla

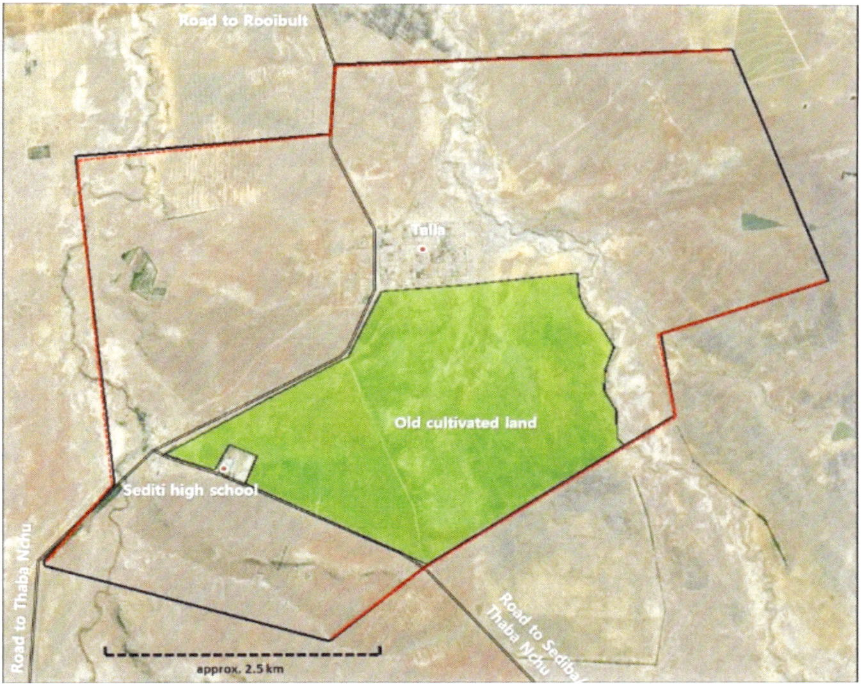

The above image shows a recent Google Earth image (2014) of the Talla area with additional geographical data collected with a GPS device. The green polygon indicates the former fields as laid out after the implementation of 'betterment' in the 1960s. In pre-'betterment' times, the area used for cultivation was slightly larger. In 2011, the former fields were used as rangelands and were no longer fenced off. The boundaries to the east (white-owned private land) and southwest (land reform farms) had fences. The boundaries to adjoining communal areas are indicated with red dashed lines and were not fenced off. Sediti high school, which is attended by many youth of the area, is located next to the road that connects Talla with Sediba and Thaba Nchu.

11.2 Map of Sediba

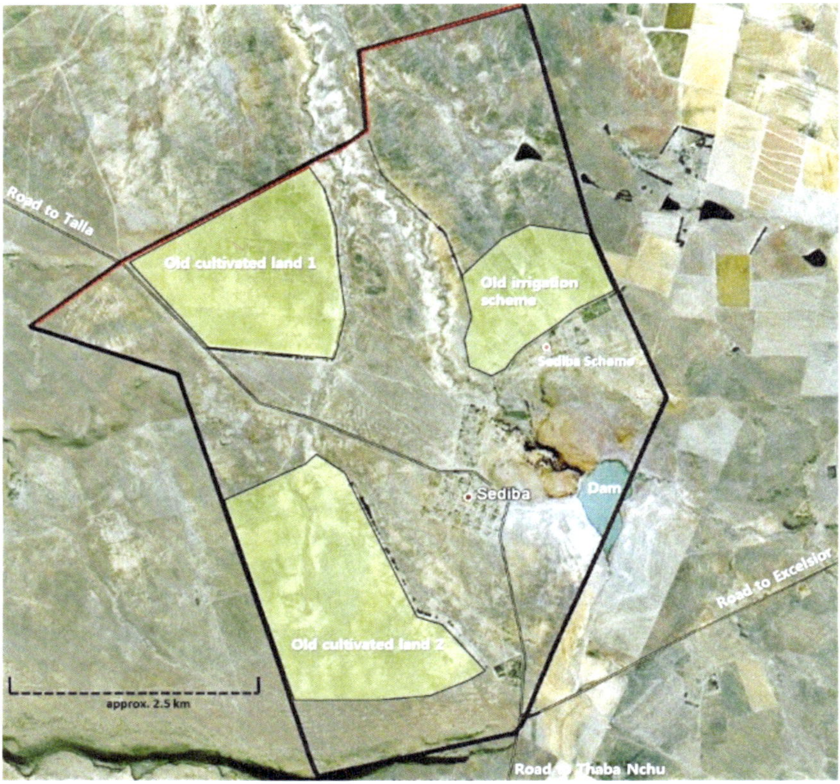

The above image shows a recent Google Earth image (2014) of the Sediba area with additional geographical data collected with a GPS device. The green polygons indicate the former fields as laid out after the implementation of 'betterment' in the 1960s. In pre-'betterment' times, the area used for cultivation was even larger. The section marked *Old cultivated land 1* was used by the *setlamo* cooperative during Bophuthatswana times (Chapter 6.4.1). In 2011, the former fields were used as rangelands and were not fenced off anymore. Except for the northern boundary adjoining the communal lands of Talla (indicated by red dashed line), the boundaries to the northeast (white-owned private land), as well as to the southeast, south and east (land reform farms) were fenced off.

11.3 Map of Middeldeel

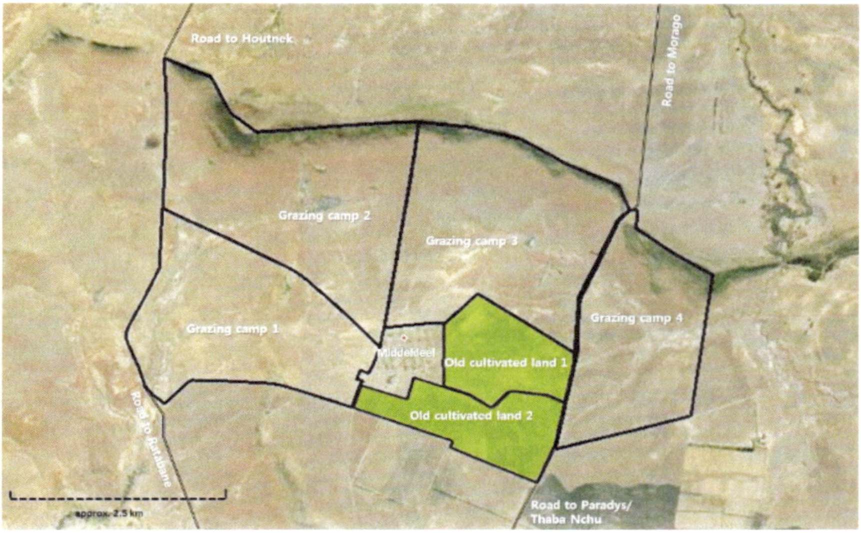

The above image shows a recent Google Earth image (2014) of the Sediba area with additional geographical data collected with a GPS device. The green polygons indicate the former fields as laid out after the implementation of 'betterment' in the 1960s. In 2011, the former fields were used as rangelands. When a sheep and wool project was initiated in Middeldeel in the mid-2000s, the village's grazing camps were fenced-off. In 2011, *Grazing camp 1* was the only one of the camps had a working wind pump. Thus, the camp was grazed by a majority of the animals. The *Old cultivated land 2* was used by the local headman. *Grazing camp 4* was out of sight of the village and farmers were reluctant to bring their livestock there because of the possibility of theft.

11.4 Names and Uses of Medicinal Plants and Their Frequency in the Free Listing Interviews (n=34)

Scientific name	Setswana / Sesotho	Frequency FL (%)	Uses/ Indication (and source of knowledge)
Aloe spp.	Lekgala	18%	Interviews: laxative, gastric derangements
Artemisia afra	Lengana	29%	Interviews: coughs, colds Watt (1932): bronchial complaints, coughs, colds, dyspepsia, loss of appetite, gastric derangements, colic, croup, whooping cough, gout
Dicoma anomala	thlwenya (also hlwenya)	47%	Interviews: coughs Pooley (1999): coughs, dysentery, toothache, sterility, gall sickness in stock
Eriocephalus sp.	segalagala (also sehalahala)	24%	Interviews: colds Thornton-Barnett (2013): High blood pressure
cf. *Eucomis* spp.	kgampupu	15%	Thornton-Barnett (2013): Banishing evil spirits.
Helichrysum sp.	phate-ya-naga	26%	Interviews: colds Pooley (1999): colds and nausea
Hypoxis hemerocallidea	African potato	15%	Thornton-Barnett (2013): Weakness in children, Prostate problems, testicular tumors, urinary tract infections.
Polygala hottentotta	lehlokoa-la-tsela	35%	Interviews: Used for ritual purposes Pooley (1999): abdominal complaints, anthrax Watt (1932): mumps, anthrax
Scabiosa columbaria	Selomi	18%	Interview: painful menstruation Pooley (1999): sterility, colic, venereal sores, painful menstruation, sore eyes and to ease childbirth
cf. *Xysmalobium undulatum*	photsetla	44%	Interviews: colds Watt (1932): fever, colic, cold

Kölner ethnologische Studien
hrsg. von Prof. Dr. Martin Rössler

Diego Augusto Menestrey Schwieger
The Pump Keeps on Running
On the emergence of water management institutions between state decentralization and local practices in northern Kunene
vol. 36, 2017, 488 pp., 34,90 €, br., ISBN-CH 978-3-643-90838-4

Silke Tönsjost
Umverteilung und Egalität
Kapital und Konsummuster bei Ovaherero-Pastoralisten in Namibia
Bd. 35, 2013, 272 S., 29,90 €, br., ISBN 978-3-643-12240-7

Katrin Schaumburg
Bonyatsi
Heirat, Status und Beziehungsstrategien in einem südafrikanischen Township
Bd. 34, 2013, 296 S., 29,90 €, br., ISBN 978-3-643-12205-6

Kerstin Hadjer
Geschlecht, Magie und Geld
Sozial eingebettete und okkulte Ökonomien im Benin, Westafrika
Bd. 33, 2009, 424 S., 39,90 €, br., ISBN 978-3-8258-1027-6

Michael Schnegg
Das Fiesta Netzwerk
Soziale Organisation einer mexikanischen Gemeinde 1679 – 2001
Bd. 31, 2005, 328 S., 34,80 €, br., ISBN 3-8258-8810-x

Susanne Berzborn
Ziegen und Diamanten
Sicherung des Lebensunterhalts im ländlichen Südafrika
Bd. 30, 2006, 480 S., 39,90 €, br., ISBN 3-8258-8469-4

Peter Finke
Nomaden im Transformationsprozess
Kasachen in der post-sozialistischen Mongolei
Bd. 29, 2004, 400 S., 29,90 €, br., ISBN 3-8258-6964-4

Clara Mayer-Himmelheber
Die Regalia des Kabaka von Buganda
Eine Biographie der Dinge
Bd. 28, 2004, 296 S., 29,90 €, br., ISBN 3-8258-6883-4

Julia Pauli
Das geplante Kind
Demographischer, wirtschaftlicher und sozialer Wandel in einer mexikanischen Gemeinde
Bd. 27, 2001, 392 S., 35,90 €, br., ISBN 3-8258-5120-6

Christoph Brumann
Die Kunst des Teilens
Eine vergleichende Untersuchung zu den Überlebensbedingungen kommunitärer Gruppen
Bd. 26, 1998, 368 S., 30,90 €, br., ISBN 3-8258-3732-7

Susanne Spülbeck
Biographie-Forschung in der Ethnologie
Bd. 25, 1998, 176 S., 12,90 €, br., ISBN 3-8258-3401-8

LIT Verlag Berlin – Münster – Wien – Zürich – London
Auslieferung Deutschland / Österreich / Schweiz: siehe Impressumsseite